A Treatise on Raisin Production, Processing and Marketing

A Treatise on Raisin Production, Processing and Marketing

Edited by
Vincent E. Petrucci, Sc.D. and
Carter D. Clary, Ph.D.

Malcolm Media Press
in association with American Vineyard Magazine
Clovis, California

First Edition 2002

Published by:
Malcolm Media Press
P.O. Box 626
Clovis, CA 93613
559-298-6675
www.malcolmmedia.com

Malcolm Media Press is an associated company with Malcolm Media Corporation and Malcolm Media • Ag Publishing, publishers of *American Vineyard*, *California Dairy*, *Central Valley Farmer*, *Pacific Nut Producer*, and *Vegetables West.*

Printed in the United States of America

This publication was written in accordance with the *American Society for Horticultual Science Publications Style Manual,* Copyright 1997, ASHS Press. The American Society for Horticultual Science may be contacted in writing at 133 South West Street, Alexandria, VA 22314-2851, by phone at 1-703-836-4606, or by email at ashapres@ashs.org.

Contents

Contents

Contents

Contents

Contents

Contents

List of Tables

Appreciation

Appreciation is hereby expressed to the many individuals who made this book possible. The editors of this Treatise recognized the authors and those who peer reviewed their manuscripts. In particular, the editors thank Barry Kriebel, President of Sun-Maid Growers of California; Vaughn Koligian, former CEO of the Raisin Bargaining Association; Terry Stark, General Manager of the Raisin Administrative Committee; Norm Engelman, Vice-chairman of the California Raisin Marketing Board; Mike Garabedian, Vice-Chairman of Valley Welding and Machine Works; Dr. David W. Ramming, Plant Breeder USDA/ARS-Fresno, California; Debbie Schmidtgall McMillian, food Technologist - Derco Foods International; Ernest Bedrosian, President, National Raisin Company, and Dr. Sanliang Gu, Ricchiuti Chair of Viticulture Research, California State University, Fresno (CSUF).

We wish to recognize Cheryl Hughes of ACG USDA/ARS-Fresno, California, for providing raisin inspection and analysis; Professor Ahai Heng of the Department of Horticulture at Shandong Agricultural University Tai'an, Shandong, P.R. China, for providing samples of raisins made in China; Vernal Amaro, Fresno raisin grower and inventor of the model DOV 2000 Grape and Raisin Harvester; Gary Pitts, inventor of the Pitts Quad-DOV raisin system; Lee Simpson, Simpson Vineyards, Madera, California; Wayne Albrecht, Albrecht Farms, Del Rey, California; William Slusser, Dept. of Geography at CSUF, for preparing the country maps illustrations shown in Chapter 1; Dr Loy Bilderback, Dept. of History at CSUF, for his consultation on text book indexing, and Alev Akman, Viticulture and Enology Research Center librarian for reference materials and general assistance.

The editors wish to recognize associate editors, Lillian Quaschnick, who prepared the original manuscript, and Randy Vaughn-Dotta at CSUF, for preparing most of the photographs in this Treatise.

Special recognition is given for the invaluable contributions of *American Vineyard* Publisher Dan Malcolm, owner of Malcolm Media Press and Malcolm Media • Ag Publishing, Clovis, California, and Malcolm Media Production Manager Mike Lawless, for the design, production and publication of this text. Recognition is also given to Elaine Fraser for page layout.

We also extend our appreciation to Curt Lynn, Farm Advisor Emeritus, University of California Cooperative Extension in Tulare County, CA, for preparing the Forward for this book.

We further wish to acknowledge our wives, Jo Petrucci and Susan Clary for their support throughout the course of this study.

Finally, the editors thank Robert Wample, Chairman of the Department of Viticulture and Enology at California State University, Fresno, events coordinator Cynthia Wood and the Assistant to the Director Helen Chrisman for providing administrative support.

Vincent E. Petrucci, Sc.D.
Carter D. Clary, Ph.D.

Foreward

Having grown up on a Fresno County raisin farm during the depression years when kids worked alongside parents to help make ends meet and during the war years when labor was scarce, I have great appreciation for progress that makes farm work easier and more productive. I recall my exuberance when we sold the mules and purchased our first tractor; also, how glad I was when we switched from wooden to paper trays because the annual job of repairing, scattering, collecting and stacking wooden trays was hard, dirty work. Later while serving as a University of California Cooperative Extension grape farm advisor, I experienced progress from the perspective of one involved in conducting applied research and extending useful information to potential users. While progress is a continuing advancement of innovation, it comes most rapidly from directed research. Such research in viticulture in California has been going on for about 75 years. This Treatise provides a concise summary and discussion of past and current research pertaining to raisins.

Currently, California produces almost half the world's raisins, of which the vast majority is produced in the southern San Joaquin Valley. California largely produces natural sun-dried raisins in contrast to most other raisin producing areas in the world where alkaline emulsion dips are used to enhance drying. Dipped raisins have a distinctively different color, texture and flavor. While many consumers in the world prefer dipped raisins, many other prefer natural sun-dried. Some consumers have no real preference, so dipped and natural sun-dried raisins often compete for sales in international markets.

The southern San Joaquin Valley's success in producing raisins may be attributed to its hot, dry summer climate that allows grapes to be naturally sundried, its abundance of sandy soils that are highly desirable for drying grapes on paper trays laid on the ground, and its available irrigation water that is absolutely necessary in an arid region. Other success factors include innovative growers who stay alert to change, a modern processing industry capable of turning out a clean, wholesome product, and successful marketing. The California raisin industry is not without problems, however. The most serious problem the industry has had to deal with throughout its history is one of oversupply, a situation common to both California and other raisin producing countries. Thus, California raisin production often exceeds sales potential, forcing lower prices and implementing some kind of supply management program. Occasional rains during the drying season have also caused serious problems at times. The industry has also suffered, on occasion, from spring frost, hail and serious pest and disease infestations. A major marketing concern is in the competition with other raisin producing areas of the world, some of which offer raisins at consistently lower prices, have lower labor costs and are supported by government programs. Currently, the California raisin industry is also becoming increasingly concerned about the availability of sufficient harvest labor, a problem that could become very serious in the future.

To a degree, California has had a competitive advantage over some of the other raisin producing areas in the world in producing and selling its raisins because of higher yields and quality. These accomplishments can largely be attributed to the adoption of modern viticultural farming practices, a direct result of research, innovation and education. The industry has been particularly well supported in such research and educational programs by the Agricultural Experiment Station and Cooperative Extension components of the University of California, California State University, Fresno and the United States Department of Agriculture. Additionally, the industry has been well served by agricultural equipment manufacturers, suppliers, financial institutions, consultants and other related agribusinesses.

Effectively transferring research and innovation to potential users is essential to the new practices are to be implemented. Publications play an important role in such transfer. There are numerous publications currently available that deal with the production of grapes and, to a lessor extent, raisins. Many of these are research papers published in scientific journals. Others include extension pamphlets that have mostly been written for producers. This Treatise adds to the list of useful publications. However, it differs from most other publications dealing with raisins in that it is more comprehensive in scope and includes all three industry components: production, processing and marketing. Written by authorities in their field of expertise, this Treatise covers efforts and findings of many researchers and experts. All information referenced is cited in accordance with standard scientific methodology so those wanting more information can readily go to the source. While this Treatise is designed to be used as a college text and reference, it is recommended for all directly involved in the California industry as well as the world.

Although most of the information presented focuses directly on the California raisin industry, considerable background information on world raisin situation and production, raisin grape varieties and production of dipped Sultana raisins are included. This information is particularly useful in understanding the competition the California industry faces and in planning for the future. This Treatise also provides an excellent discussion of current trends, including the development and introduction of new raisin grape varieties, drying grapes on the vine (DOV), larger trellising systems and mechanized harvesting.

Grape production in California has changed significantly in the past 75 years, particularly in vine training and pruning, trellising, fertilization, irrigation, pest and disease control and mechanization. Raisin drying practices, on the other hand, have changed little. Natural sun-dried raisins are still largely dried on trays laid on the ground between the vine rows. However, this is about to change as significant progress is being made in drying grapes on the vine. The advantage of the DOV system is that the dried grapes can be harvested mechanically, thereby greatly reducing labor requirements. Also DOV grapes are less subject to rain damage. Additionally, when DOV is coupled with larger trellising systems, raisin yields may be increased significantly. The DOV system, including trellising configuration and mechanized harvesting, is thoroughly discussed in this Treatise.

The objective of the California raisin industry is to produce and sell raisins at prices favorable to all involved in the process. Achieving this is an increasingly competitive market requires a highly efficient and effective production, processing and marketing system. Raisin growers, for instance, must produce and dry a quality product at the lowest price possible while still remaining viable. Processors, in turn, must efficiently de-stem, clean and package the raisins to assure they meet consumer satisfaction. To complete the process, sellers must successfully compete in an international market where consumers have a wide selection of foods to choose from and are very price-conscious.

Raisins will continue to be a particularly note-worthy agricultural crop in the world and especially in California. After all, they represent one of the first fruit crops to be preserved and stored for later consumption. Additionally, they are one of the first dried fruit crops commercially traded, dating back to biblical times. Economically, raisins will remain an important crop in several countries. Promoters of family farms recognize raisins as a specialty crop that is well suited for small-scale family operations. Consumers identify raisins as a tasty, nutritious snack food, salad and desert topping and as a bakery goods ingredient.

In California the raisin industry will face new challenges as it moves into DOV systems and utilizes greater amounts of mechanization. Some fear DOV, with its higher production potential, will only add to world raisin surpluses. Others see opportunities to sell more raisins by offering lower prices made possible by greater production efficiency. It would be foolish to believe there will not be problems of one kind or another regardless of what happens in the future. However, by working together for a common gain, California raisin producers, processors and marketers can deal with the problems in a way the industry will maintain world leadership.

Curtis D. Lynn,
Farm Advisor Emeritus, University of California Cooperative Extension,
Tulare County, California.

Dedication

On behalf of the California State University, Fresno and the authors of this Treatise, the editors wish to dedicate this work to the Bertha and John Garabedian Charitable Foundation, whose financial support made it possible to prepare and publish this Treatise on Raisin Production, Processing and Marketing. We thank Individual Trustees Silvestre Arias, Glenn E. Rose, C.P.A., Malcolm H. Stewart, Esq., H. Tookoian, M.D. and the Corporate Trustee, Bank of America, N.A. for their confidence and support of this Treatise.

Vincent E. Petrucci, Sc.D.
Carter D. Clary, Ph.D.

John Garabedian

Contributors

Eddie Wayne Albrecht, A.A.
Chapter 8
Partner, Albrecht Farms,
Del Rey, California

Gregory T. Berg, M.S.
Chapter 5
Consultant, Viticulture Solutions,
Fresno, California

James E. Casey, Jr., Ph.D.
Chapter 6
Lecturer, Department of Agricultural Economics, California State University, Fresno,
Fresno, California

Carter D. Clary, Ph.D.
Chapters 8 & 9, Co-Editor
Assist. Prof., Ag. Tech.and Mgmt, Depart. of Biological System Engineering, Washington State University and former Director of the Dried Foods Tech. Lab., VERC, CSUF,
Fresno, California

Peter R. Clingeleffer, B.S.
Chapter 7
Principal Research Scientist, CSIRO, Division of Horticulture,
Merbein, Australia.

Michael J. Costello, Ph.D.
Chapter 10
Assistant Professor of Horticulture and Crop Science Department, California Polytechnic University
San Luis Obispo, California

Bernadine B. Ferguson
Chapter 14
Foods and Culinary Consultant,
Fresno, California

Kip R. Green, M.S.
Chapter 10
Horticulural Crops Advisor,
Fresno, California

Z. Jo Harper, B.S.
Chapter 10
André Tehelistcheff Honoree,
California State University, Fresno
Fresno, California

Curtis D. Lynn, M.S.
Author of Forward
Farm Advisor Emeritus, University of California Cooperative Extension,
Tulare County, California

Dan Malcolm
Publisher
CEO, Malcolm Media Corporation
Clovis, California

Mark A. Mayse, Ph.D.
Chapter 10
Professor Emeritus of Entomology (posthumously), Department of Plant Science, CSUF
Fresno, California

Clyde Nef, M.S.
Chapter 13
Retired, Former Manager of the California Raisin Advisory Board & Manager of the Raisin Administrative Committee, Fresno, California

Thomas J. Payne, M.A.
Chapter 14
Food Industry Consultant, San Mateo, California

Vincent E. Petrucci, Sc.D.
Chapters 1, 3, & 8, Editor
Professor Emeritus of Viticulture and Director Emeritus of the Viticulture and Enology Research Center, CSUF, Fresno, California

Lillian Quaschnick, B.A.
Chapter 2, Associate Editor
Teacher, Hayward High School Hayward, California

David W. Ramming, Ph.D.
Chapter 4
Research Horticulturist, USDA/ARS, San Joaquin Valley Agricultural Sciences Center, Parlier, California

Sarah A. Shepard, M.B.A.
Chapter 6
Instructor, Department of Agriculture, West Hills College, Coalinga, California

Allan Shields, Ph.D.
Chapter 2 Feature
Professor Emeritus of Philosophy, San Diego State University, San Diego, California

R. Keith Striegler Ph.D.
Chapter 5
Extension Horticulturist, Fruit, University of Arkansas,Fayettville, Arkansas & Former Director of VERC, Fresno, California

Richard Van Diest, B.A.
Chapter 12
Marketing Specialist, USDA, Fresno, California

Randy Vaughn-Dotta
Associate Editor
Campus Photographer, Academic Innovation Center, CSUF, Fresno, California

Julian Whaley, Ph.D.
Chapter 11
Professor Emeritus of Plant Pathology, Department of Plant Science, CSUF, Fresno, California

R. Lynn Williams Ph.D.
Chapter 6
Assistant Professor, Department of Agricultural Economics, CSUF, Fresno, California

CHAPTER 1.

Overview Of World Dried Grape Production

Vincent E. Petrucci, Sc.D.

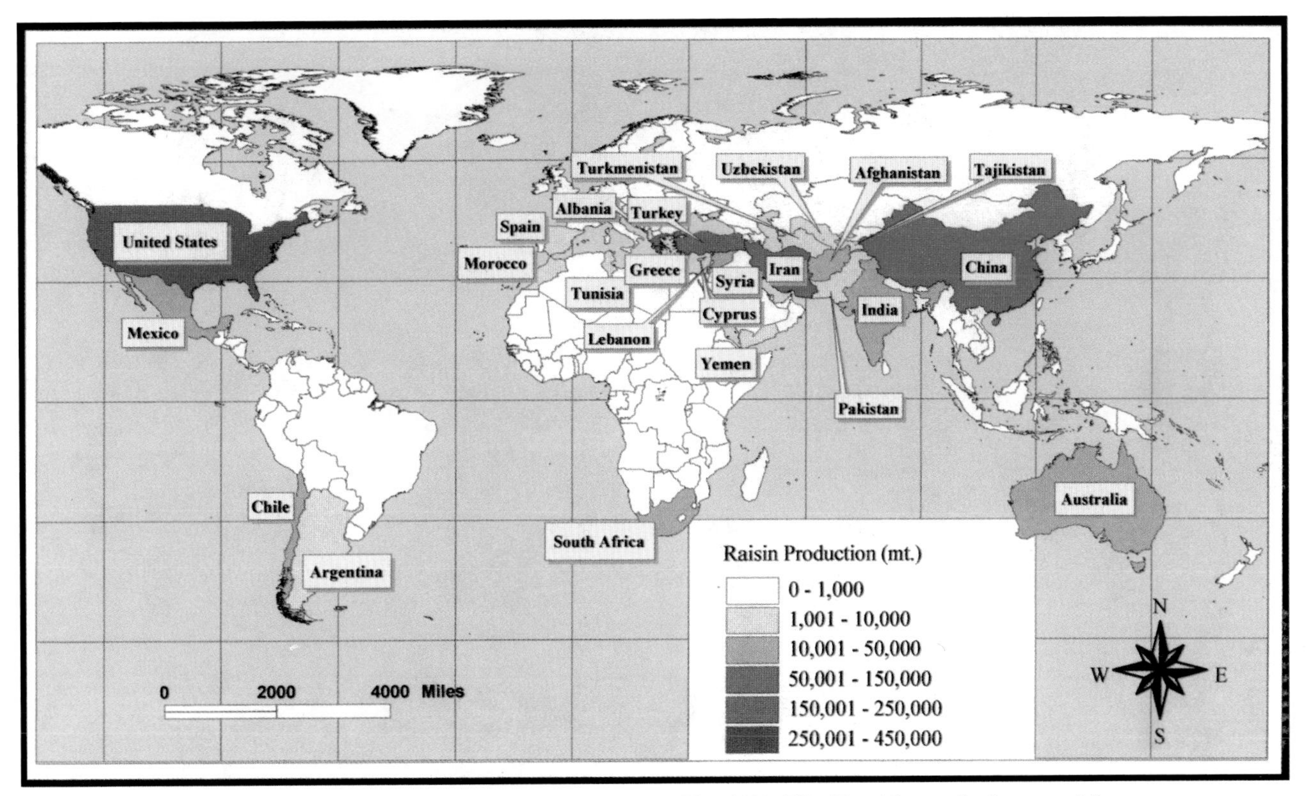

Fig. 1.I.0: World raisin producing countries.

Introduction

Grapes produced for use as wine, raisins, table grapes and other uses continues to be the world's number one deciduous fruit crop. Grapes are also the number one deciduous fruit crop in the United States and the leading fruit crop in California. Grapes are grown on six continents in 62 different countries, each with a minimum of 12,000 acres (5,000 ha) in production supplying grapes to commercial trade (**Table 1.I.1**). In 1997, the world vineyard surfaces reached 18,915,744 acres (7,651,000 ha). **Table 1.I.2** illustrates the distribution of this acreage with Europe as the leading continent at 62%, followed by Asia, South America, North and Central America, Africa and Oceania. North and Central America (Canada, Guatemala, Mexico and the U.S.) lead the world in yield per acre, averaging 8.1 tons per acre (18.2 t/ha) compared to the world average of 3.4 tons per acre (7.62t/ha). This may be attributed to the research and development programs at U.S. universities, United States Department of Agriculture and other research agencies.

Vineyard planting showed steady growth up to 1980 when vineyard areas of the world started to decrease (**Table 1.I.3**). This is related to economic measures undertaken by the European Union (EU) to encourage the removal of wine grape vineyards to reduce wine grape surpluses. The former USSR has also removed vineyards, however, at the same time, efficiency of production has increased. This is due in part to the selection of higher producing clones, disease free plants, more effective grape pest management, and advanced viticulture technology, including more productive trellising systems, row spacing, canopy management, vine nutrition, and irrigation innovations. Another factor contributing to increased yields is the increase in acreage in viticultural regions, such as North and Central America, where yields generally have been high.

When viewing world utilization of grapes by class, wine grapes rank first, followed by table grapes and raisin grapes. Raisin grapes account for 7.59% of the total world grape production (**Table 1.I.4**).

Table 1.I.1
World Raisin and Sultana Production by Country.

Country	Tons	Percent
United States[1]	442,200	35.8%
Turkey[1]	275,000	22.2%
Iran[1]	112,200	9.1%
China[6]	77,000	6.2%
Greece[1]	58,300	4.7%
Australia[1]	41,800	3.4%
South Africa[1]	38,500	3.1%
Chile[1]	35,200	2.8%
Afghanistan[2]	30,800	2.5%
India[4]	30,800	2.5%
Mexico[1]	19,250	1.6%
Syria[2]	17,600	1.4%
Argentina[5]	10,780	0.9%
Lebanon[2]	8,800	0.7%
Spain[5]	8,250	0.7%
Cyprus[3]	6,600	0.5%
Morocco[2]	5,500	0.4%
Tunisia[3]	4,400	0.4%
Pakistan[2]	4,400	0.4%
Albania[2]	3,300	0.3%
Tajikistan[2]	3,300	0.3%
Uzbekistan[2]	3,300	0.3%
Turkmenistan	1,210	0.1%
Yeman[2]	1,210	0.1%
World	**1,139,700**	**100.0%**

Sources:
[1]International Conference of Sultana and Raisin Producing Countries, Estimates. San Francisco, CA. 1998.
[2]FAO Production Yearbook, Vol. 51-1997. Series No 142, Rome, Italy.
[3]O.I.V. Bulletin Supplement 803-804. January and February, 1998. Paris, France.
[4]S.D. Shikhamany, National Research Center, Pume, India.
[5]Agriculture Attaché Query Detail, 1998. U.S.D.A./F.A.S., Washington, DC
[6]Editorial Board for China Agriculture Yearbook, China Agriculture Yearbook (1993-1996), Beijing: China Agriculture Press, 1993, 1994, 1995, 1996.

Table 1.I.2
World Grape Production by Continent, 1997.

Continent	Acres	Percent World Total	Total Tons	Tons/Acre
World	18,915,744	100	64,313,700	3.4
Europe	11,722,224	62	32,935,100	2.8
Asia	4,185,096	22	14,105,300	3.4
S. America	1,033,296	5.5	5,395,500	5.2
N.C. America	902,280	4.8	7,337,000	8.1
Africa	880,032	4.7	3,433,100	3.9
Oceania	192,816	1	1,107,700	5.7

Source: FAO Production Yearbook, Vol.51, 1997

Table 1.I.3
Total World Grape Acreage Production and Yield per Acre, 1971-1997.

Years	Acres	Tons	Tons/Acre
1971-75	24,613,631	62,089,328	2.52
1976-80	25,236,323	37,827,424	2.69
1981-85	24,272,633	70,347,408	2.90
1986-90	21,534,765	67,701,760	3.14
1991-95	19,980,506	68,788,720	3.44
1996	19,130,482	65,722,944	3.44
1997	18,915,744	64,313,700	3.40

Source: O.I.V. Bulletin, 1998

Table 1.I.4
World Grape Utilization by Class, 1997.

Class	Share (%)
Wine Grapes	78.70
Table Grapes	13.71
Raisin Grapes	7.59
Total	**100.00**

Source: O.I.V. Bulletin, 1998
F.A.O. Production Yearbook, Vol. 52, 1998

Some Definitions

The terms "raisins" and "sultanas" will be used interchangeably in this text. Nef (1999) states that "all raisins are dried grapes, but not all dried grapes are raisins." There are two major grape varieties used to produce grapes that are dried into raisins and/or sultanas. These varieties, which are synonymous, are the *Vitis vinifera Thompson* Seedless and the *Vitis vinifera* Sultana. The terms "Thompson Seedless," "Sultana," "raisins" and "sultanas" are used inconsistently, and sometimes interchangeably, from country to country. Furthermore, ampelography texts seem to be inconsistent in defining these terms. Galet (1979) defines Thompson Seedless as Sultanine Blanche, Sultanina, Sultana (Australia and South Africa). Galet defines no separate heading for Sultana. Perold (1974) defines Sultana as Sultanina, Thompson's Seedless (California), Sultanieh, Sultan, and Sultani and gives no separate heading for Thompson (or Thompson's) Seedless. Winkler (1974) defines Sultanina as a synonym for Thompson Seedless, stating that the Thompson Seedless is also called Sultana in Australia and South Africa. Winkler does not separate Sultana as a different variety. Kerridge and Antcliff (1996) state that in California, the Sultana is called Thompson Seedless to distinguish it from another variety introduced erroneously as Sultana.

Yet there is a difference between the Thompson Seedless and the Sultana in many people's minds. In many parts of the world, packages labeled "raisins" are sold next to packages labeled "sultanas." For the purpose of this text, the term "raisins" will be used to refer to dried grapes unless otherwise specified. In California, all the dried fruit produced from these two similar varieties and some other minor varieties are called "raisins." The rest of the world, however, makes a distinction between "raisins" and "sultanas," generally using the former to refer to the natural sun-dried California product and the latter to refer to the product treated with various dipping solutions and then dried. With the exception of golden raisins, California raisins are darker in color than sultanas produced in other countries.

The Food and Agriculture Organization of the United Nations (F.A.O.) and the Office International de la Vigne at du vin (O.I.V.) refer to the natural sun-dried product produced in California as raisins. Dried grapes produced in other parts of the world are called sultanas. This may account for some of the statistical inconsistencies encountered in providing production information later in this chapter.

It should be noted that the countries of Afghanistan, Australia, Greece, Iran and Turkey prefer to use the name "sultana" both in the fresh and dried form. Most other coun-

Table 1.I.5
Utilization of Raisins and per Capita Consumption by the Leading Raisin Producing Countries, 1997-98.

Country	Population (m)	Production (1,000 tons)	Domestic (tons)	Export (tons)	Import (tons)	Per Capita (lbs.)
United States	271.6	402.0[1]	242,000	160,000	12,100.0[3]	1.90
Turkey	63.5	250.0[1]	40,000	210,000	5,000.0[5]	1.42
Iran	71.5	102.0[1]	25,500	76,500	0.0	0.71
China	1243.7	70.0[6]	63,000	7,000	5,200.0[3]	0.11
Greece	10.5	53.0[1]	7,950	45,050	3,000.0[3]	2.05
Australia	18.3	38.0[1]	7,600	30,400	9,600.0[3]	1.90
South Africa	43.3	35.0[1]	5,250	29,750	0.0	0.24
Chile	14.6	32.0[1]	4,800	27,200	0.0	0.66
Afghanistan	22.1	28.0[2]	8,000	20,000	0.0	0.72
India	960.2	28.0[4]	28,000	0	6,700.0[3]	0.58
Mexico	94.3	17.5[1]	1,750	15,750	1,600.0[3]	0.41
Syria	14.9	16.0[3]	16,000	0	0.0	2.14
World Total	**5848.7**	**1,099.0**				**0.38**

Sources:
[1]International Conference of Sultana and Raisin Producing Countries, Estimates. San Francisco, CA. 1998.
[2]FAO Production Yearbook, Vol. 51-1997.
[3]O.I.V. Bulletin Supplement 803-804. January and February, 1998.
[4]S.D. Shikhamany, National Research Center, Pume, India.
[5]Agriculture Attaché Query Detail, 1998.
[6]China Agricultural Yearbook, 1998.

tries prefer to use the term raisin when referring to the dried Thompson Seedless.

Raisin Production By Country

Raisins have been produced and used since biblical times originating in the Middle East. Production has spread to other parts of the world (**Figure 1.I.0** and **Table 1.I.1**), mostly to the United States, Turkey, Iran, China, Greece, Australia, South Africa, Chile, Afghanistan, India, and Mexico. These countries account for 92% of the world's total raisin production. Other countries of smaller production are Syria, Argentina, Spain, Cyprus, Morocco and Tunisia.

The major and minor raisin producing countries are discussed in this chapter in order of volume of production from largest to smallest.

While every effort was made to obtain information on all raisin-producing countries, we were not successful in every case. Unless noted otherwise, an expert, a representative or an attaché query supplied information for each country. The sources for each country are cited at the end of each section. Raisin producing countries where detailed information was not available include Pakistan, 4,000 tons (3,636 t); Albania, 3,000 tons (2,727 t); Tajikastan, 3,000 tons (2,727 t); Uzbekistan, 3,000 tons (2,727 t); Turkmenistan, 1,100 tons (1,000 t); and Yemen, 1,100 tons (1,000 t).

The world's leading raisin producing countries use the typical trade channels for exporting raisins. Per capita consumption by each country is based on domestic supply plus imports (**Table 1.I.5**).

1. United States

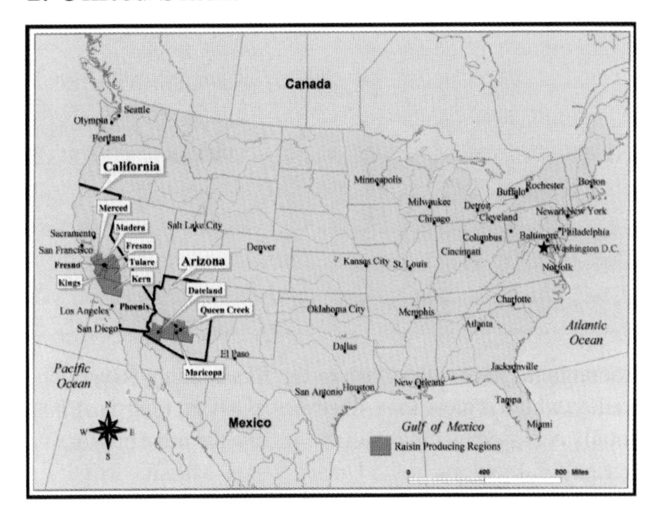

Fig. 1.1.0: United States raisin producing areas.

History

The United States (**Table 1.I.1**) is the world's leading raisin producer, averaging 350,000 to 400,000 tons (318,182 to 363,636 t) annually. This represents 36% of the world's production. All aspects of raisin production in the United States are discussed in subsequent chapters. California accounts for over 90% of total U.S. grape production (**Table 1.1.1**), using almost 3 million tons of grapes to produce approximately 400,000 tons (363,636 t) of raisins annually. Arizona is the only other state producing raisins at about 2,000 tons (1,818 t) per year.

Average area in raisin grape vines, 2000: 200,000 acres (80,940 ha), based on average yield of 2 tons/acre.

Average raisin production, 2000: 400,000 tons (363,636 t)

The Central Valley of California reaches north to the

Table 1.1.1
Total Grape Production by Type and State for the United States – 2000 - 2001.

State & Type	Total Production		Bearing Acres		Yield Per Acre[1]	
	2000 Tons	2001 Tons	2000 Acres	2001 Acres	2000 Tons	2001 Tons
Arizona	20,008	15,488	4,100	3,200	4.88	4.84
Arkansas	4,200	2,700	1,400	1,500	3.00	1.80
California	7,062,580	5,931,470	827,000	851,000	8.54	6.97
Wine	3,361,720	3,100,800	458,000	480,000	7.34	6.46
Table	774,300	697,060	89,000	91,000	8.70	7.66
Raisin	2,912,000	2,133,000	280,000	280,000	10.40	7.62
Georgia	3,504	3,201	1,200	1,100	6.98	2.35
Michigan	87,250	28,905	12,500	12,300	4.96	6.23
Missouri	2,950	2,297	850	870	3.47	2.64
New York	154,035	148,995	31,500	31,500	4.89	4.73
North Carolina	2,298	2,002	600	700	3.83	2.86
Ohio	7,700	6,000	2,000	2,000	3.85	3.00
Oregon	18,630	23,520	8,100	8,400	2.30	2.80
Pennsylvania	62,976	61,440	12,800	12,800	4.92	4.80
South Carolina	520	N/A	400	N/A	1.30	1.2
Washington(all)	264,880	283,200	44,000	48,000	6.02	5.90
Wine	90,000	100,080	20,000	24,000	4.50	4.17
Juice	174,960	183,120	24,000	24,000	7.29	7.73
US	**7,695,628**	**6,518,730**	**946,450**	**977,970**	**8.12**	**6.67**

[1]Yield is based on total production

Source: USDA Non Citrus and Nuts 2001Preliminary Summary, January, 2001.

Table 1.1.2
California Raisin Production by County, tons, 1994-1998.

Year	Fresno	Madera	Tulare	Kern	Kings	Merced
1994	317,000	60,581	34,300	9,360	5,800	2,860
1995	277,000	57,582	31,900	16,000	5,600	2,334
1996	233,000	61,957	34,200	2,500	6,500	1,244
1997	321,000	90,938	33,700	6,280	5,500	2,771
1998	206,000	52,569	24,600	3,150	6,500	1,574
Average	270,800	64,785	31,740	7,458	5,980	2,156
Percent	71.0	17.0	8.0	2.0	1.5	0.5

Source: County Agriculture Commissioner, Annual Reports from each county.

Sacramento River and south to the San Joaquin River. The valley, which is more than 400 miles (640 km) long, is almost totally enclosed by mountains and is one the most productive agricultural regions in the United States. Massive irrigation projects transfer water from the rainy northern half to the drier southern half of the valley. The heart of the California raisin industry is located in the San Joaquin Valley. Mountains to the east contain a watershed managed by a system of dams and reservoirs used to irrigate the crops on the valley floor.

Raisins were first produced in this region in 1873. **Chapter 2** provides a complete history of the development of the California raisin industry. Raisin production is centered in six counties in the central San Joaquin Valley (**Table 1.1.2**). Rank by county from 1994 to 1998 is Fresno, Madera, Tulare, Kern, Kings, and Merced (**Table 1.1.2**). Fresno and Madera Counties account for 71 and 17% of the production, respectively. Overall, the raisin industry is comprised of about 5,000 growers concentrated within a 75-mile (121 km) radius of the city of Fresno, California.

There are 21 raisin processing plants and 27 raisin dehydrators located in Fresno and Madera counties (**Appendix B and C**). Sun-Maid Growers of California is the world's largest raisin packer, processing 35% of the total California raisin crop. Ninety-percent of California raisins are naturally sun dried with the remaining portion being mechanically dried by gas-fired dehydrators. California's Central San Joaquin Valley is ideally suited for the production of natural sun-dried raisins because of its climate. The drying season extends from mid-August to mid-October. **Tables 1.1.3 a, b** and **c** illustrate an ideal weather pattern for natural sun drying. Warm temperatures prevail throughout the growing season and harvest, averaging 4,500 to 5,000 degree-days[1]. This, coupled with high sunshine intensity, low humidity, and minimal summer rainfall gives California raisin growers the weather conditions needed to dry the crop either on the ground or on the vine. These two methods of natural sun drying are described in detail in **Chapter 5**.

Characteristically, the California raisin grower carefully watches the ripening of the grapes from veraison to a point where sufficient sugar has been reached to begin harvest. It is

[1]Amerine and Winkler, 1944, established climate regions of California in degree-days which measures the length of time the thermometer remains over 50 °F between April 1 and October 31. Thus, if the mean temperature over a five day period was 70 °F, the "summation" of heat would be (70-50=20)x5=100 degree-days.

Table 1.1.3
Climatic Data for Fresno, California, 1961 – 1990.

Table 1.1.3a: Temperature and Daylight Data

Date	Mean Temp (°F)	Maximum Temp (°F)	Hours of Sunshine per day
August 15	80.5°	97.0°	13.57
September 1	78.0°	94.0°	12.83
September 15	74.5°	90.0°	12.42
October 1	70.5°	86.0°	11.80
October 15	65.5°	80.0°	11.28

Table 1.1.3b: Precipitation During the Drying Season

Date	Average Precipitation
July 1 to August 15	.01"
July 1 to August 31	.03"
July 1 to September 1	.04"
July 1 to September 15	.13"
July 1 to October 1	.29"
July 1 to October 15	.43"
July 1 to October 18	.49"
July 1 to October 19	.51"

Source: Brad Adams, Hydrometeorological Technician, Fresno, California.
Data from National Climatic Data Center, Asheville, North Carolina.

Table 1.1.3c: Relative Humidity

	Relative Humidity at 1600 hrs (4:00 p.m.)	Normal Rainfall
August	25%	0.03 inches
September	28%	0.24 inches
October	35%	0.53 inches

*Average annual rainfall July 31 through June 30 is 10.6 inches

always the compromise between fruit maturity and weather forecasts that influences the decision to harvest.

From August 15 to October 15, the average mean temperature decreases by 15 °F (9.4 °C); the hours of daily sunshine decrease by 2.17 hours; and the relative humidity increases by 10% based on 40 years of weather records. The average rainfall increases significantly from 0.13 inches on September 15 to 0.5 inches (3.3 mm to 12.7 mm) on October 19. Therefore, the later the harvest, the shorter the drying season. Using the Thompson Seedless variety as an example, most growers will harvest no later than September 15, a date when the fruit usually has reached 22% sugar and when there is sufficient time to sun dry the crop. Growers who purchase crop insurance must have their crop harvested no later than September 20 or their contracts will be invalid.

Those growers choosing to go to dehydrators will normally harvest their crops as late as they can to obtain maximum sugar content to increase yield. Varieties recently released by the USDA such as DOVine and Summer Muscat, and the present grower favorite, the Fiesta, all ripen 10 to 20 days earlier than Thompson Seedless (see **Chapter 3**). These new varieties are particularly adapted to drying on the vine (DOV) which virtually eliminates losses due to rain damage. Another added feature of the DOV process is that it lends itself to almost complete mechanization. This is a valuable economic factor considering the ever-increasing labor shortage in California. There are predictions that in the near future new varieties suitable for raisins that will dry naturally on the vine without cutting the fruit canes or requiring chemical drying aids will be produced (Ramming, personal communication).

While most of the raisins produced in California are from the six counties already mentioned, it must be noted that a small amount is produced in the very hot Coachella Valley, California's earliest ripening table grape district where temperatures soar to as high as 120 °F (49 °C). Because of the extremely high daytime temperatures, the table grape strippings left on the vine dry into raisins. It is not feasible to dry this fruit in the direct sunlight as the raisins will burn and have an undesirable caramelized flavor. This can also be a hazard in the San Joaquin Valley when temperatures on occasion reach 110 °F (43 °C).

As stated earlier, raisins have been produced in Arizona for many years, primarily as a secondary product from table grape vineyards. Currently raisins are produced on only about 4,000 acres (1,620 ha) in Arizona, but there are only about 750 acres (300 ha) dedicated to raisin production. Raisin production on the remaining acreage is from table grape strippings remaining after the table grape harvest. The total Arizona production in 1999 was about 2,000 tons (1,818 t) per year with one grower producing about 1,200 tons (1,100 t) on 700 acres (283 ha). Four growers produce most of the raisins with only two having vines dedicated to raisins and the others simply producing raisins by dry stripping table grapes after harvest. Most raisins are produced near Maricopa, Arizona, with some production near Phoenix and Dateland.

The climate where raisins are produced in Arizona can be classified as arid with an annual rainfall ranging from 5 inches (12.7 cm) at Dateland to 8 inches (20.3 cm) in the Phoenix and Maricopa areas. Rainfall is typified by a "monsoon" season, which generally begins in early July and produces about one-half of the annual precipitation between July 1 and September 15. Maximum temperatures in July run between 106 and 107 °F (41.1 and 41.6 °C). Minimum temperatures in December run between 34 and 36 °F (1.1 and 2.2 °C).

The primary varieties for raisin production are Thompson Seedless, Zante Currant, Flame Seedless and Perlette. Thompson Seedless and Zante Currant may be planted specifically for raisin production and Flame Seedless and Perlette are primarily planted as table grapes. Some new plantings of DOVine are being made near Maricopa. Vineyards that are dedicated to raisin production dry the raisins on paper trays in the same fashion as is practiced in California.

Sun Raisins is the only raisin processing plant in Arizona located near Queen Creek.

Credits and Sources

Information on Arizona provided by:

Dr. Donald C. Slack, Department of Agricultural and Biosystems Engineering, and Dr. Michael Kilby, both of the University of Arizona, and Kalem Barserian of American Vineyards, Arizona.

2. Turkey

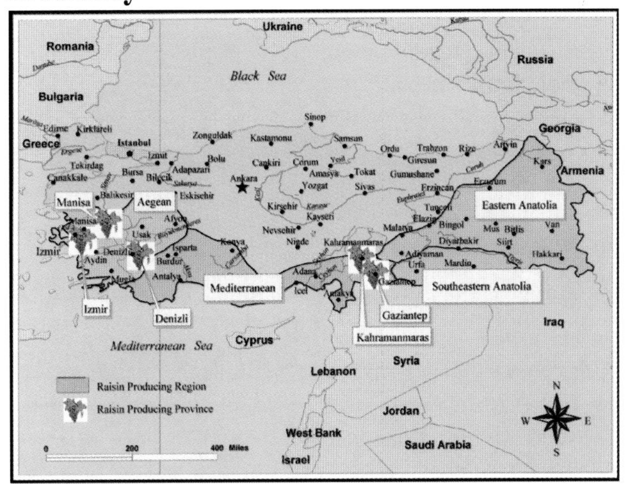

Fig. 1.2.0: Turkey raisin producing areas.

History

Archeological excavations show evidence of viticultural activity in Anatolia as early as 3500 B.C. and that viticulture was well developed by 1800 to 1500 B.C. when grapes and wine were offered as a votive to the Gods by the Hittites.

Today, Turkey is 2nd only to the United States in raisin production. The primary areas for raisin production are the provinces of Izmir, Manisa and Danizli in Turkey's Aegean region, and to a lesser extent in the regions of southeastern Anatolia, Eastern Anatolia and the Mediterranean region. The raisin grape areas are growing due to demand. The seedless grape growing area in particular continues to increase and constitutes approximately one-third of a total grape crop averaging about 3.4 million tons (3.1 t) annually. About 10% of the seedless grape crop is consumed fresh and the remainder is dried (Eris and Barut, 1994).

Average area in raisin grapevines, 1994-98: 163,020 acres (66,000 ha).

Average raisin production, 1994-98: 223,520 tons; (203,200 t). ***Production in 1998-99:*** 275,000 tons; ***Estimate for 1999-2000:*** 250,000 t. 209,000 tons, 190,000 t (1999 Sultana Conference, London, England).

Raisin Varieties - Aegean Region

The Aegean is an area where mountains descend abruptly into the sea, with a coastline of 1,750 miles (2,800 km). In this coastal region, the summers are hot and dry and the winters are moderately warm with an average rainfall of 25.5 inches (65 cm). The Aegean coastal region reaches 104 °F (40 °C) during the summer and can have winter minimum temperatures as low as -18.6 °F (-28.1 °C). Four major raisin grape varieties are grown in this region.

Yuvarlak Cekirdeksiz (Round Seedless) - Figure 1.2.1

This variety bears fruit that is white to light golden color. The cluster is large weighing 12.4 to 15.9 oz (350 to 450 g), shouldered, long cylindrical, and compact. The berries are small (1 to 3 g) and round with a neutral flavor, ripening early August. More than 90% raisins are made in the Aegean region come from this variety. It is Turkey's favored raisin grape and has an attractive bright color, tender, soft texture, and excellent quality. It is also used as a table grape. The vines are very vigorous and productive and cane pruning is required.

Fig.1.2.1: Yuvarlak Cekirdeksiz.

Sultani Cekirdeksiz (Thompson Seedless) - Figure 1.2.2

This variety is the same variety grown in California. It has the same descriptive characteristics as the Yuvarlak Cekirdeksiz except for smaller berry size (1 to 2 g). The berry shape is ellipsoidal elongated with a cluster density described as well filled to compact. In Turkey, it is known as a variation of Yuvarlak Cekirdeksiz and is acknowledged as the best known and extensively grown raisin variety in the world.

Fig. 1.2.2: Sultani Cekirdeksiz.

Cal Karasi - Figure 1.2.3

This is a black, seeded (two seeds) variety with ellipsoidal berries of medium size (2 to 3 g). It has a moderately tough skin and a neutral flavor. The clusters are large, 10.6 to 17.6 oz (300-500 g) and compact, and the fruit ripens mid-season. This variety produces well in the Denizli area of the Aegean region. Its' intense red color makes it suitable for bulk red wine. Some of the production is made into raisins or consumed as a table grape.

Fig. 1.2.3: Cal Karasi.

Iskenderiye Misketi (Muscat of Alexandria) - Figure 1.2.4

This is the same variety grown in California and described in **Chapter 3**. It is particularly adapted to the hot areas of Istanbul, Izmir and Manisa of the Aegean region. It is used commercially as a table grape and for raisins. As in other parts of the world, this variety sometimes experiences poor flower set with many shot berries, resulting in straggly clusters.

Fig. 1.2.4: Iskenderiye Misketi.

Raisin Varieties - Mediterranean Region

The Mediterranean region has mountains that extend to the coast in parallel formation with heights up to 9,961 ft (3,036 m). The length of the shoreline is 932 miles (1,500 km). Typical Mediterranean climate prevails in the coastal belt, with moderately warm winters and precipitation of 30.6 inches (776.8 mm) and hot, dry summers. Frost is very rare. Three major raisin grape varieties are grown in this region.

Sergi Karasi (Karalik, Milreri Siyahi and Lanlanrkara) - Figure 1.2.5

This variety is grown in the G. Antep and K. Maras districts of Turkey's Mediterranean region. It is a reddish-purple to black seeded variety with very large berries weighing 6 to 7 g each. It has a tough skin, is sweet, and is high in tannin. It is used commercially as a table grape, wine grape, and raisin grape. Other varieties grown to a lesser extent in the Mediterranean region are Ag Besni and Besni, described below.

Fig. 1.2.5: Sergi Karasi.

Raisin Varieties - Eastern Anatolia Region

This is the largest agricultural producing region in Turkey located at altitudes of 4,921to 8,202 ft (1,500 to 2,500 m) in 75% of the region. Land formation in Eastern Anatolia consists of mountain chains, large plateaus, plains and valleys. The Tigris and Euphrates rivers are vital water sources. The average winter low temperature is 49 °F (9.5 °C) and the average summer temperature is 72.5 °F (22.5 °C). Large temperature differences exist with maximums at 107.6 °F (42 °C) and minimums of -45.4 °F (-43 °C). Average rainfall is 22 inches (560 mm).

Besni (Peygamber and Bamba) - Figure 1.2.6

Besni is grown in southeast and eastern Anatolia. It is the most important seeded variety grown in southeastern Anatolia. Berries are white and large and weigh 0.21 to 0.25 oz (6 to 7 g) with an elongated ellipsoidal shape. The berries have a thin skin, neutral flavor and ripen mid-season (late August). The dried grapes are used for the production of distilled spirits or are consumed as a table grape.

Fig. 1.2.6: Besni.

Raisin Varieties - Southeastern Anatolia Region

This region is Turkey's smallest raisin producing area and has mountain ranges ranging from 1,640 to 2,625 ft (500 to 800 m). The eastern section of the region is more undulating and the terrain is rougher. The Tigris and Euphrates rivers are the most important rivers in the region and the South Anatolia water project, Turkey's largest, serves this region. The average low temperature is 61.5 °F (16.5 °C). The maximum temperature reaches 117 °F (47.6 °C) and minimums reach – 11.6 °F (-24.2 °C). Average total rainfall is 22.7 inches (576 mm). Three major varieties are cultivated in this region.

Ag Besni - Figure 1.2.7

Fruit produced by this variety has large white (5 to 6 g) berries that are of an elongated oval shape with moderately tough skin and neutral flavor. It ripens in late August to early September. It is used as a table grape and for raisins.

Fig. 1.2.7: Ag Besni.

Rumi - (Urumu) Figure 1.2.8

Rumi is grown in the G. Antep district. Its fruit is greenish-yellow, seeded (three to four) and the berries are round, medium-sized with a neutral flavor. The fruit ripens in early September. Rumi is used commercially as a wine grape, for unfermented grape juice, food products and raisins. The wines are of fair quality. Some of the crop is dried for distillation to produce wine spirits. Cane pruning is required.

Fig. 1.2.8: Rumi.

The previously described Sergi Karasi and Besni varieties are also grown for raisins in the southeastern region of Anatolia, but their production is of minor importance.

Of the eight varieties described, the two varieties that stand out are the Yuvarlak Cekirdeksiz (Round Seedless) and to a lesser extent, the Sultani Cekirdeksiz (Thompson Seedless). The Turkish people prefer the Round Seedless for the production of their sultanas, which are recognized worldwide. The other varieties mentioned, when dried, are distilled to produce grape spirits.

The production of seedless raisins/sultanas is reported by TEKEL (the state liquor monopoly) Deputy Director General Dr. Niyazi Adali. **Table 1.2.1** clearly shows the popularity of Turkish seedless raisins with an increase in production during the last five years (1994-99) of 93,500 tons (85,000 t). Turkey continues to lead the world in exports as shown in **Table 1.2.2.** Turkey's primary markets are Germany and England.

Table 1.2.1

Production of Seedless Raisins in Turkey, 1994-95 through 1998-99.

Year	US Tons
1994-95	181,881
1995-96	220,462
1996-97	242,508
1997-98	256,839
1998-99	275,578

Source: Dr. Niyazi Adali, Deputy Director General, Tekel, Unkapani, Istanbul, Turkey.

Table 1.2.2

Seedless Raisins Exports (World and Turkey), 1994-95 through 1998-99.

Years	World Exports	Turkish Exports	Turkey's Share (%)
1994-95	492,733	172,653	35.0
1995-96	524,700	188,159	35.8
1996-97	503,756	187,611	37.2
1997-98	498,245	213,107	42.8
1998-99	550,263[1]	163,358[2]	29.7

[1]Estimated
[2]First 5 months
Turkeys export is mostly to Germany and England.

Source: Niyazi Adal, Deputy Director General, TEKEL, Istanbul, Turkey.

Viticultural Aspects

Most Turkish vineyards are trained to the vase system (**Figure 1.2.9**) (Eris and Barut, 1993). High training systems (**Figure 1.2.10 a** and **b**) however are being used in the vineyards of the Aegean and Marmara regions and in the larger vineyards established in recent years in other regions (Eris and Barut, 1993).

Fig. 1.2.9: Vase Training System.

Fig. 1.2.10a: High "T" trellis (young vineyard).

Most vineyards need no irrigation due to sufficient rainfall. Advances in fertilization techniques have been made in recent years, especially in new vineyards. Fertilizers are applied based upon the results of foliar and soil analysis. Common diseases and pests in Turkish vineyards include Plasmopora viticola, Uncinila Necator, Botrytris cinerea, phylloxera, nematodes and Lobesia botrana. Turkey is deal-

Fig. 1.2.10b: High "T" trellis (Mature vineyard). Insert: Sultani Cekirdeksiz cluster.

Fig. 1.2.11: Turkish dipping facility

ing with phylloxera since it has proliferated in recent years, except in the Mediterranean region where resistant rootstocks are used. This pest is a hazard in all other regions, especially in central Anatolia and in the regions to its east. In recent years, the use of American rootstocks has increased due to the demonstration of their value to growers. The most commonly used American vine rootstocks are Rupestris du lot, 5BB, 420A, 99R, 110R, 140R, and 41B. The use of rootstocks such as Dogridge and 1103 Paulsen are also becoming common (Eris, 1992). As a result of more intensive cultivation through better irrigation and trellising, Turkey has experienced higher yields in its seedless grape production in recent years. Private processors have financed much-needed improvements in the industry. The industry is seeking to improve the quality of its raisins through improved growing, harvesting and drying techniques with the goal of producing a cleaner product. These techniques include trellising the vines, replacing sacks with small plastic harvest crates, and the use of concrete drying beds or plastic sheeting.

Dehydration and Processing

Harvest begins when the grapes are between 22 and 23% sugar, which occurs during late August through September. The grapes are harvested by hand, placed on a tractor drawn trailer and hauled to the dipping area. When the grapes reach the dipping area (**Figure 1.2.11**), they are dumped into the dipping solution. The dipping solution consists of 5 to 6% potassium carbonate (K_2CO_3) mixed in water. This solution is heated and mixed with olive oil ranging from 0.88 to 1.54 lbs (0.4-0.7 kg) per 26.4 gallons (100 L) of water. After dipping, the grapes are spread onto concrete slabs covered with paper, fine canvas, or a similar material (**Figure 1.2.12**). Unacceptable fruit is removed to insure quality. In some instances, a clean, compacted soil area is used instead of concrete. The grapes dry into raisins in five to 10 days depending on climatic conditions. At this point, they are brown to yellow-amber in color and have a moisture content of 10 to 13%. In recent years, growers have tried rack drying similar to the Australian method. This requires 10 to 20 days drying time. There is some interest among Turkish producers in mechanical dehydration to use gas or oil-fired dehydrators, but this process has not been found economical in Turkey.

Domestic Utilization

40,100 tons (37,000 t), 1.42 lbs (064 k) per capita, 1998.

Fig. 1.2.12: Grapes drying on concrete slabs covered with fine canvas.

About half of the domestic utilization is for confectionery consumption and the remainder is used for distillation. Domestic consumption fluctuates based on supply and quality according to the quantity TEKEL consumes for distillation each year.

Exports, t, 1998:		*Imports:*	
United States	518	Iran	4,500
United Kingdom	36,679	Not listed	500
Netherlands	29,712		
Germany	36,118		
Italy	16,576		
France	11,367		
Belgium	10,730		
Australia	6,891		
Spain	6,180		
Ireland	4,215		
Brazil	4,427		
Not listed	29,587		

Marketing

The private sector handles the majority of marketing in both domestic and import markets. The Izmir Commodity Exchange is the main cash market for raisins. There are numerous firms dealing in dried fruits in Turkey, but the top 10 control about 85% of the raisin market. Recently, there have been policy reforms to reduce the government's role in agriculture. The Turkish government no longer sets a procurement price for certain crops, including grapes. There is no direct export subsidy for raisins, although there is government subsidized credit given to producers and packers at lower than commercial interest rates by the Agricultural Bank of Turkey. To protect the domestic industry, the government levies a 58.5% import duty on the CIF value of raisin imports. Virtually all Turkish exports are shipped in bulk for industrial uses.

Information of Interest

Seedless grapes are generally grown on small farms that produce a variety of crops. Sources estimate that as many as 60,000 farmers produce seedless grapes, which equals an average farm size of about 3 acres (1.2 ha). There is high demand for Turkish raisins in both the export and domestic markets. Turkey has expanded the cultivation of raisin grapes, therefore land available for other purposes is decreasing.

Credits and Sources

Information on Turkey provided by:

Niyazi Adali, Department Director General, TEKEL, Istanbul, Turkey.

Sources:

Attila Eris and Erdogan Barut, "Grape and Small Fruit Growing in Turkey," Journal of Small Fruit & viticulture, Vol. 2(2) 1994; Susan Shayes, Agriculture Counselor, USDA Foreign Agricultural Service, GAIN Report, U S Embassy, Ankara, Turkey; A. Unal Sarigedik, Ag Specialist, Dried Fruit Annual Report, attaché query, American Embassy, Ankara, Turkey, 1996-1998; Aysun Aloc, administrative assistant, U S Embassy, Ankara, Turkey, 1998.

Photo Credits:

Figures 9 & 10a: P.R. Clingeleffer

3. Iran

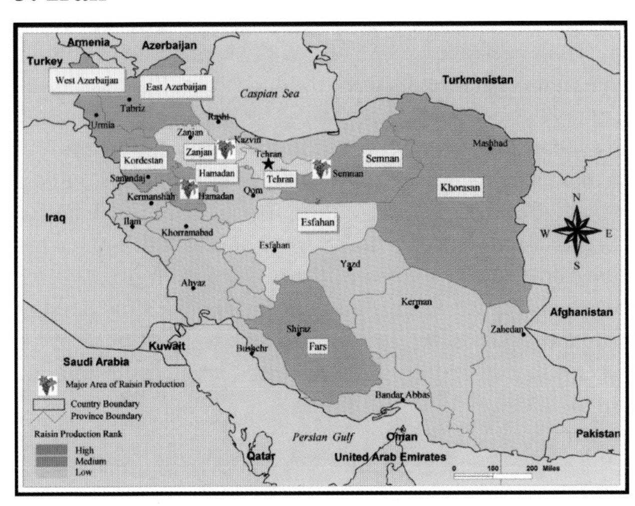

Fig.1.3.0: Iran raisin producing areas.

History

Iran produced 102,000 tons (92,727 t) in 1998. Raisin production in Iran dates back to Ancient Persia. There is no recorded date of the first production of raisins, however for many centuries raisins have been one of Iran's staple foods.

Grapes are cultivated throughout the country. The state of Fars ranks first in grape production accounting for 19.7% of this country's production. Khorasan, with 16.8%, ranks second. The other five states or provinces, in ranking order, are Kordistan, Kazvin, West Azerbaijan, East Azerbaijan and Hamadan. The major area of raisin production is centered around Semnon in the Khorasan province and Hamadan in the Kordistan province. These six provinces produce a total of 68.4% of the country's grapes. The remainder of the grape production is in the states of Tehran, Central Zanjan, Esfahan and Kordistan.

Area in raisin grape vines, 1996-1998 average: 61,750 acres (25,000 ha, estimated). ***Estimated for 1999 - 2000:*** 132,000 tons (120,000 t) (1999 International Conference of Sultana and Raisin Producing Countries, London, England).

Average raisin tonnage, 1996-1998: 102,000 tons; 92,727 t; 78% is exported, 22% is for domestic consumption.

The average estimated raisin production in Iran from 1996 to 1998 was 107,030 tons (97,300 t) (ICSRPC, conference minutes, 1996, 1997, 1998-99). Of this amount, 78% was estimated for export and the remaining 22% for domestic consumption.

Raisin varietals grown for commercial production in descending order of importance: Asgarry is a round to oval, white seedless, thin-skinned variety similar to Thompson Seedless (**Figure 1.3.1**); Shahamie is a black seeded variety similar to Monukka; Hossini Fakheri is red-seeded with medium to large berries; Pykannie, Sahebie, Kandahar and Rish Baba have large oval white-seeded berries; Yaghotie has small, round, red berries similar to Black Corinth, but is slightly larger (**Figure 1.3.2**). Most of the known varieties are named in the Farsi language, but it's possible for a variety to have several names throughout the country.

Viticultural Aspects

Most of the grapes are grown on the ground without any support from trellising (**Figure 1.3.3**), depending upon the weather conditions of the area. In the cold regions, the vines are planted on the bottom of a ridge. The ridge is formed from soil about 2 to 3 ft high against the wind to protect the plant from freezing. This also keeps the fruit from coming in contact with irrigation water.

Water is delivered in underground canals called "ghanuts". This method of subterranean water delivery has been practiced since ancient times in China, Iran, and some of the Asian countries. Due to adverse land and weather conditions, and

Fig. 1.3.1: Non-trellised Asgarry white seedless.

Fig. 1.3.2: Yaghotie grape cluster.

Fig. 1.3.3: Typical vineyard site in the Kazvin area.

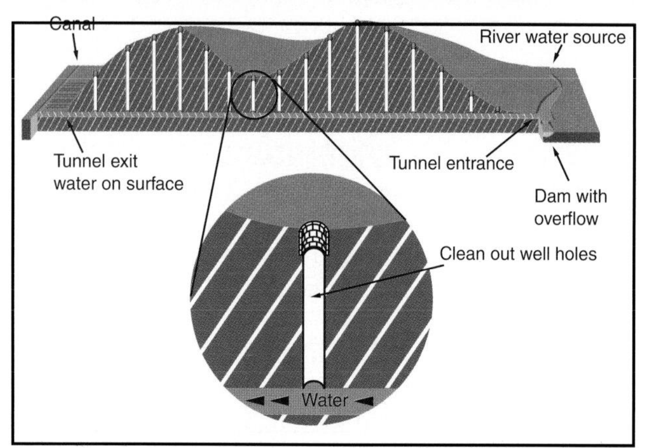

Fig. 1.3.4: Ghanut irrigation system.

Fig. 1.3.5: Ghanut clean out well hole.

Fig. 1.3.6: Vine in an individual irrigation basin. Parviz Aminian, right, and Dr. Mehdi Aminian, left.

the lack of heavy equipment, long distance above ground water delivery was not practiced. This system starts at a water source such as a river or aquifer. A series of closely spaced wells are dug in a path toward the area to be irrigated. The wells are then joined together by tunnels and the water flows by gravity to the targeted area, appearing at the surface (**Figure 1.3.4**). **Figure 1.3.5** shows a rock-lined hole for canal cleaning.

In warmer areas, grapes are planted on open, flat land. They are planted in individual basins for irrigation (**Figure 1.3.6**), have short trunks and are head pruned. Grapes are grown under both irrigated and non-irrigated conditions, depending on availability of irrigation water, topography and climate. The average yield from irrigated vineyards is 4.8 tons (4.36 t) of grapes per hectare compared to 1.7 tons (1.5 t) of fresh fruit per hectare in non-irrigated vineyards.

Dehydration and Processing

There are several methods of producing raisins in Iran. The fruit is harvested, taken to a dry-yard and spread on the ground to sun dry. Another method is to pick the fruit, dip it in a soda ash solution, spread the fruit on trays and dry the fruit in dehydrators. Some raisins are dried on racks in rain protected areas. Some raisins are processed by hand, but most are processed mechanically and packed in bulk using conventional processing equipment imported mainly from the United States. There are 43 raisin processing and packing facilities with a capacity of 81,730 tons (74,300 t). Most of them are located in East and West Aserbaijan, Tehran, and Khorasan.

Domestic Utilization:

The per capita consumption of fresh grapes and raisins combined is about 64 lbs of the fruit produced annually.

Exports, metric tons, 1998:	
Azerbaijan	3,700
Pakistan	6,016
Poland	6,848
Russia	6,848
Ukraine	16,227
United Arab Emirate	12,395
Other	27,627

Information of Interest

It's reported that 35% of Iran's total raisin production is not marketable because of the perishability of the fruit, the inaccessibility of some vineyards and both pre- and post-harvesting methods. There are no special organizations that focus on grape production in the various provinces. Hence, there is a great need for cooperatives and advisory organizations for grape and raisin production in Iran.

Credits and Sources

Information on Iran provided by:
Parviz Aminian, M.S., Viticulture, CSU, Fresno
Photo Credits:
Figures 1-5 : V. E. Petrucci

4. China

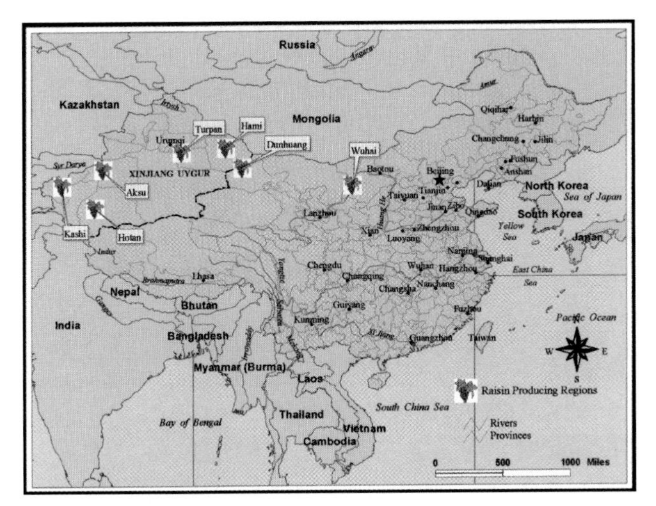

Fig. 1.4.0: China raisin producing areas.

History

China may very well be the 4th leading raisin producing country in the world; statistical data varies. Raisin production in China can be traced back to about 2,000 B.C. to the Han Dynasty and to the Xinjiang Uygur Autonomous Region, China's largest province with a total area of 637,560 square miles (1,650,000 square kilometers). Xinjiang enjoys a very diverse topography.

It has mountains such as Chogori peak with elevations of over 24,600 ft (7500 m), the famous mountain lake of Tianchi in the Tianshan Mountains, as well as valley lakes. The Taklamakan Desert, China's largest, is in Xinjiang as is the Turpan Oasis, the lowest elevation in China at 260 ft (79 m) below sea level.

China's grape producing area is located in the Turpan Depression at the southern foot of the Tianshan Mountains. Traditionally, the main production areas are in the Turpan and Hami basins and since the 1970s have been extending into the edge of Talimu oasis, in areas such as Hotan, Kashi, and Aksu. Raisins are also produced in Dunhuang, and Wuhai, Inner Mongolia.

Area in raisin grape vines: 32,000 acres (12,955 ha), estimated.

Raisin tonnage: 77,000 tons (70,000 t), estimated, is produced in mainland China of which 66,000 tons (60,000 t) are being produced in the Turpan Basin (China Agricultural Yearbook, 1998).

Raisin varietals grown for commercial production in descending order of importance: Thompson Seedless (Wuhebai) (**Figure 1.4.1** and **1.4.2**) accounts for 90% of raisin production. Other cultivars used for raisins include Jingzaojin, Wuzihei, Hetianhuong, and Hetianhuang. Varieties such as Longuan (Dragon's Eye), Nunai (Cow's Teat) and Suosuo, a variety resembling Black Corinth (Zante Currant) are also used in the Turpan region (Peter Dry, 1999).

Viticultural Aspects

Xinjiang has a continental climate in its temperate zones. The Turpan Valley (**Figure 1.4.3**) is located in a sterile, sandy desert but is irrigated by a 2,000 year-old irrigation system invented in ancient Persia (**Figure 1.3.3**). Water is brought by this system of underground channels or water tunnels from the northern Tianshan Mountains. These systems are called karez (**Figure 1.4.4**) in China and ghanut in Iran. There are about 1,600 karez wells in Xinjiang of which about 1,000 are in Turpan.

Temperatures can reach 120 °F (49 °C) in the summer months and as low as -40 °F (-40 °C) in the winter requiring protection in most areas. The vines are covered with soil during winter and uncovered at the beginning of spring. This task is simple where vines are grown on or close to the ground, but it becomes very labor intensive where the vines have been trained to an overhead trellis system (**Figure 1.4.5**). The vines must be detached from the trellis and buried during the winter and uncovered and reattached to the trellis in the spring. The light level reaches 3,000 light hours per year and rainfall is minimal, averaging 0.71 inches (18 mm) per year.

Many vineyards employ the multiple trunk system (**Figure 1.4.6**) which is practiced to a degree on the East Coast of the United States. When the main trunk becomes so rigid it can't be bent down to ground level, it is

Fig. 1.4.1: (below) Thompson Seedless (Wuhebai) in Hotan area.

Fig 1.4.2: (right) Typical clusters of Thompson Seedless (Wuhebai) in Kaski area.

Fig. 1.4.3: Turpan Valley vineyard plantings.

Fig. 1.4.4: Karez underground tunnel irrigation system.

replaced with a younger flexible trunk, normally rising from the base of the older, larger main trunk. Vines are often trained into small arbor systems and pruned to spurs (two to three buds or canes eight to 15 buds) depending on the fruiting habit of the variety. Other training methods include growing the vines on the ground where the vine trunks are basically grown in trenches (**Figure 1.4.7**) with the upper structure (arms, canes or spurs) grown slightly 2 to 3 ft (0.75 to 1 m) above the soil surface.

Dehydration and Processing

A number of drying methods are used. The most popular method is shade drying in earthen brick chambers called Qunje (**Figure 1.4.8**), having a roof and walls constructed with every other brick missing to allow warm breezes to flow freely through the chambers. The fruit is hung on racks within the drying chamber (**Figure 1.4.9**). This process takes from four to six weeks and produces the green raisins preferred by the Chinese market (**Figure 1.4.10**).

Fig. 1.4.5: Overhead trellis system.

Fig. 1.4.6: Multiple Trunk System.

Fig. 1.4.7: Vines growing in trenches.

Fig. 1.4.8: Qunje earthen brick drying chambers.

A second method (**Figure 1.4.11**) is similar to the Australian rack-dried system detailed in **Chapter 7**. A third method involves spreading the grapes on cement slabs or plastic sheets to dry in the sun, producing golden brown raisins (**Figure 1.4.12**).

Mechanically heated air dehydration is only used occasionally. It has been reported that chemical drying aids, which can reduce drying time by 50%, have been developed and patented by the Xinjiang Grape Research and Development Center.

Processing after drying consists of mechanical destemming, removal of all debris such as sand and leaves, washing and re-drying. In some instances, a 0.5% olive oil emulsion is applied before packaging. Hand sorting for color uniformity is done prior to packaging.

Fig. 1.4.9: Fruit being placed on Qunje drying racks.

Domestic Utilization

In China 80% of the raisins produced are consumed on the mainland and 20% are shipped to Hong Kong and Macao.

Credits and Sources

Information on China

Fig. 1.4.10: Qunje dried green raisins.

Fig. 1.4.11: Rack drying Thompson Seedless (Wuhebai).

Fig. 1.4.12: Sun drying of grapes on plastic sheets.

provided by:

Dr. Sanliang Gu, Ricchiuti Chair of Viticulture Research, Viticulture and Enology Research Center, CSU, Fresno and Zhichao Li, professor, Xinjiang Grape Research and Development Center, Xinjiang, China

Source:

Peter Dry, Australian *Wine Industry Journal*, vol. 14, #2, 1999

Photo Credits:

Figures 1, 2, 8, 9, 12: Zhichao Li; Figures 3, 4: Sino - Overseas Grape Vine & Wine Issn-7360; Figure 1.4.10: Randy Vaugh-Dotta; Figures 5, 6, 7: V.E. Petrucci

5. Greece

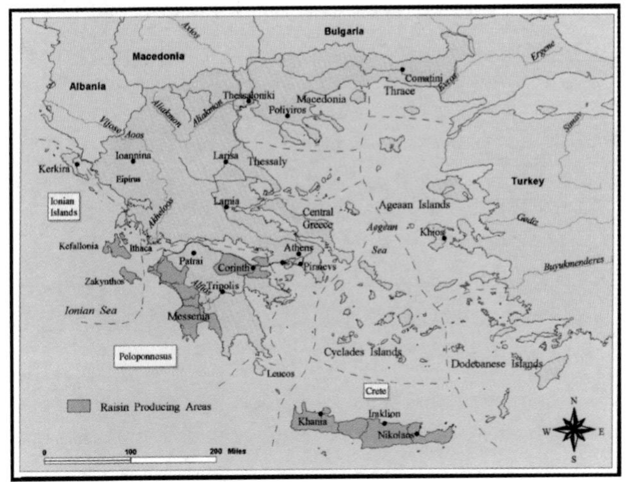

Fig 1.5.0: Greece raisin producing areas.

History

For many years, Greece was the 3rd leading country. However, a widespread outbreak of phylloxera several years ago sent Greek production of Sultanas into a rapid decline. Currently, Greek production of Sultanas is on the rise, and it is estimated that by the year 2000 between 85 and 90% of existing vines will be replaced with rootstock resistant to phylloxera. This should boost yields to an estimated 66,000 tons (60,000 t).

Greece may very well be the oldest grape producing country in Europe, second only to Turkey. Aristotle mentioned Corinthian grapes and Corinthian raisins were cultivated and traded in the northwest Peloponnesus in the 14th century. They were also grown in the Ionian Islands and in Morea, an area above Peloponnesus. In 1516, Corinthian raisins were introduced to Zakynthos and then to Kefallonia. Naming the variety Corinthian became an issue since it was also grown in other areas of Messina besides Corinth.

The first law regarding Corinthian raisins was passed in July of 1892 by the new Greek government to control the commercialization of this variety. Additional regulations were adopted in 1944 to protect Corinthian raisin production under the administration and order of the Greek Autonomous Raisin Organization. Today the growing and trade of raisins comes under the KOA (Public Trade Organization) and under the regulations of the European communities.

Greece's grapes are primarily for raisin production, the majority of which are Zante Currants produced in the Peloponnesus. Almost half of the remaining raisins/sultanas are produced on Crete.

Area in raisin grape vines, 1996: 101,073 acres (40,920 ha) (O.I.V. Bulletin, 1997).

Raisin tonnage, 1994-1997 average: 107,804 tons (98,004 t).

Raisin varietals grown for commercial production in descending order of importance: Black Corinth (Zante Currant), Thompson Seedless (Sultana) and Muscat of Alexandria.

Viticultural Aspects

Crete, an island located in the Aegean Sea 160 miles

Fig. 1.5.1: Drip irrigation on two-wire trellised vines.

Fig. 1.5.2: Three-cross arm "T" trellis.

(257.5 km) south of Athens, is Greece's major Sultana producing area. A chain of limestone mountains runs intermittently through the island from east to west. The northern side of Crete slopes gradually to a coastal plain and on the southern side drops steeply to a rocky shore. The large Sultana producing area of Iraklion lies in a relatively wide coastal plain on the north side of the mountains, roughly midway between the two other Sultana producing centers of Khania on the west and Sitea on the east.

The weather is well suited to raisin production characterized by hot, dry summers and mild, moist winters. The growing season is long, with a cumulative heat summation of 4,707 degree-days calculated from April 1 through October 31. Rainfall during the dry season, April to September is 3.24 inches (8.2 cm) and during the rainy season, October to March, totals 17.05 inches (43.3 cm). The total annual rainfall is 20.29 inches (51.5 cm).

About 85% of the soils in Crete are on limestone. The soils of the Iraklion area are also of calcareous origin. Most Iraklion soils are rich in potassium with varying amounts of phosphorus, but nitrogen content is low. Sultanas are grown on loamy soils, which predominate in the district. In Khania and Sitea, stony soils containing lime are used. Supplemental irrigation is required for optimum production. Drip irrigation is being practiced in the most modern vineyards (**Figure 1.5.1**). The average yield of Sultana raisins in Crete is about 2 tons (1.8 t) per acre, and is the same for the all-important Iraklion region. Lower yields of 1.5 tons (1.4 t) /acre are found in the less important western producing area around Khania. In eastern Crete, or Sitea, production of Sultanas is minor but because of irrigation, yields are 2.75 tons (6.2 t/ha) /acre.

About 70% of Greece's Sultanas are produced on Crete with the remaining 30% grown on the peninsula of Peloponnesus. The total number of vineyards is large and the average size vineyard is about 2 acres (0.81 ha).

In some vineyards, vines are spaced 6.56 x 6.56 ft (2.0 m x 2.0 m), which equals 1,012 vines per acre (2,500 vines/ha). In other vineyards, vines are spaced 8.5 x 4.6 ft (2.6 m x 1.4 m), which equals 1,117 vines per acre (2,759 vines/ha) (Costas Athanisiades, personal communication, April 2000).

Sultanas, because of their fruiting habit, are cane pruned with four to six canes averaging 10 to 12 buds and for each cane a two-bud renewal spur is left. The number of fruiting canes and renewal spurs is based on each individual vine's vigor and capacity. Trellising systems range from a simple two-wire vertical (**Figure 1.5.1**) to a more elaborate system (**Figure 1.5.2**) utilizing a three cross arm "T" trellis system. In the simple two-wire system, the first wire at a height of 24 inches (60 cm) and a second catch wire at 10 to 12 inches (25 to 30 cm).

The Peloponnesus produces about 30% of Greece's Sultanas and virtually all of the Greek currants (Zante Currant is also known as Black Corinth). The Peloponnesus is the large peninsula that juts out from the Greek mainland west of Athens and is separated from the mainland by the Gulf of Corinth on the north and the Corinth Canal on the east. This region is mountainous with only 20% cultivatable land. The climate is Mediterranean and rainfall is limited except in the higher elevations. The coastal plains receive 15 to 20 inches (38 to 51 cm) of rainfall in some areas and as much as 30 inches (76 cm) in others. Summers are hot and dry; winters are mild. The main area of production is the province of Corinth. Currants are also produced on the Ionian islands of Ithaca, Leucos, Zante and Cephalonia (Winkler et al., 1977). Sultanas are grown mostly on the plains, and currants are grown in the hills.

Vine spacing for Black Corinth (Zante Currant) is 5.6 x 5.6 ft (1.7 m x 1.7 m) which equals 1,389 vines per acre (3,430 vines/ha) (Costas Athanisiades, personal communication, April 2000). Cultivation of the older Currant vineyards is the same ancient practice of head or vase training, leaving three to four arms (Greek literature refers to arms as *cordons*) with spurs of two to three buds. The new plantings, however, follow a linear design similar to that of the Sultana, except that winter pruning consists of short spurs of two to three buds arising from a bilaterally cordon-trained vine. Since the currant yields are less than Sultana yields, larger acreages are required to supply market demands. Most farmers produce one or the other, but not both.

Dehydration and Processing

Black Corinth (Zante Currant): The fresh grapes are harvested by hand and transported to the drying area. Here they are dipped in an alkaline solution consisting of potassium carbonate and water at a concentration varying from 0.3 to 0.6%. The fruit is spread either on plastic sheets, concrete slabs or packed soil. Seventy-five percent of the Corinthian crop is sun dried in this manner. Another practice consists of

Fig. 1.5.3: Fruit being dipped prior to placement on drying rack.

drying the fruit in the shade on wire racks similar to the Australian method (**Figure 1.5.3**). Drying time is approximately 20 days. The raisins are then removed from the wire racks and exposed to the sun for about two days. A third method being practiced more recently is to cut the bunches from the canes and allow them to dry in the shade of the vine canopy. These raisins are of exceptional quality.

Thompson Seedless: The Thompson Seedless (Sultana) came to Greece from Asia Minor early in the 19th century. Cultivation spread to all parts of Peloponnesus but concentrated mostly in Crete.

The raisin and sultana processing is done with standard equipment, pre-cleaning, removal, washing and just enough drying to control moisture to satisfy the requirements for export. The raisins are then graded and standardized to meet Greek regulations. The raisins are normally packed into 5 to 15 kg containers.

Domestic Utilization

3,850 to 4,400 tons (3,500 – 4,000 t) are consumed annually. Sultanas are consumed as a snack with some used by bakeries and confectioneries.

Exports, metric tons, 1997:		***Imports:***	
European Union	25,349	United States	20
Australia	480	>European Union	161
Germany	15,093	United Kingdom	100
Serbia	236	Germany	61
France	3,511	Turkey	1,138
United Kingdom	4,781	Unlisted	585
Italy	1,149		
Other E U	815		

Marketing

In Crete, the Central Confederation of Sultana Cooperatives (KSOS) operates a $1.6 million EU funded program to improve raisin quality and sales. Introduced in 1997, this program includes a new packing system using plastic containers. In place of sacks, 35, 45 and 450 kg containers contributed to a better dried fruit quality. Over 100,000 of these containers have been distributed to farmers. Other program activities include training producers and traders, advertisement methods and exploring new production methods.

Information of interest

In 1997, direct payments to farmers replaced the guaranteed price system, according to the Ministry of Agriculture. Farms with a minimum yield of 2800 kg/ha were paid $3,791 and farms with a minimum yield of 1,300 kg/ha received $2,577 - $2,650. In 1998, the income supports were set at $3,716 and $2,679 to $2,813, respectively. Vineyards with low yield in remote/mountainous areas are being abandoned. Most acreage still in production is being replanted with healthy rootstock and irrigated. The most recent estimate by the KSOS is that 75% of the total crop area will return to normal production after the replanting and will result in higher yields and better quality.

Credits and Sources

Information on Greece provided by:

N.P. Psarros, Director Ministry of Agriculture, Athens, Greece. Interpreted by Eli Skofis, Harry Costis and Themis J. Michailides; Costas Athanisiades, Agriculture assistant, American Embassy, Athens, Greece; D.M. Rubel, *The Greek Raisin Industry.* Foreign Agriculture Service M-75. Dec, 1959

Sources:

Clay Hamilton, U.S. Embassy, Greece, Dried Fruit Annual Report, attaché query, 1998; Elizabeth Berry, USDA Foreign Agricultural Service GAIN Report, U.S. Embassy, Athens, Greece; A.J. Winkler, James A. Cook, W.M. Kliewer and Lloyd A. Lider. *General Viticulture.* 1974

Photo Credits:

Figures 1 and 2: E. Skofis; Figure 3: Carter Clary

6. Australia

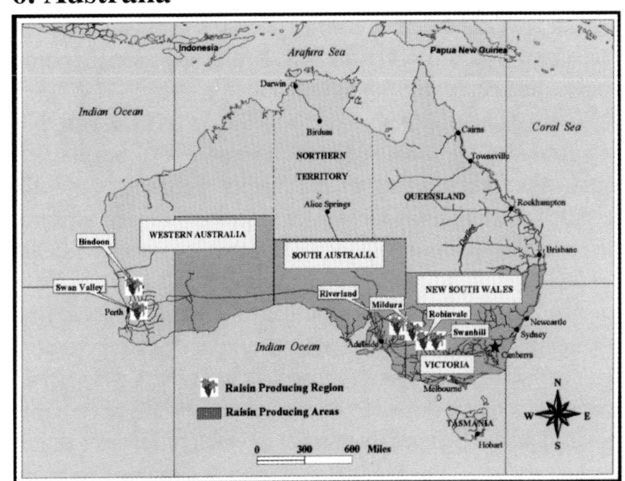

Fig 1.6.0 Australia raisin producing areas

History

Australia is the world's 6th largest raisin producer. Australia's raisin industry began in 1813 and is located mostly in the southeastern part of the continent in the Sunraysia region of NW Victoria and SW New South Wales, and the Riverland region of South Australia. A few raisins are grown in the Swan Valley and Bindoon region of Western Australia. The bulk of Australian raisins and half the currants, however, are produced in Sunraysia.

Fig 1.6.1: Rack drying Australian Sultanas, as pointed out by Vince P. Petrucci, 1969.

The Sultana vines originally planted in Australia came from vineyards established in the Cape of Good Hope, near Capetown in South Africa, which in turn came from Greenbryser, England. Known as Lady de Coverley's grape, Sultanas from this source were planted in Mildura for the first time in 1890. During this same period, Sultana cuttings were brought to Sunraysia from California. Growers, however, were advised not to plant the California material.

Average area in raisin grape vines, 1994-97: 20,000 acres (8,000 ha).

Average raisin tonnage, 1994-97: 47,400 tons (43,090 t)

Raisin varietals grown for commercial production in descending order of importance:

Sultana (41,060 tons, 37,327 t) is one of the first varieties to be planted in irrigated areas in 1888.

Zante Currant, also known as *Black Corinth* (4,168 tons; 3,790 t), is a very early maturing variety with very small black berries. They are widely planted in the Sunraysia region and also in Riverland and Western Australia. Production is not consistent due to poor fruit set, despite the use of setting sprays, and susceptibility to rain damage.

Carina, a CSIRO bred cross of Shiraz and Sultana, is a vigorous variety producing long narrow loose bunches of small, black berries that are not readily damaged by rain. It requires a setting spray for satisfactory crop yield and uniform berry size. Because of its loose bunches it is a good candidate for DOV.

Muscat Gordo Blanco (Muscat of Alexandria) was first planted in Mildura in 1888 and is one of Australia's most widely planted varieties. It is used for wine, table grapes and raisins. It must be de-seeded to be a commercially acceptable raisin. This is an expensive process, which produces a rather sticky product, subject to deterioration in storage. The main advantage of this variety is large size and Muscat flavor.

Walthan Cross, also known internationally as Rosaki, is a variety with large, white berries and a neutral flavor. In Australia it is marketed as a raisin along with Muscat Gordo Blanco. Approximately 2,200 tons (2,000 t) of these two varieties of raisins are produced annually (1994-98) (Ross Skinner, personal communication, Jan. 1999).

Merbein Seedless, a Sultana type, is also dried, and other wine and table grape varieties have been dried from time to time.

Fig 1.6.2: Curtains on drying racks to aid in drying completion of Sultanas.

Viticultural Aspects

In 1957, the CSIRO began an aggressive program to identify and evaluate the most desirable (superior) clones of the Sultana variety. Those clones showing leaf roll virus and other problems were eliminated and higher yielding virus-free clones became the standard clones used in Australia. About 20% of the Australian Sultana production is from grafted vines which provide higher yields. Rootstock effect on increased vine vigor gave Sultana raisin growers the ability to use taller and wider trellis systems, thus producing yields in excess of 3.9 tons per acre (9.6 t/ha).

Australia raisin and sultana production methods continue to improve due to the various canopy management systems, based on the taller, wider trellises, selection of the more vigorous clones, use of rootstocks, improved fertilization, irrigation and soil management techniques.

Australia, like most grape producing countries, has problems with grape diseases and insect pests. Growers who prefer natural control agents for aboveground pests practice chemical control on a limited basis. Rootstocks are used extensively where nematodes are present in damaging numbers in the sandy soils. Phylloxera has not been found in the dried fruit producing regions but remains a threat.

Dehydration and Processing

Rains during the ripening and drying season, often damage the grape crop. Rack drying (**Figure 1.6.1**) and the newer trellis drying systems continue to make sultana production a successful industry in Australia. In some instances, curtains are placed on the sides of the drying racks to channel hot air supplied by a gas-fired burner blower to complete the drying process (**Figure 1.6.2**). The dried fruit is then removed from the drying racks by a tractor-mounted shaker (**Figure 1.6.3**). Australia is world-famous for the production of light colored sultanas that are a distinctly different product from California's sun-dried Thompson Seedless. Clingeleffer describes the techniques used in Australia for producing light-colored, amber raisins in **Chapter 7**.

Domestic Utilization

Domestic consumption in 1997 was 40,000 tons (36,300

Fig 1.6.3: Sultanas being removed from rack by tractor-mounted shaker.

t) in a population of 19 million. While total dried fruit consumption has shown a slight increase in recent years, per capita consumption of dried sultanas has seen a slight drop.

Exports, metric tons, 1997:	
Portugal	62
Iran	3,111
Italy	83
Germany	1,741
Canada	2,010

Imports:	
United Kingdom	1,466
Turkey	6,587
New Zealand	1,198
Iran	3,111
Japan	457
Greece	1,249
Belgium	207
Chile	432
Netherlands	83
United States	96
France	41
Taiwan	83
Malaysia	113

Marketing

In 1997, Australia experienced the smallest sultana crop since 1928. This was attributed to several factors including limited sunlight and cool temperatures in November/December of 1996, a freeze in September of 1996, vine stress after a high yield 1996 crop, and diversion to winery use. Additionally, in 1996 some sultana vineyards were replaced with wine varieties. Three marketers and four packers serve the industry. The Australian Dried Fruit Board (ADFB) and the Overseas Trade Publicity Committee (OTPC) help promote and advertise Australian dried fruits overseas. Tariffs dropped from 23% in 1988 to the current 5% and no tariffs were assessed for less developed countries on January 7, 1996.

Information of Interest

Australia has been working on improving the quality of sultanas by tightening standards, increasing storage capacity of unprocessed sultanas, holding minimum stock levels, installing laser sorters and bin tipping of grower deliveries. The majority of Australia's imported sultanas is inferior to the Australian Sultana and competes at the low end of the market. They are used in the baking and cooking industries. However, imports depress the price of Australian sultanas sold in the local market.

Credits and Sources

Information on Australia provided by:

Ross Skinner, Dried Fruits Research and Development Council/Australian Dried Fruits Board, Mildura, Victoria, Australia; Scott Turner, Agricultural Specialist, Embassy of the United States of America, Office of the Agricultural Counselor, Canberra, Australia; James Truran, Dried Fruit Annual Report, attaché query, 1998

Photo Credits

Figures 1-3: V.E. Petrucci

7. South Africa

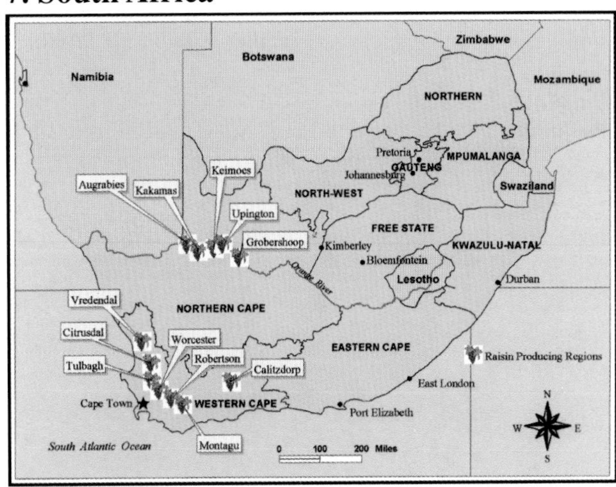

Fig 1.7.0: South Africa raisin producing areas.

History

South Africa is the 7th largest raisin producer in the world. Founder of the Dutch colony, Jan van Riebeeck, introduced grapes to South Africa in 1655. Three years after Riebeeck landed at the Cape of Good Hope, he planted 1,200 vines, among them the Muscat of Alexandria. Although Cape raisins were popular at the court of Queen Victoria, there was no organized export or marketing of dried fruit. Krone and Company erected the first packing house for dried fruit near Worcester in 1875.

Near the end of the 19th century, Harry Pickstone of California influenced people, like grower Piet Cilliers of Wellington, to visit California to gain expertise. However, Senator A.S. Brink is seen as the founder of the grape industry along the Orange River, since he planted the first Sultanina (Thompson Seedless) vines at Keimoes in 1919.

The main production area is along the Orange River which stretches about 106 miles (170 km) to the east and 99 miles (160 km) to the west of Upington. Total Sultanina plantings in this area are about 29,640 acres (12,000 ha), but this includes wine and table grapes (**Table 1.7.1**).

Area in raisin grape vines, 1998: 38,883 acres (15,742 ha).

Average raisin tonnage, 1994-1998: 37,802 tons (34,365 t).

Raisin varietals grown for commercial production in descending order of importance (**Table 1.7.2**):

Sultanas: Sultanina (Thompson Seedless), Merbein Seed-

Table 1.7.1
Production of Raisin Grapes by Region, dry tons, 1996.

Region	Area	Sultanas	Currants	Raisins
Western Cape	Vredendal	657.1	1,589.9	249.3
	Citrusdal	9.0		4.2
	Tulbagh/Worcester		15.5	
	Robertson/Montagu	0.7	2.8	
	Calitzdorp	0.4	4.9	2.7
	Other		1.4	
Northern Cape	Groblershoop	1,390.2	7.6	0.5
	Karos	2,815.0	1.2	1.2
	Upington	6,113.5	0.0	34.7
	Neilersdrift	2,863.2		
	Kelmoes	6,573.9	0.3	
	Kakamas	4,647.7	0.1	4.8
	Namibia	147.7		1.2
	Augrabies	6,013.2	0.0	0.6
Total		**31,231.6**	**1,623.6**	**299.1**

Table 1.7.2
Production of Raisin Grapes in South Africa, dry tons, 1992-98.

Year	Thompson Seedless	Unbleached Sultanas[1]	Golden Sultanas	Currants	Muscat & Monukka	Total
1994	20,904	5,649	4,882	1,263	307	33,005
1995	25,951	5,890	6,731	1,117	328	40,017
1996	19,586	3,593	5,154	1,473	271	30,077
1997	26,321	11,240	6,518	1,543	330	45,952
1998	13,347	2,934	6,508	1,130	208	24,127
Mean	20,950	5,861	5,959	1,306	289	34,365

[1]OR and WP unbleached sultanas

less, Fiesta. Sultanina clones H4, H5 and M12 from Australia and TS 66 from the United States as well as a local selection (14/2) are generally used. Sultanina comprises 95% of the total raisin grape production in South Africa.

Monukka type: Black Monukka, Flame Seedless.

Currant type: Zante Currants, Cape Currants.

Others: Muscat of Alexandria and Datal.

Hanepoot grapes produce a dark brown raisin, which accounts for less than 1,100 tons (1,000 t).

Viticultural Aspects

Many of the older plantings are own-rooted. Sultanina is planted on a 9.8 x 6.6 ft (3 m x 2 m) spacing and trained to a T-trellis. A larger gable system is being used to increase production. The vines are flood or furrow irrigated. New plantings are on 143 B, Ramsey and 99 Richter rootstocks because of nematode and phylloxera infestation. All harvesting and pruning is still done manually, but increasing production costs have stimulated grower interest in new trellis systems and mechanical harvesting.

Dehydration and Processing

Thompson Seedless grapes are harvested at 22 to 23% sugar and laid out on natural stone or concrete slopes in full sun until they are dry. No chemicals are used and the final product is very dark. Orange River unbleached Sultanas are

Fig 1.7.1: Bleached Sultanas.

harvested when the sugar reaches 18 to 20%. They are dipped in a cold lye solution of Sultanol oil and potassium carbonate and placed on wire racks in full or semi-shade. The final product is light brown. Western Cape Sultanas are harvested when the sugar reaches between 18 and 20% and treated with a hot lye mixture before drying in semi-shade. Golden or bleached Sultanas (**Figure 1.7.1**) are dipped the same way as unbleached (Orange River) Sultanas, placed on drying racks, bleached with SO_2, and dried in the shade.

All the raisins/sultanas are processed in Upington. Raisins are delivered in half-ton bins or stored at one of its depots. Processing is continuous all year. The raisins are washed using equipment similar to that used in California. Laser scanning equipment does screening for foreign material and color. The factories can process up to 1,100 tons (1,000 t) of raisins weekly (**Figure 1.7.2**). Packaging varies from 2.1 oz (60 g) containers to 4.6 lbs (2 kg) containers for retail sale to bulk packaging ranging from 6.6 to 33 lbs (3 to 15 kg) cartons (**Figure 1.7.3**).

Domestic Utilization

Utilization is 40% domestic and 60% export. In 1998, domestic consumption was 9,928 tons (9,025 t). The majority of domestically consumed raisins are consumed as ingredients in foods.

Exports, metric tons, 1997: ***Imports:* -0-**

United States	1,142
United Kingdom	6,099
Singapore	765
Japan	1,912
Austria	332
Netherlands	2,557
Canada	2,421
Norway	546
Germany	7,642
New Zealand	528
Other	6,239

Marketing

In 1907, the farmers in the Western Cape established a cooperative and the following year the South African Dried Fruit Company was founded. In 1962, the company's name was changed to the South African Dried Fruit Cooperative

Fig 1.7.2: Raisin packaging plant.

Fig 1.7.3: Raisin packaging line using two-ounce containers.

LTD (SAD). The South African Dried Fruit Board, founded in 1938, appointed the SAD as its marketing agent. This insured a uniform standard for all grades of dried fruit and, because only one packing plant handled the dried fruit, it could be done at a lower unit cost. The Board's main functions were the creation of stabilization funds, promotion, inspection and quality control, and supervision of distribution. Changes in the Marketing Act mandated the formation of a non-profit company called the Dried Fruit Technical Services, which replaced SAD. Under the New Agricultural Products Marketing Act, South African domestic dried fruit marketing has been liberalized. This has encouraged farmers to market their products through institutions of their own choosing and has widened domestic marketing. The United States may become more interested in entering the South African market since raisins imported into South Africa are duty free.

Information of Interest

In 1981, the industry founded an experimental/demonstration farm at Upington. New cultural practices are being developed and demonstrated to producers and research findings are tested on a large scale before implementation by the industry. New varieties and clones are evaluated on the experimental farm and computer controlled micro-irrigation was pioneered here. The farm also has a nursery, which supplies producers with the best available planting material. The raisin grape industry assists in identifying research priorities and sponsors research projects conducted by the Institute for Fruit, Vine and Wine, as well as by universities.

Credits and Sources

Information on South Africa provided by:

J. T. Loubser, Ph.D, ARC Fruit and Vine and Wine Research Institute, S. A.

Source: Joseph Lopez, Dried Fruit Annual Report, Attaché Query, 1996-98

Photo Credits:

Figures 1-3: V.E. Petrucci

8. Chile

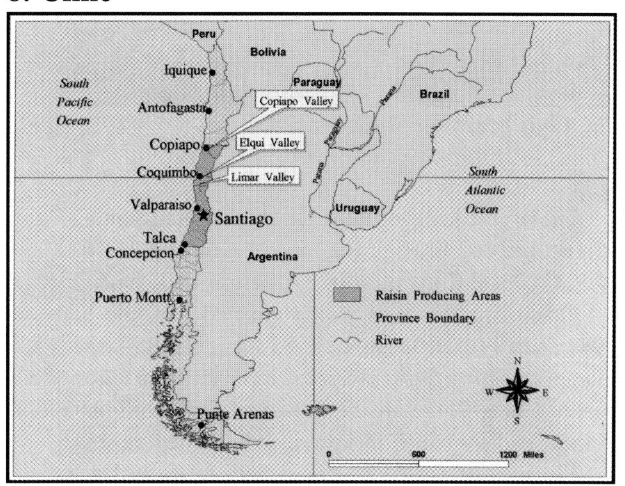

Fig. 1.8.0: Chile raisin producing areas.

History

Chile is the 8th largest producer. The majority of Chile's vineyards are located between 33° and 38° south latitude in the valleys dividing the coastal hills from the Andes. Raisins are produced in the warmer, northern districts. Raisin grape production in Chile began more than 200 years ago in the northern valleys of Elqui, Copiapo and Limar where a sun dried fruit made from Muscat type varieties were produced. Chile began exporting raisins about 50 years ago and the exports grew along with the table grape exports. In 1983, Chile exported about 2,750 tons (2,500 t) of raisins. By 1998, Chile's export production increased to 31,900 tons (27,500 t). The production level for export can vary considerably due to fluctuating diversions to the winery or table grape markets. Although Chile's participation in the world raisin market is minor, its advantageous climatic conditions, excellent soil and water resources and the geographic isolation of its vineyards gives this country great potential in raisin production.

Most of the total dried fruit production in Chile occurs in six regions in the northern and central parts of Chile. Table grapes and raisins are produced from Copiapo, 528 miles (850 k) north of Santiago, to Talca 186 miles (300k) south of Santiago.

Area in raisin grape vines, 1998: 81,543 acres (33,000 ha).

Average raisin tonnage, 1994-1997: 37,015 tons (33,650 t).

Raisin varietals grown for commercial production in descending order of importance:

Thompson Seedless represents 43% of the table grape acreage of 52,223 acres (21,143 ha) from which an average of 37,015 tons (33,650 t) of raisins are produced as a secondary

Fig. 1.8.1: Furrow irrigation in Chile.

Fig. 1.8.2: Overhead trellis supporting pruned Thompson Seedless.

crop.

The largest acreage of this variety is around Santiago with 15,166 acres (6,140 ha). Region six cultivates 11,167 acres (4,521 ha) and Coquimbo 10,878 acres (4,404 ha).

Flame Seedless was widely planted in Chile between 1984 and 1994. Between the 1983-84 and 1985-86 seasons, Flame Seedless exports increased by 874% from the northern part of Chile. There are 22,050 acres (8,927 ha) planted to Flame Seedless which make very good brown raisins.

Fiesta is harvested in November and early December, earlier than the other seedless varieties. The last fruit comes from Curico in March.

Other table grape varieties such as Perlette and Superior Seedless are also used for raisins.

Viticultural Aspects

In general, the grape growing regions of Chile have mild to moderate winters, rainfall varying from 2 inches (50 mm) at Copiapo to 19.7 inches (500 mm) at Talca. The climate, which resembles that of California, is very dry with a long growing season, warm to hot summers, very low spring rainfall and practically rainless summers. There is high effective heat summation during the growing season and disease is easy to control.

Soils are alluvial and irrigation systems are fed by gravity from the streams in the Andes. Vineyards are irrigated using furrows or basins (**Figure 1.8.1**) or by drippers or micro-sprinklers. The furrow system, which requires less investment, is the most common. Drip irrigation has expanded rapidly in the last decade mainly in the northern desert areas where water and good soils are scarce. Many drip-irrigated vineyards are planted on steep terrain where furrow methods of irrigation are impractical.

The multiple Guyot is the standard pruning system for cane pruned varieties like Thompson Seedless and Superior Seedless. Some fruitful cultivars like Ribier and Flame Seedless are either spur or cane pruned. Thompson Seedless typically has four arms and four diagonals that come out of these arms (**Figure 1.8.2**). From each of these, one to three canes of 12 to 24 buds are left at pruning. In northern Chile, where the vines are less vigorous, 12 canes of 12 buds are left giving 144 buds plus 12 buds from the spurs. In central Chile

Fig. 1.8.3: Overhead trellis supporting pre-harvest Thompson Seedless.

where the vines are more vigorous, twice as many buds per vine are left. Flame Seedless is pruned to 256 buds. Most growers prune according to crop forecasts made by a dissection of the buds and a microscopic examination that estimates bud fertility. The summer pruning of table grape and raisin grapevines is frequently carried out and consists of removing shoots, water sprouts and leaves. Trunk suckering is performed in early spring after bud break and before bloom. Excess nitrogen combined with an abundant water supply induces growth of water sprouts from the base of the trunk of young and vigorous vines. This can also occur as a result of a very short pruning in vines with a high percentage of blind buds or bud necrosis. Head suckering, or the removal of shoots that arise from the buds in older wood in the permanent parts of the vine, the arms and the place where the arms separate or head of the vine, is also done before flowering. The removal of undesirable succulent shoots in a radius of 15.7 inches to 23.6 inches (40 to 60 cm) around the head permits a better sunlight penetration and better light exposure to the shoots that are going to be the fruiting canes the following season.

The overhead trellis, similar to the parral of Argentina and the parronal of Spain, is by far the most popular trellis system (**Figure 1.8.3**). This system is suitable for moderate to high vigor varieties. When properly managed, it provides for good light distribution over a large leaf area within the canopy,

Fig. 1.8.4: Grapes sundrying on concrete slabs.

Fig. 1.8.5: Grapes sundrying on plastic film on hillsides.

which is very efficient for photosynthesis under these conditions. It can produce a high yield and quality of both table and raisin grapes. The system permits the clusters to hang free so that the berries are free from wind scaring, the bloom is not rubbed away, and the berries are easy to manipulate and spray. Canopy density can be manipulated. Vigorous cultivars such as Thompson, Flame Seedless, and Superior Seedless grown in central Chile, and trained on the parronal trellis system use a 13.1 x 13.1 ft vine spacing with 253 vines per acre (4 x 4 m with 625 vines/ha). Recently, vines have been planted at 10.8 x 9.8 ft (3.3 x 3.0 m), even in good soils. With varieties of less vigor, like Ribier, Black Seedless or Red Globe, the recommendation is 11.5 x 8.2 ft with 462 vines per acre (3.5 x 2.5 m with 1,142 vines/ha). In the northern districts where good soils and water quality are limited and vine vigor is low, the distance is 11.5 x 6.5 ft (3.5 x 2.0 m) for Thompson Seedless and Flame Seedless.

Dehydration and Processing

Currently, about 75% of Chile's raisin production is sun dried. Grapes are placed on concrete or clay slabs (**Figure 1.8.4**) or on plastic film on hillsides (**Figure 1.8.5**). When sun dried no additives are used. Normal drying time is 15 days. Tunnel drying is also practiced with and without SO_2. Two types of raisins are produced based on the color; golden raisins are called "rubias," (**Figure 1.8.6**) and brown raisins are called "morenas" (**Figure 1.8.7**). After drying is complete, the raisins are re-hydrated slightly and washed at 176 °F (80 °C). The raisins are then mechanically destemmed. State

Fig. 1.8.6: Golden raisins (Rubias).

Fig. 1.8.7: Brown raisins (Morenas).

of the art processing machinery purchased from the United States is utilized. In some cases mineral or vegetable oil is applied to the raisins before packaging. The processed raisins are packaged in 22 lbs (10 kg) cardboard boxes. Raisins are classified according to size ranging from "corinto," <0.28 inches (<7 mm), small 0.31 to 0.35 inches (8-9 mm), medium 0.36 to 0.43 inches (9.1-11 mm) and jumbo >0.43 inches (>11 mm). According to Juan Eduardo Laso M of Del Monte Fresh Produce in Santiago, Chile, over 50% of Chilean raisins are over 11.5 mm in size.

Domestic Utilization

Twenty percent of the production is used domestically. Statistics on per capita consumption are not available. Chilean raisin production is mainly oriented to satisfy the fruit cocktail market.

Exports, metric tons, 1997:

United States	2,888
Netherlands	1,234
Brazil	4,043
Ecuador	11,116
Mexico	3,905
United Kingdom	1,101
Colombia	3,783
Canada	740
Peru	3,710
France	629
Venezuela	2,222

Marketing

Over half of Chile's raisins are made from large grapes, which are low in demand and price on the world market. Most exporters try to maintain close to zero stock levels. A flat 11% import tariff is charged for raisins. There is also an 18% value added tax on all consumer items, both domestic and imported.

Credits and Sources

Information on Chile provided by:

Roger P. Kerneur, Farrior Farms Chowchilla, California and Jorge Perez-Harvey, Ph.D. Facultad de Agronomia e Ingenieria Forestal, Pontificia Universidad Catolica, Chile.

Sources: Luis Hennicke, Dried Fruit Annual Report. Attaché Queries, 1996-1998; Jerold Rebensdorf, Fresno Raisin Cooperative, Fresno, California; Juan Eduardo Laso M., Del Monte Fresh Produce (Chile) S.A., Santiago, Chile.

Photo Credits: Figures 1-3: V.E. Petrucci, Figures 4, 5: J. E. Lasso M.; Figures 6, 7: Randy Vaughn-Dotta.

9. Afghanistan

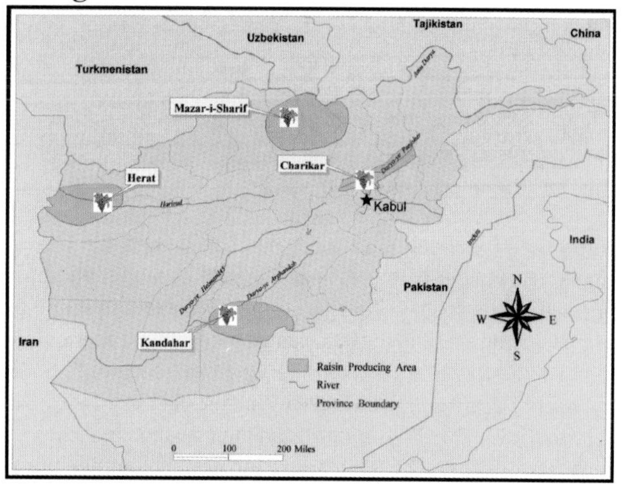

Fig. 1.9.0: Afghanistan raisin producing areas.

History

In 1980, Afghanistan was the world's 5th ranked producer of raisins, producing over 66,200 tons (60,200 t) (O.I.V. 1996), and raisins were the country's leading source of foreign exchange and trade credits. The principal export markets were the former Soviet Union, Pakistan and India, representing a total sales of 59,620 tons (54,200 t). Afghanistan still ranks 9th in world raisin production with a total production of 30,800 tons (28,000 t) and is exporting 22,000 tons (20,000 t).

Grapes for all uses are widely cultivated, as indicated by a bearing acreage of 118,560 acres (48,000 ha) (O.I.V., 1996). Grapes for raisins account for about 15% of the acreage. The main centers of production are located in Kandahar in the southeast, Herat in the northwest, Mazar-i-sharif in the north, and in a belt stretching south from Charikar at the mouth of the Panjshir Valley to Mir-bacha-kot, north of Kabul. These are also the main population centers due to the availability of water and cultivatable land.

Area in raisin grape vines: 17,290 acres (7000 ha) (O.I.V. 1998).

Raisin tonnage: 30,800 tons (28,000 t) (O.I.V., 1998).

Raisin varietals grown for commercial production in descending order of importance: Thompson Seedless accounts for the majority of the raisin production.

Viticultural Aspects

Training methods vary in response to limiting climatic factors including sandstorms in the region of Herat and cold winter temperatures in Mazar-i-sharif and the area north of Kabul. Trellising is not used and vine support is found only in the Herat region where the vines are trained to a vertical cordon against the side of earthen brick supports some 78 to 120 inches (2 to 3 m) in height and approximately 39 inches (1 m) in width. The walls are oriented to shield the vines from sandstorms.

Elsewhere, vines are planted on berms 56 inches (142 cm) high and about 25 inches (63.5 cm) across created by deep ditches dug down on either side of the berm (**Figure 1.9.1**). The ditches are cross-diked about every four to six plants. These ditches are filled two-thirds to three-fourths full to irrigate the vines. Mature vines are 3 to 4 ft tall (0.9 to 1.2 m) and are not trellised, but are elevated about 20 inches (0.5 m)

Fig. 1.9.1: Trenched soil berm "trellis".

Fig. 1.9.2: Sun drying grapes on packed earth.

Fig. 1.9.3: Adobe brick drying chamber with perforated walls.

by weaving the dormant canes into a basket shape about 3 ft (0.9 m) in diameter. This low style of training facilitates burying canes along the row in autumn to protect the vines from winter injury. The vines are pruned in spring to six to eight bud spurs, which protrude tangentially from the perimeter of the basket.

Irrigation is accomplished by diversion from streams. Water is circulated among the growers of a district on a periodic basis. The seasonal nature of the water supply has created a vineyard architecture dominated by water storage considerations. Grapes are not generally dried in the field, so a row spacing of about 10 to 13 ft (3 to 4 m) is used to accommodate the irrigation basins. The high relief of the root zone has the deleterious effect of concentrating salts.

Dehydration and Processing

At harvest, grapes are either sun dried or shade dried. In either case, they are brought from the vineyard to a central location where they can be guarded to avoid pilfering during the drying process. Sun-dried raisins are generally dried on packed earth and occasionally on mats (**Figure 1.9.2**).

Afghanistan's shade-dried raisin is a unique product. Drying is accomplished by hanging the clusters on cane scaffolding in a structure built for this purpose. These structures are built of adobe brick with walls perforated by leaving out alternate bricks from each course (**Figure 1.9.3**). The structure is roofed similar to the Chinese earthen brick chambers called Qunje (**Figure 1.4.6**). The resulting raisin is green to greenish gold in color and retains the flavors of the fresh fruit. These are generally packed on the cluster in sawdust. The majority is exported to India.

Processing machinery is standard and much of it has been imported from California.

Credits and Sources

Information on Afghanistan provided by:

Robert B. Siegfried, Assistant Engineer, Santa Clara Valley Water District.

10. India

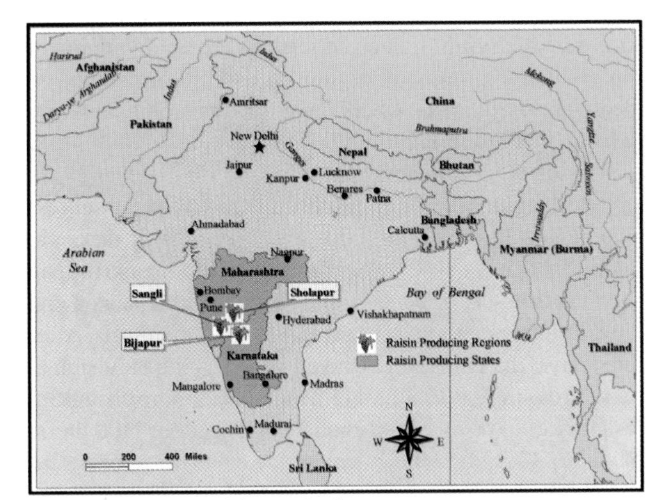

Fig. 1.10.0: India raisin producing areas.

History

India ranks 10th in world raisin production. Grapes are grown over an area of 85,000 acres (34,413 ha) in India with an annual production of one million fresh tons. Seedless grapes account for about 60% of the area and 50% of the production. Approximately 30% of the seedless grapes and less than 1% of seeded grapes are dried for raisins; the rest is for table use. Raisin production is confined to very hot and dry tropical regions of the country including the Snagli and Solapur districts of Maharashtra and Bijapur in Karnataka, where the maximum temperature is 107 °F (42 °C). The relative humidity is less than 15% during harvest and drying.

Tonnage produced, 1998: 31,020 tons (28,200 t).

Raisin varietals grown for commercial production in descending order of importance: Thompson Seedless and its local clones Tas-A-Ganesh, Sonaka and Manik Chaman. No raisins are made from Perlette.

Viticultural Aspects

Small, round, green and pulpy raisins with a coating of natural bloom and thin skin are preferred in India. To produce loose clusters to facilitate the drying process, a 10 ppm gibberellic acid (GA3) spray is applied at 132 to 158 gallons per acre (1,220-1,344 L/ha) 15 days after budbreak to elongate the clusters. Berry thinning either by GA3 or manually is not practiced. To obtain spherical berries, GA3 is not sprayed at bloom or after berry set to avoid elongation of the berries. To produce berries with thin skin, a post bloom application of GA3 is avoided. Berry thinning and girdling are also avoided in order to restrict the size of berries and the thickness of skin.

Adequate leaf/fruit ratio is maintained to achieve increased sugar content in the berries. Clusters are tipped by one-quarter to one-third or sometimes by one-half of their length depending on the number of leaves available for nourishing the cluster. Generally one-fourth of the cluster is clipped when 15 leaves are available, one-third is clipped if 12 and one-half if eight to nine.

Approximately 165 to 220 lbs (75-100 kg) of potash is applied per acre depending upon the petiole level of potassium at bloom. Half the dose is applied at 8 mm size of the berries and the remaining half at berry softening. Clusters are retained on the vine until the total soluble solids content of berries reaches 22 °Brix.

Clusters are protected from direct sunlight to avoid the destruction of chlorophyll to maintain green berries. Sunburn of berries is not a problem in the bower (tall trellis) trained vines with good canopy coverage (**Figure 1.10.1**). In case of sparse coverage of the canopy in either bower T or Y trellises, shoots are positioned manually to protect the clusters from sunburn.

Dehydration and Processing

Golden and green raisins are made in India. Green raisins constitute about 80% of the total production. Pre-drying treatments for the two types differ, but the drying process, destemming and grading is the same.

Golden raisins are bright yellow or golden yellow in color,

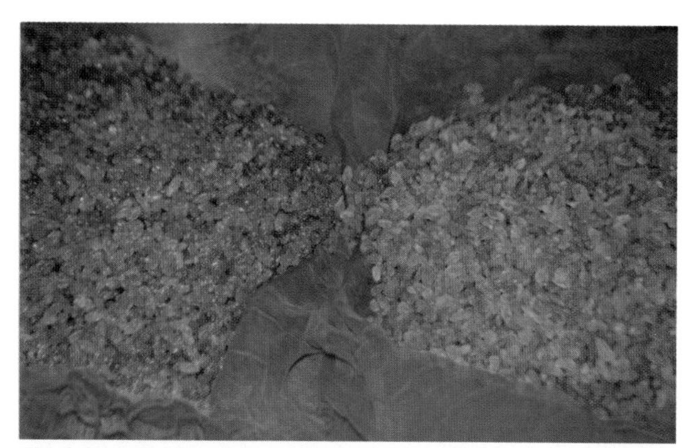

Fig. 1.10.1: Thompson Seedless in "T" trellis showing full canopy.

Fig. 1.10.2: Grapes being dipped into a hot lye solution.

shiny, translucent, moderately tender and sometimes sticky). Dipping the cleaned and trimmed clusters in a boiling solution of 0.2 to 0.3% sodium hydroxide (lye solution) for two or three seconds produces cracks in the skins to help the grapes dry (**Figure 1.10.2**). The grapes are rinsed in cold water immediately after dipping to cool them sufficiently.

The dipped grapes are examined to determine the effectiveness of dipping the grapes to crack the berries. If cracking is inadequate, the concentration of lye solution and/or the duration of dipping time are increased. The dipped and washed clusters are exposed to sulfur fumes for two to four hours by burning sulfur powder in an air-tight room. About 2.2 to 4.4 lbs (1 to 2 kg) of sulfur is burned to bleach one ton of grapes. After sulfuring, the grapes are dried on racks in sheds with covered roof and open sides. Berries are dried to about 15% moisture in 12 to 15 days depending on the temperature in the drying sheds.

Green raisins are produced from Thompson Seedless clones following the Australian method of making Sultanina. The commercial "dipping oil" is imported from Australia for dipping the grapes before drying. The dipping oil contains 60% ethyl esters of C16 and C18 fatty acids with free oleic acid and emulsifiers. A dipping emulsion of 26 gallons (100 L) is prepared by the addition of 5.3 lbs (2.4 kg) of potassium carbonate and 0.4 gallons (1.5 L) of dipping oil to 26 gallons (100 L) of water.

Cleaned and trimmed clusters are placed in perforated plastic crates and washed in clean water to remove the extraneous dirt. After draining the water completely, the crates are dipped for five minutes in cement tanks of dipping emulsion. The dipped clusters are spread on nylon net racks in the drying sheds.

Fig. 1.10.3: Drying sheds with open sides.

Fig. 1.10.4: Spraying fruit on drying racks with a drying emulsion.

Fig. 1.10.5: Worker removing fruit from drying rack.

Drying sheds are built using angle iron posts and galvanized iron/asbestos sheet roofing. The sides are partially covered with black nylon net, allowing only 15% light when fully covered. These sheds are built in the vineyard where the maximum temperature is above 104 °F (40 °C) and where the relative humidity does not exceed 15% at any time during drying. If such conditions are not available in the vineyard, groups of drying sheds are constructed in clean elevated areas where such drying conditions are favorable (**Figure 1.10.3**).

Drying racks are fabricated using 0.4 x 0.4 inches (1 x 1 cm) mesh nylon nets. Vertical clearance between the successive tiers is 12 inches (30 cm). About 12 tiers of racks are made. The width of each rack is about 5 ft (1.5 m) and the gap between two sets of racks is also 5 ft (1.5 m). Cleaned and dipped clusters are spread in a single layer on these nets. After three days, the grapes are sprayed with dipping emulsion at two-thirds strength (0.32 gal [1.2 liters] of dipping oil and 3.5 lbs [1.6 kg] of potassium carbonate in 26 gallons [100 liters] of water). If drying is not adequate, an emulsion at one-third strength (0.16 gals [600 ml] of dipping oil and 3.5 lbs [1.6 kg] of potassium carbonate in 26 gallons [100 L] of water) is

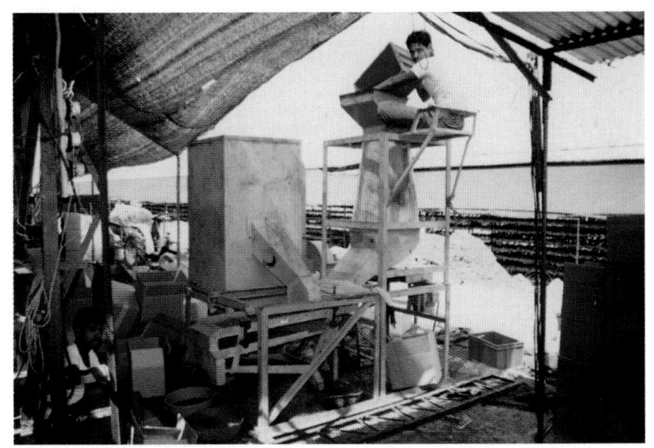

Fig. 1.10.6: Mechanical de-stemming and size grading.

Fig. 1.10.7:Worker manually grading raisins for color.

sprayed on the fifth day (**Figure 1.10.4**).

The drying takes 10 to 12 days depending on the ambient temperature, humidity and air circulation in the sheds. When adequate drying is achieved, the dried grapes are detached from the rachis by passing two iron pipes, one above and one below the rack (**Figure 1.10.5**). The raisins are collected from the bottom most rack, destemmed, cleaned, graded and packed.

Destemming is done either manually or mechanically. In mechanical destemming, separation of raisins from stems and grading is done simultaneously. The dried pieces of clusters collected from the racks are poured into the hopper of the machine. The raisins are destemmed while the pieces of clusters pass between two soft rollers, both moving inwards. While falling into a sieve tray, a fan blows off the trash and the raisins pass through a series of sieve trays, each with different size perforations shaking horizontally in an inclined plane. The largest raisins are collected from the top tray, the medium from the middle and the small from the bottom (**Figure 1.10.6**).

The large and medium sizes are graded for green and brown raisins manually. The dried grapes collected from the bottom-most rack are rubbed gently by hand against a perforated iron sheet fixed on a raised platform to remove finer particles. The mass containing the large broken pieces of rachis and separated berries is sifted through a forced air stream, which removes the trash as well as lightweight berries away from the sound berries. The sound raisins are then sorted by size through sieving. Each grade is manually sorted for color (**Figure 1.10.7**).

The graded raisins are weighed and packed into 2.2 lbs (1 kg) or 1.1 lbs (0.5 kg) plastic bags. The plastic bags are imprinted with the trademark and details about the raisins, such as type, grade, weight, date of pack, etc., packed in cardboard boxes and stored. Normally 26.5 lbs (12 kg) of raisins are packed in each box. Sometimes the entire 26.5 lbs (12 kg) container is lined with thick plastic bags and packed.

Raisins are stored and released for sale according to market demand. Adequate care is taken to prevent insect and rodent infestation.

Marketing

All raisins produced in India are consumed domestically. Until the late 1980s, 5,500 tons (5,000 t) of raisins valued at $5 million were imported from Afghanistan and Iran each year. Raisins are consumed mainly as a dried fruit in winter and given to the convalescent patients. The indigenous systems of medicine, namely Ayurveda and Unani, recommend the use of raisins for treatment of many ailments. The popular tonic "Chyavanprash" contains raisins. Raisins are also used in the preparation of a variety of desserts, sweets, and bakery products.

Credits and Sources

Information on India provided by:

Ahmed Ahmedullah, Prof. Emeritus, Horticulture, Washington State University; and S.D. Shikhamany, Director, National Research Center for Grapes, Pune, India

Sources: The Maharashtra State Grape Growers Association, National Horticulture Board, Pune, India

Photo Credits:

Figures 1-7: S.D. Shikamany

11. Mexico

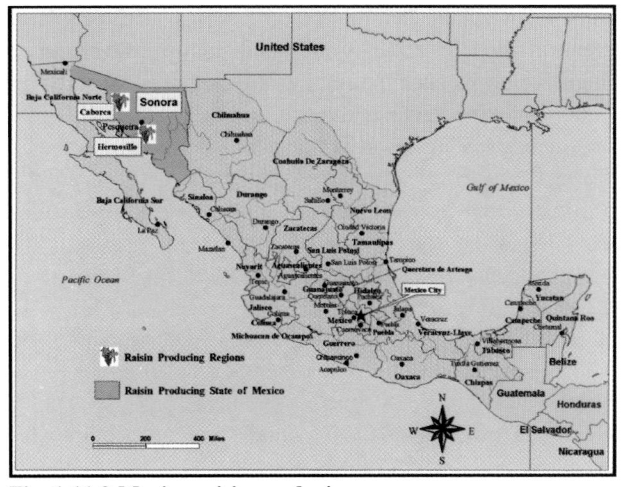

Fig. 1.11.0:Mexico raisin producing areas.

History

Mexico ranks 11th in world raisin production and is the oldest grape producing country in the Americas, producing grapes since 1518. The primary raisin producing areas are the Costa de Hermosillo and Caborca in the state of Sonora.

Table 1.11.1
Raisin Production in Caborca, Mexico, 1985-1998.

Year	Acres	Tons
1985	7,729	7,357
1986	7,324	6,403
1987	14,423	10,423
1988	11,396	11,471
1989	13,344	12,247
1990	10,045	8,925
1991	7,611	6,736
1992	6,178	5,534
1993	9,761	8,708
1994	15,459	16,550
1995	15,271	16,204
1996	14,300	14,991
1997	14,730	20,434
1998	11,367	17,844
Mean	**11,353**	**11,702**

Source: District of Rural Development No. 139, Dept of Agriculture and Rural Development, Caborca, State of Sonora, Mexico

Table 1.11.2
Raisin Production in Hermosillo, Mexico, 1998.

Variety	Tons
Flame Seedless	521
Perlette	684
Superior Seedless	39
Ruby Seedless	
Thompson Seedless	46
Other	
Total	**1,290**

Fig. 1.11.1: Laying grapes on paper trays.

Fig. 1.11.2: "Parrones" overhead trellis.

Fig. 1.11.3: Third-leaf vine on parrones trellis.

Grapes were introduced to the Sonora area in the 1950s by colonizers from Europe, mainly Italians who brought grapes for making wine for personal use. In time, a small acreage of wine grapes was planted, including Thompson Seedless. In the 1960s, the region developed water problems and the government imposed rationing of deep well water. This resulted in a change in crop patterns, with extensive areas planted to cereal grains and cotton converting to intensive crops such as grapes. Viticulture, with its three options of wine, table, and raisin grapes, became an important industry. Thompson Seedless was introduced to the La Costa de Hermosillo area. It proved unsuccessful because of a climate that did not provide enough chilling to break dormancy. The variety was relocated to the Caborca area with excellent results in production and quality of table grapes and raisins.

During the 1970s, Costa de Hermosillo, Pesqueira and Caborca were studied to determine adaptable classes of grapes including wine, table or raisin grapes. Costa de Hermosillo was found to be suitable for wine grapes for brandies, excellent for table grapes such as Flame Seedless, Superior Seedless, and Perlette. These grapes could also be made into raisins, but of only fair quality. Caborca is also suitable for wine grapes for brandy, produces excellent quality table grapes, and it is Mexico's best area for raisin production. Thompson Seedless produces high sugar (22-24 °Brix) resulting in sweet, plump and meaty raisins. Pesqueira is the earliest area for table grapes of excellent quality, however raisins are not produced due to untimely summer rains.

During the 1980s, the suitability of each area was proven and the grape industry expanded through the 1990s; wineries grew in number, table grape facilities increased, and eight raisin processing plants were constructed in Caborca.

Area in raisin grape vines: 11,367 acres (4600 ha).

Raisin tonnage, 1998: 19,250 tons (17,500 t); 17,807 tons (16,188 t) from Caborca; 1,443 tons (1,312 t) from Hermosillo. (**Tables 1.11.1** and **1.11.2**).

Raisin varietals grown for commercial production in descending order of importance: Thompson Seedless (90%), Perlette, Flame Seedless and Superior Seedless.

Viticultural Aspects

All irrigation water is pumped from wells, making electricity the most expensive factor in Mexican raisin production. Due to limited water supplies, drip irrigation is widely used with good results.

New fields are being planted at a density of 11.8 x 2.6 ft

Fig. 1.11.4: Double cordon structure.

Fig. 1.11.6: Raisin processing line in Caborca.

Fig. 1.11.5: Raisins ready for field collection.

with 1,420 vines/acre (3.6 x 0.8 m, 3,507 vines/ha) in African (gable) trellising trained in a single or double cordon with a double crossarm in the cordon area. Trellising and training of older plantings of raisin vineyards as well as the practice of drying on paper trays down vineyard rows are similar to that of California (**Figure 1.11.1**).

The latest innovation is the planting of Chilean "parrones" (**Figures 1.11.2** and **1.11.3**) with a density of 11.8 x 4.9 ft with 753 vines/acre (3.6 x 1.5 m, 1,860 vines/ha). Training has been modified to develop a double cordon, which follows a transversal line to the plants (**Figure 1.11.4**). Test plots indicate that this system will triple the production of grapes trained on the conventional trellis.

Dehydration and Processing

Natural sun drying is the principal method of dehydration, taking advantage of the very hot and dry desert conditions (**Figure 1.11.5**). All other methods of drying are not considered cost effective.

Domestic Utilization

Domestic consumption is estimated at 13,200 tons (12,000 t). Christmas and Holy Week are the most important weeks for raisin consumption, but raisins are usually not available because of high storage costs, and most raisins are exported as soon as possible. Because of this, the raisin processing plants are at their peak season (**Figure 1.11.6**). Since facilities for storage are inadequate, the domestic market is usually saturated right after harvest and shortages occur thereafter.

Exports, metric tons, 1996:		**Imports:**	
United States	9,185	United States	1,122
Brazil	288	Chile	3,679
Canada	1,086	Other	138
Other	219		
Guatemala	322		

Marketing

Seedless raisins are packed in cartons of 22 lbs (10 kg) for domestic use and in 30 lbs (14 kg) cartons for export use. Under NAFTA, there are no tariffs.

Information of Interest

Processors sell their best product for export right after harvest, even if the domestic market price is higher because of the overriding benefits of the export market, such as payment on delivery. Lower quality raisins are then imported to meet domestic needs.

Credits and Sources

Information on Mexico provided by:

Ing. Hector Inacio Tapia Contreras, Department of Agriculture and Rural Development, Sonora, Mexico, District of Rural Development No. 139, Caborca, Mexico.

Source:

D. Flores, Dried Fruit Annual Report, U.S. Embassy, Mexico City, Mexico, attaché query, 1998

Photo Credits: Figures 1-6:

Hector Inacio Tapia Contreas

12. Syria

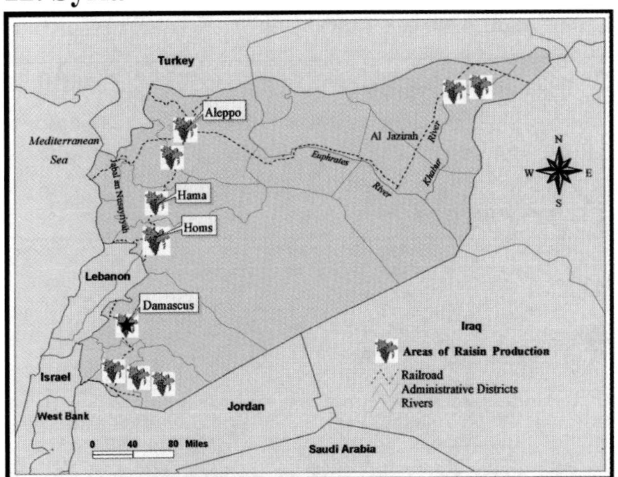

Fig. 1.12.0: Syria raisin producing areas.

History

Raisin production in Syria ranks 12th. Grape and raisin production in Syria, as in most of the Middle East, date back many centuries. It is estimated that one-third of the land is arable, but only one-half is actually cultivated. The best farmland may be found along the coast and in the Jabal and Nusayriyah, around Aleppo in the regions between Hama and Homs. Raisins are produced in the Damascus area and in the land between the Euphrates and Khabur rivers, known as the Al Jazirah (in Arabic Al Jazirah is "the island"). Grapes are also grown for fresh table use.

Area in grape vines: 165,490 acres (67,000 ha) (O.I.V. Bulletin, 1998).

Area in raisin grape vines: 7,410 acres (3,000 ha).

Average raisin tonnage, 1993-97: 14,439 tons (13,126 t).

Raisin varietals grown for commercial production: The two primary varieties are Asgarry (White Seedless) and Maviz (Black Seedless) which are similar to the varieties Thompson Seedless and Black Monukka, respectively. Some seeded table grape varieties are also dried, as well as the

Muscat of Alexandria.

Dehydration and Processing

Most of the fruit is dipped in a solution containing soda-ash and olive oil and spread out in the sun on a smooth soil surface or cement slabs.

Marketing

All of the raisins produced are consumed domestically with per capita consumption of 1.71 lbs. No imports or exports have been reported.

Syria Raisin Production by Year – 1993-1997:

Year	Production (metric tons)
1993	8,453
1994	11,391
1995	13,590
1996	19,445
1997	12,746
Average	13,126

Credits and Sources

Information on Syria provided by:

Jaber Dalati, American Embassy, Damascus, Syria and Fred Nury, Ph.D., Department of Food Science and Enology, CSU, Fresno.

13. Argentina

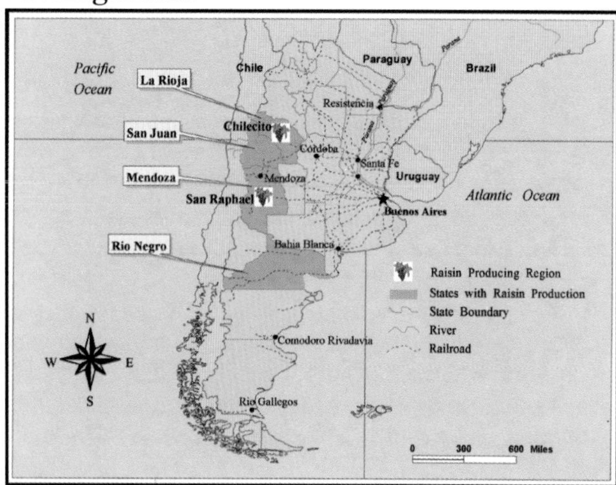

Fig. 1.13.0: Argentina raisin producing areas.

History

Argentina ranks 13th in world raisin production. Jesuit fathers introduced grapes to Argentina in 1650. Today, vineyards are mainly located on the high plateau bordering the Andes in the states of Mendoza and San Juan and to a lesser extent in Rio Negro. The climate on the plateau is temperate with cool winters and warm to hot summers. There are occasional hailstorms in San Raphael where vines are protected with overhead screens (**Figure 1.13.1**). Raisins are also produced in Chilecito in the province of La Rioja, in a valley in the foothills of the Andes, at an altitude of 3937 ft (1200 m) above sea level. This area has a mountain climate with very long, sunny summers.

Area in raisin grape vines, 1996: 10,770 acres (4,360 ha).

Average raisin tonnage, 1994-1998: 10,780 tons (9,800 t).

Raisin varietals grown for commercial production in descending order of importance: Thompson Seedless, Arizul and Perlette.

Viticultural Aspects

Since the production regions are semi-arid, flood irrigation is used from a plentiful water supply in the nearby Andes Mountains (**Figure 1.13.2**). Various trellis systems are used, ranging from a two- to three-wire vertical system to the commonly used overhead pergola systems (**Figure 1.13.3a** and **b**). Some grape growers in the province of La Rioja are practicing organic raisin production and exporting to the United Kingdom and Germany.

Dehydration and Processing

When the grapes have reached optimum maturity (22 °Brix), the fruit is harvested into containers and hauled to the drying area. The drying areas consist of rock slabs, which absorb heat from the sun. Grapes are spread uniformly onto wooden trays and placed on the slabs (**Figure 1.13.4**) to optimize exposure to the sun. After a few days, the grape clusters are turned to expose the other side to the sun. This is done as many times as needed to achieve uniform drying. The raisins are dark purple to brown with the characteristic taste of a natural sun-dried raisin (**Figure 1.13.4**). Some of the grapes are sulfured and dried in the same manner. These

Fig. 1.13.1: Screen to protect against hail damage.

Fig. 1.13.2: Flood irrigation.

Fig. 1.13.3a: Overhead pergola trellis system.

Fig. 1.13.3b: Pergola trellis system with wide avenue for turning space.

Fig. 1.13.4: Sulfured and non-sulfured grapes drying on wooden trays set on rock cobblestone.

Fig. 1.13.5: Lug box container of sulfured raisins.

assume a light reddish to yellow color (**Figures 1.13.4** and **1.13.5**), with a lingering taste of sulfur. Some grapes are dried in gas fired dehydrators in San Juan without any pretreatment.

The raisins are stored in disinfecting rooms to maximize storage time prior to processing. Processing equipment, similar to lines designed in California, is used to destem, size grade, and remove sub-standard raisins. The raisins then pass through a water bath followed by a centrifuge to remove excess water. Workers inspect the raisins on moving belts, separating out those of inferior quality. A bath of protective oil (medicinal Vaseline) is then applied to prevent stickiness and to give the raisins a shiny appearance. Packaging consists of placing the raisins in 7 oz (200 gm) bags which are then placed in cardboard boxes.

Domestic Utilization

An estimated 3,300 tons (3,000 t) is consumed annually by a population of 35.5 million people. Per capita consumption is estimated at 0.19 lbs (0.105 kg) per year.

Exports, metric tons, 1997:	
Brazil	3,827
Dominica	148
Peru	48
Spain	420
Portugal	90
Canada	19
Paraguay	294
Costa Rica	49
Ecuador	19
Dominican Republic	243
United States	38
Imports:	*50 - 100 t*

Marketing

Raisins are marketed by the private sector. There are no subsidies for either production or exportation, although the government does have a program to assist exporters with up to one-half the cost of promoting their products in overseas trade programs and shows.

Credits and Sources

Information on Argentina provided by:

Dr. Viviana Michael of the Cooperativa Vitivinifruticola, Chilecito, La Rioja, Argentina

Source: Gary Groves, Office of Agricultural Affairs, U.S. Embassy, Buenos Aires

Photo Credits:

Figures 1-5: V.E. Petrucci

14. Lebanon

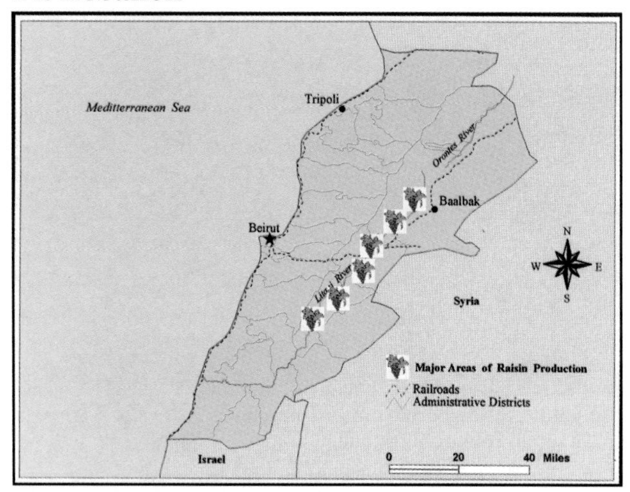

Fig. 1.14.0: Lebanon raisin producing areas.

The 14th ranked raisin producer is Lebanon. It is estimated that approximately 12,000 acres (4,856 ha) are used for production of raisins. Lebanon produces about 8,000 tons of raisin (F.A.O. Production Yearbook, Vol 51, 1997). They consume all domestic production and there is no record of raisin imports.

Lebanon's major fruit production areas are centered on the eastern border of the Anfi-Lebanon Range. Between these two mountain ranges lies the fertile valley of Al Biga. The Orontes in the north and the Litani in the south are the main Rivers. Major crops supported by this fertile valley are grapes (raisins), citrus, sugar beets and potatoes.

Total grape acreage: 68,694 acres (27,800 ha).

15. Spain

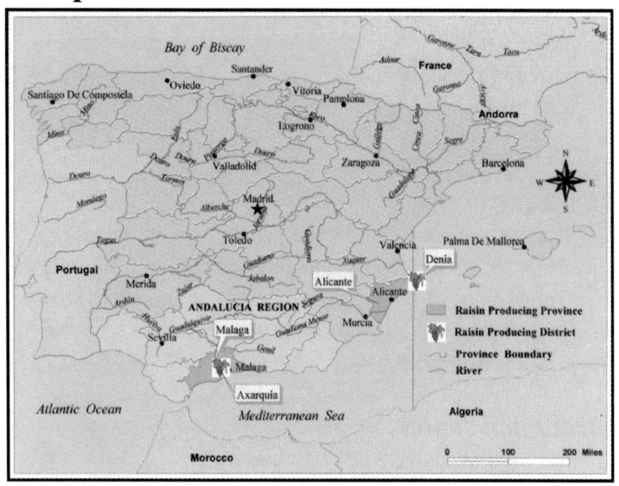

Fig. 1.15.0: Spain raisin producing areas.

History

The 15th ranked raisin producer is Spain. It is difficult to determine when raisins were first made in Spain, but it is known that raisins were being exported in the 13th century. The primary production area is Andalucia, with 70% of total area in the province of Malaga in the Axarquia region. The remaining 30% are produced in the Alicante province.

Spain was once the world leader in the production of "Malagas" which is made from the Muscat of Alexandria grape and exported worldwide. At the beginning of this century, there were 227,875 acres (112,500 ha) dedicated to raisin production in the Malaga province. Phylloxera has destroyed most of the Spanish vineyards and raisin production has been declining. The last removal of about 250 acres (101 ha) took place in 1994. Raisin grape acreage is expected to continue to decline due to the age of the vines, difficulty of cultivation, costs, their location in areas difficult to mechanize, low yields of about 500 lbs per acre (560 kg/ha) and demand for seedless varieties and less expensive imports.

Fig. 1.15.1: Non-trellised Pedro Ximenez (for raisins).

Fig. 1.15.2: Closely spaced Muscat of Alexandria.

Area in raisin grape vines, 1995: 9,927 acres (4,019 ha).

Average raisin tonnage, 1994-95: 6,985 tons (6,350 t).

Raisin varietals grown for commercial production in descending order of importance: Muscat of Alexandria, Pedro Ximenez (**Figure 1.15.1**), Lairen, Cogadera, Mantuo and Sultanina (Thompson Seedless).

The Muscat of Alexandria accounts for an estimated 90% of total raisin production.

Viticultural Aspects

The raisin vineyards of Spain are not irrigated. The vine spacing in general is 5.24 x 5.24 ft (1.6 x 1.6 m) (**Figure 1.15.2**), and the vines are trained to the head system with three to five spurs, one to three buds in length (**Figure 1.15.3**). Most of the surviving vineyards are on phylloxera resistant rootstock consisting mainly of Rupestris St. George and 110 R. In the raisin vineyards of Malaga and Alicante, the traditional method of cultivation in the vineyards is done manually and

Fig. 1.15.3: Head trained Muscat of Alexandria.

Fig. 1.15.4: Use of animal power for vineyard cultivation.

Fig. 1.15.5: Packaging of Muscat raisins in Malaga.

Fig. 1.15.6: Spain's world-famous Muscat layered cluster.

with the use of animals (**Figure 1.15.4**).

Dehydration and Processing

In the Malaga province, Muscat of Alexandria grapes are harvested and transported from the vineyard to a drying area where the fruit is spread out on cement or paved slabs to dry in the sun. In the Alicante province (Denia), Muscat grapes are sun dried after they have been dipped into a lye solution. Through this process the bloom is removed giving the Muscat raisins of Alicante a shiny appearance.

Standard processing and packaging equipment for free-flowing individual Muscat raisins is used (**Figure 1.15.5**). However, hand processing is still used, particularly with the world famous Spanish layered clusters, which are dried as whole clusters. They are then carefully collected to keep the raisins attached to the rachis (**Figure 1.15.6**). The individual raisins in the cluster are carefully cleaned by hand, and the clusters are packed in layers with each layer being of the same depth. Packaging varies from 5.5 to 45 lbs (2.5 to 20 kg).

The single berries of Muscat raisins are packaged in cartons of many assorted sizes. The most popular retail package weighs 1 lb (454 g) and commercial packages range from 30 to 45 lbs (14 to 20 kg).

Domestic Utilization

Approximately 5,620 tons (6,182 t) of Spain's raisins are consumed domestically. This accounts for about 90% of Spain's production. Per capita consumption is 0.7 lbs (318 g). The rest of the crop is exported to Portugal, England, France, Germany and Switzerland. Spain imports about 8,250 tons (7,500 t).

Marketing

Imported raisins are cheaper than those produced domestically and Malagas, or Muscat raisins, have seeds and the market demands seedless raisins. The aid from EU for the cultivation of dried grapes to the farmers is also limited for production of Spanish Muscat raisins. EU regulation 1594/98 establishes 3,290 ECU/ha for sultanas, 3,080 ECU/ha for currants and 880 ECU/ha for Spanish Muscat seeded raisins.

Information of Interest

In 1920, Spain exported 14,275 tons (12,977 t) of dried Muscat grapes while in 1997 exports were only 560 tons (508 t).

Credits and Sources

Information on Spain provided by:

Alberto Garcia de Lujan, Director de la Estacion Experimental, Apartado 589, Rancho de la Merced, Jerez de la Frontera, Spain.

Source:

Christopher Rittgers, Agricultural Attaché, American Embassy, Madrid, Spain.

Photo Credits:

Figures 1-4: V.E. Petrucci; Figures 5 and 6: Alberto Garcia de Lujan

16. Cyprus

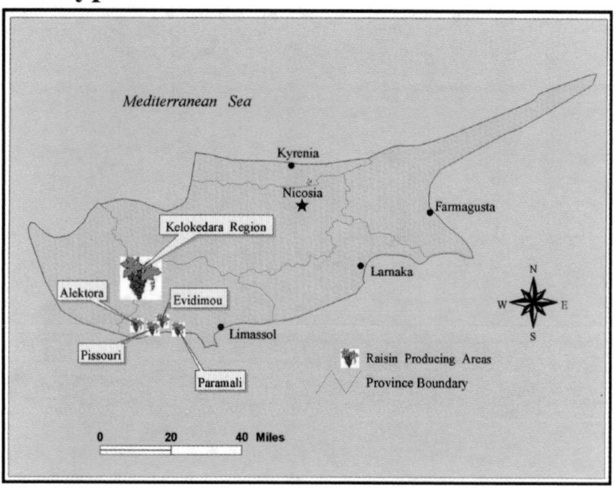

Fig. 1.16.0: Cyprus raisin producing areas.

History

Cyprus is the world's 16th largest raisin producer. Archaeological evidence indicates that the cultivation of grapes started in Cyprus sometime early in the second millennium B.C. The earliest evidence for the existence of "wild grapes" in Cyprus was found in 1973 during archaeological excavations at a Neolithic site. Though we cannot be certain how this wild species (*Vitis vinifera* ssp. *Silvestris*) was used, there is no doubt that it was harvested. The grapes may have been dried to provide food for the winter months.

Today, sultana raisins are produced mostly in the coastal area of Pissouri, Alektora, Paramali, and Evdimou villages, situated between Limassol and Paphol.

Average area in raisin grape vines: 2,534 acres (1,026 ha) (estimate, 1993 - 1995).

Average raisin tonnage: Approximately 4,276 tons (3,887 t) of black scalded raisins are produced yearly. Sultana production averaged 590 tons (536 t) from 1993 to 1997.

Raisin varietals grown for commercial production: Two types of raisins are produced; the black scalded raisins with seeds and Sultana (seedless) raisins, made from the Mavro, a large black seeded grape, and the Sultana table grape (Thompson Seedless), respectively (**Figure 1.16.1**). Until recently, the black scalded raisins constituted about 90% of the total production of all types of raisins produced and are grown mainly in the Kelokedara region of the Paphos district. Dipping Mavro grapes in boiling water produces Black scalded raisins.

Fig. 1.16.1: Thompson Seedless grape cluster.

Fig. 1.16.2: Thompson Seedless supported by overhead trellis in the Village of Pissouri.

Viticultural Aspects

Vine densities of the Mavro variety range from 890 to 1,245 vines per acre (2,200 - 3,075 vines per ha). The vines are trained without trellising to a "goblet" which is the head bush system. Average fresh fruit production is 3 tons per acre (6.7 t per ha). The vineyards usually are not irrigated, relying mostly on the area's rainfall.

The density for vineyards planted to the Sultana grape variety (Sultana raisins) varies from 680 to 690 vines per acre (1,680 to 1,704 vines per ha). The Sultana vineyards are usually trellised (**Figure 1.16.2**). The average production is 0.61 tons of raisins per acre (1.4 t/ha). The island's good climate and soil create a favorable situation for raisin production and the quality of wine and raisins is excellent.

Dehydration and Processing

All raisins produced are sun dried. The most common method of preparing the grapes for sun drying is to harvest the grapes into baskets or perforated drums and dip them in a solution of potassium carbonate and 1% of good quality olive oil for about two minutes. The grapes are drained and spread in thin layers on special plastic to dry in the sun. After three or four days the grapes are turned to allow for uniform drying. The raisins are usually ready for harvest after eight to 10 days. Black scalded raisins have a drying ratio of three to one. Sultana raisins have a dry ratio of 3.75 to one. Processing machinery imported from California is used for cleaning the raisins. This equipment employs a feeder hopper and cluster breaker with conveyors, a destemming machine with oiling unit, shaker-grading screens, vacuums, conveyor belts for hand sorting, and bag/carton filling devices. Black scalded raisins have been packed in 50 kg hessian (burlap) bags for shipment to the former Soviet Union as well as in 15 kg cartons, destined for the rest of the world. Sultana raisins are packaged in 12.5 kg cartons.

Domestic and Export Utilization

The last year statistics were available was 1989. At that time, the annual domestic consumption of black scalded raisins was 11 to 16.5 tons (10 to 15 t). The production of black scalded raisins from 1970 to 1980 averaged 9,900 tons (9,000 t) per year and was exported primarily to the former Soviet Union with some raisins exported to Europe. The annual domestic consumption of sultanas was 330 to 385 tons (300 to 350 t). The major markets used to be Belgium, Germany, Switzerland, Romania, Egypt, Sri-Lanka and Russia.

Information of Interest

The Vine Products Scheme (V.P.S.) was established by the government in 1949 for the protection and assistance to vine growers who had been for some time suffering economically due to problems of disposing of their vine products. The primary objective was to concentrate raw grape alcohol and raisins (black scalded raisins) and dispose of them. The delivery of both products to V.P.S. was compulsory.

The Vine Products Commission (V.P.C.) was formed in 1968 to take over the work carried out by the V.P.S. and it continued collecting and exporting raisins. The V.P.C. also started collecting and exporting sultana raisins. These arrangements continued until 1989 when the V.P.C. stopped collecting sultana raisins.

Credits and Sources

Information on Cyprus provided by:

Dr. Nearchos Roumbas, Department of Agriculture, Viticulture and Oenology, Limassol, Cyprus

Photo Credits:

Figures 1 and 2: V.E. Petrucci

17. Morocco

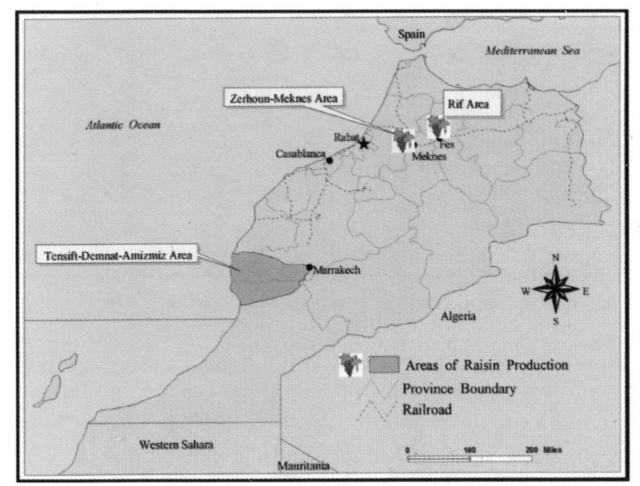

Fig. 1.17.0: Morocco raisin producing areas.

History

Morocco ranks 17th in world production. It is likely that the cultivation of the wild vines began in the Mediterranean area and was spread to Asia and North Africa by the Phoenicians. In Morocco, grapevines were cultivated through the natural sowing of seeds of cultivated vines. It appears that some Muscat genotypes were imported from the Orient by the Phoenicians in about 2000 B.C.

Morocco cultivates about 114,114 acres (46,200 ha) of grapevines. Of these, 24,700 acres (10,000 ha) are wine grapes, 76,570 acres (31,000 ha) are table grapes, and 12,844 acres (5,200 ha) are raisin grapes.

Area in raisin grape vines: 12,844 acres (5,200 ha).

Average raisin tonnage, 1993-1998: 9,625 tons (8,750 t).

Raisin varietals grown for commercial production: Grape varieties used for raisins are Taferialt, Zerhouni, Koukkali, Thompson Seedless, Kings Ruby and Muscat of Alexandria. Of particular interest to Moroccan grape and raisin producers is the cultivar Taferialt (**Figure 1.17.1**) which can be grown as a table or raisin grape. The clusters can remain on the vine for some time without loss of quality. The cluster is medium to large, loose to well filled, with a large, round berry. It has a juicy texture, a thin skin, and seems to be resistant to downy mildew (*plasmopara viticola*), powdery mildew (*uncinula necator*) and has an affinity with Richter 99 rootstocks in the Zerhoun area.

Viticultural Aspects

The influence of the vineyard site in terms of climate, microclimate, soil aspects, etc. on vine growth and productivity in Morocco is most interesting. Climate and soil are decisive factors in determining the growth potential of the vine and dictates vine spacing and the type of trellis system needed to accommodate this growth potential.

In the modern vineyards, the classical spacing of trellises on espalier (double guyot) pruning is 5 x 10 ft (1.5 m x 3 m). The traditional vineyards with goblet (head) pruning are planted on the square system 13 x 13 ft (4 x 4 m) to 16.5 x 16.5 ft (5 x 5 m). For vines trained on an overhead trellis (pergola), the spacing ranges 10 x 10 ft (3 x 3 m) to 8 x 8 ft (2.5 x 2.5 m).

Fig. 1.17.1: Taferialt raisin variety with seeds.

Fig. 1.17.2: Taferialt vine supported by pergollete trellis system.

The goblet form of head training has been the standard system for traditional and modern vineyards. On sites where new vineyards were established beginning in 1966, various types of trellising have been used. The arbor system employs two variations including the pergola at a height of 6 ft (1.8 to 2.0 m) accommodating four or more canes or the smaller version called the pergolette at a height of 4.3 to 4.6 ft (1.3 to 1.4 m) (**Figure 1.17.2**). Other trellises include the double royat supporting two cordons with spurs at regular intervals and the guyot which supports a pruning system carrying one cane and one spur on a unilateral cordon which are renewed each year.

Irrigation systems are dependent on water availability, rainfall and ground water. The standard systems are flood, drip and micro-jets. Smaller vineyards practice the flood system and some vineyards are not irrigated and completely dependent on rainfall. Drip irrigation is used in more modern vineyards.

Fertilization practices are adapted according to the fertility of the soil and the vigor of the vineyard. The injection of fertilizer into the irrigation system is the usual practice in modern vineyards. Recommendations for fertilizer use are distributed among small growers and extension technicians.

Morocco is in tune with modern practices to enhance fruit quality, such as canopy management, pulling leaves, cluster thinning, girdling and use of growth regulators.

Vineyards are located within an area of 435 miles (700 km) from north to south. The variations in climate within this area affect the maturation time, which varies from July to November.

Dehydration and Processing

Morocco relies on two basic methods of dehydration. The first is the old practice of sun drying method where the harvested grapes are dipped in a solution containing carbonate, ash of bean straw and olive oil. The clusters are placed on straw mats or clean cemented soil. Drying takes 10 to 15 days. The second method uses hot air dehydrators using gas or oil. The fruit is hand picked, dipped in a hot alkaline solution (caustic soda 0.3% to 0.4%) for 20 to 60 seconds then exposed to sulfur dioxide for 1.5 to two hours. The resulting product is similar to the California golden raisins.

At most packing houses, raisins are hand sorted and cleaned for defects and are re-hydrated before packaging. Packaging type depends on the market. Bags of 8.8 oz (250 g), 17.7 oz (500 g), and 2.2 lbs (1 kg) are used for retail markets and packages of 11 lbs (5 kg) and 22 lbs (10 kg) are used for wholesale.

Domestic Utilization

Morocco consumes all the raisins it produces and according to the latest O.I.V. Report in 1996, it imported about 1,540 tons (1,400 t).

Information of Interest

The promotion of raisin consumption focuses on the nutritional value of raisins and targets schools, hospitals, and sports organizations. Morocco also has ongoing programs sponsoring the development of technology needed to develop acreage for high quality raisin production. It also has a dynamic marketing policy, which is aimed at protecting Moroccan producers against competition.

Credits and Sources

Information on Morocco provided by:

Dr. Houcine El Miniai, Commission Nationale Vitivincule Du Maroc - Rabat Instituts, Morocco.

Sources:

Ministere de l'Agriculture et de la Reforme Agraire, DPVCTRF (Maroc): Synthese des monographies varietes et porte-greffer de vigne Decembre 1998; Societe de Developpements Agricoles: Presentation de la vigne marocaine des raisins et raisins secs. 1985.

Photo Credits:

Figures 1 and 2: Houcin El Miniai

18. Tunisia

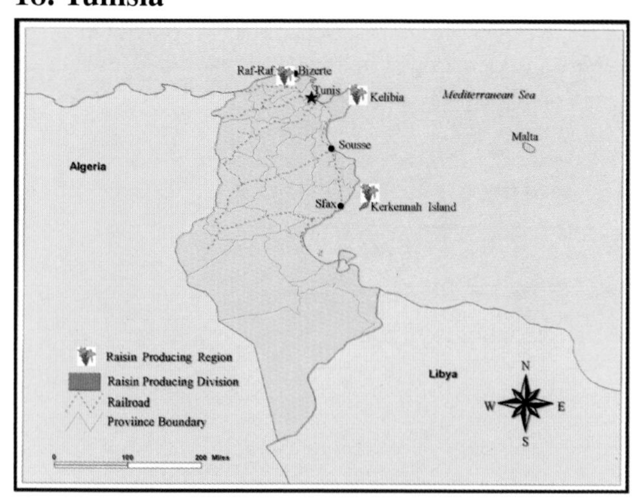

Fig. 1.18.0: Tunisia raisin producing areas.

History

Tunisia ranks 18th in world raisin production. Raisins have been produced in Tunisia for centuries and have always been an important part of the Tunisian diet, both in cooking and as a snack food. The growing regions include the Grombalia and Sousse areas.

Total area in grape vines: 685,746 acres (277,630 ha).

Area in raisin grape vine: 9,880 acres(4,000 ha) (O.I.V. Bulletin, 1996).

Average raisin tonnage, 1990-96: 5,280 tons (4,800 t) (O.I.V. Bulletin, 1996).

Raisin varietals grown for commercial production in descending order of importance: Asli, Rezzagui, Muscat of Alexandria and Sultanina (Thompson Seedless).

Dehydration and Processing

The island region of Kerkenah situated near the southern Tunisian city of Sfax cultivates the varieties Asli and Rezzagui. The fruit is partially dried on the vine, dipped in seawater and spread on the sand to dry. In the Region of Raf-Raf, situated on the northeast coast near the city of Bizerte, the primary variety is the Muscat of Alexandria (meski raf raf) (**Figure 1.18.1**). The grapes are dried using the ancient method of dipping the fruit into a boiling solution of soda ash and lime. The grapes are spread on vegetable debris or soil to dry in the sun (**Figure 1.18.2**). After drying, the raisins are deseeded. In the region of Kelible, situated on the east coastal area of Cap Bon, Muscat of Alexandria is the main variety and is dried as in Raf-Raf. The cured and processed raisins are packed in shallow containers (**Figure 1.18.3**).

Fig. 1.18.1: Meski Raf Raf (Muscat of Alexandria).

Fig. 1.18.2: Drying grapes on soil covered with a canvas at Raf Raf.

Fig. 1.18.3: Raisins in shallow containers in Kelible region.

Marketing

All of the raisins are consumed by the domestic market. An additional 630 tons (570 t) are imported (O.I.V. Bulletin, 1998).

Credits and Sources

Information on Tunisia provided by:

Fethi Askri, Director General, Office National de la Vigne, Tunis, Tunisia

Photo Credits:

Figures 1-3: Fethi Askri

Summary

The seven leading countries in sultana and raisin production entering commercial channels are the United States, Turkey, Iran, Australia, South Africa, Greece and Chile (International Sultana and Raisin Conference, 1999, London; 2000 Australia). As shown below, the world's export market for sultanas and raisins entering commercial trade, Turkey supplies the most raisins to exports, followed by the United States, Iran, Greece, Chile, South Africa and Australia.

	World Rank	***Export Rank***
United States	36%	20%
Turkey	31%	44%
Iran	13%	16%
Australia	7%	4%
South Africa	6%	5%
Greece	4%	6%
Chile	4%	5%

***Figures are for 1999-2000 and do not include Golden Seedless or Zante Currants.*

The United States continues to lead the world in total raisin production. The San Joaquin Valley of California has an abundance of fertile soils, which are irrigated and support the highest yield and quality in the world. With an ideal climate for making natural sun-dried (dark) raisins and with the current trend of producing raisins on elaborate trellis systems, including overhead arbors, California continues to have problems with overproduction. In 2000, the United States produced the largest raisin crop in the nation's history with 427,000 tons (388,182 t). The goal of the United States is to seek ways to increase raisin sales in the domestic and international markets while continuing to be competitive.

California raisins are produced mostly from Thompson Seedless, however this acreage is expected to decline over the next several years. California wineries have utilized the Thompson Seedless for about 25% of their total crush. Winery demand for Thompson Seedless grapes since declined sharply due to increases in plantings of other white wine varieties. In the year 2000 only 12% of Thompson Seedless was crushed.

Turkish production of Sultanas is expected to remain fairly stable over the next several years at an estimated 250,000 tons (227,273 t) annually. Turkey produces primarily for the export market and in 1999 exported 212,047 tons (192,770 t) or about 84% of its total production. The raisin producing areas of seedless grapes decreased slightly in 1999 after high demand from both the international markets for Sultanas and the domestic market for fresh seedless grapes promoted increased plantings in recent years (GAIN Report #TU9043 FAS, USDA, 1999). The Turkish industry is focusing on improving quality rather than increasing quantity. Packaging will be a likely future trend in an attempt to increase consumer-pack for exports.

Iran continues as the world's third largest Sultana and raisin producing country. Of its total production, 78% is exported. Iran supplies 16% of all the raisins entering the export market. Iran uses unique irrigation called a "ghanut" underground canal delivery system. More recent plantings use traditional means of pumping water from wells. Iran doesn't have any special programs for raisin production and marketing, and there is a great need for cooperatives and advisory organizations for grape and raisin production in Iran.

The Australia Sultana and raisin industry is on a continual decline. Due to an increase in export demand for Australian wines, plantings of premium wine grapes is replacing multipurpose grapes like Sultana and Muscat. The Australian Dried Fruit industry further states "that in past years multipurpose grapes have accounted for around one-third of the overall wine grape intakes. This is expected to fall to less than one-fifth by the 2001 season" (Attaché query, 2,000).

South Africa's raisin industry produces primarily for the export trade, which accounts for about 84% of its total production. South Africa produces three kinds of raisins from the Thompson Seedless grape. Grapes are laid out on cement slabs upon reaching 23% sugar and are dried in the sun, producing a dark brownish raisin. A second type is produced by dipping the fruit in a unheated lye solution of sultanol and potassium carbonate for a short time. The fruit is dried in the shade on wire shelves and produces raisins of a light brown color. Golden raisins are the third major product and the most expensive. This fruit is dipped as described above, sulfured and then dried in the shade producing a golden raisin. South African marketing policy has made an abrupt change since the Dried Fruit Board in 1997 closed and freed the industry from all controls. There are no subsidies or export incentive programs, however the industry is protected by a 16% import duty.

According to the Central Confederation of Sultana Cooperatives, Greece's major Sultana vineyard plantings in Crete are expected to remain the same. Greece produces primarily for the export market supplying about 4% of the world market. About 95% of Greece's total production of Sultanas is exported. Greek producers expect new regulations regarding their Sultana production, processing and trades. There will be more emphasis on improving packaging and quality control.

Chile's raisin production is dependent on the country's table grape industry and to a lesser extent on the wine grape industry. When prices for table grape culls become higher than usual, wineries purchase less culls. Since Chile's raisins are produced from large size table grapes culls and stripings, they have a lesser demand and command lower prices in the world markets. However, some of the large size, high quality raisins are used in snack foods may receive price premiums. Over 90% of Chile's raisin production is exported.

Other raisin producing countries that have the potential for greater participation in the world trade are China and Argentina. Eighty percent of China's reported 70,000 tons (60,000 t) is consumed domestically. The remaining 20% is sent to Hong Kong and Macao.

Argentina continues to increase production and presently consumes about 35% of its production domestically with 65% going to the export market. Argentina is planning to expand its export market.

References

Adams, Brad. Climatic data for Fresno, California, 1961-1990. National Climatic Center, Asheville, North Carolina. Personal communication.

Amerine, M. A. and A. J. Winkler. 1944. Composition and quality of musts and wines of California grapes. Hilgardia 15:493-695.

California Agricultural Statistics Service. California fruit & nut review 1998. 18(9).Sacramento, CA.

California Agricultural Statistics Service. Summary of county agricultural commissioners reports. Gross value by commodity groups. California. 1966-67. Sacramento, CA.

California Agricultural Statistics Service. 1990. California's 50 major agriculture crops. 1990 dot maps. Sacramento, CA.

California Agricultural Statistics Service. County agriculture commissioners reports 1998. Sacramento, CA.

Raisin production by counties 1994-1998. County agricultural commissioner annual reports for Fresno, Madera, Tulare, Kern, Kings and Merced counties.

Eris, Atila and Erdogan Barut. May 1993. Grape and Small fruit Growing in Turkey.

Eris, A. 1992. Special Viticulture (Turk). Uludag Univ. FAC. of Agric. Pub. No. 52.

FAO Production Yearbook. 1997. 51. United Nations, Rome, Italy.

GAIN Reports. 1999. FAS/USDA. #TU9043, AS9018, SF9007, GR9015, CI9018.

Galet, Pierre. 1979. A practical ampelography. Cornell University Press, Ithaca, New York.

International Sultana and Raisin Conference, 1999, London and 2000, Australia.

Kerridge, G.H. and A.J. Antcliff. 1996. Wine grape varieties of Australia. CSIRO Pub. Australia.

Nef, Clyde. 1998. A history of the California raisin industry under federal and state marketing orders. Malcolm Media Press. Clovis, CA.

Perold, A. I. 1927. A Treatise on Viticulture. McMillan and Company Ltd, London.

Winkler, A. J., A. J. Cook, W. M. Kliewer and L. A. Lider. 1974. General Viticulture. University of California Press, Berkeley, CA.

CHAPTER 2.

The History of the California Raisin Industry

Lillian Quaschnick, B.A.

Fig. 2.1: Horse-drawn wagon of grapes.

Early History

The grapevine has been part of human history, wherever people have put down roots and established a society; they have taken the grape with them to put down to its roots. As people adapted to new lands and new civilizations evolved, so did the grape, evolving to meet the needs and tastes of the new civilizations. The earliest evidence of grape leaves and seeds have been found in fossilized remains of the Tertiary period, roughly 12 million years ago (Winkler, 1965). These remains indicate that the vine flourished in what is now Germany, France, England, Iceland, Greenland, North America and Japan. *Vitis vinifera,* the species that was the source of all cultivated grapes before the discovery of the New World, originated and was first cultivated in Asia Minor. Remains of grape seeds were found in the refuse mounds of Bronze Age lake dwellers of south-central Europe indicate grapes were part of these cultures. The Egyptians were the first to produce and appreciate wine around 2800 B.C. (Bonner, 1983). The pharaohs were entombed with jugs of wine inscribed with the vintage, the name of the wine maker, the name of the grape used and the location of the vineyard from which the grapes came. Stuller and Martin (1989) point out that although the Egyptians are well known for the pyramids, they were the first to define the sciences of viticulture and enology. Wine is mentioned in the Old Testament and grapevines are used figuratively in the New Testament as a metaphor for Christ and the Christian life.

The Bible also mentions raisins in Samuel, 30:12:

> "And they gave him a piece of cake of figs and two clusters of raisins, and when he had eaten, his spirit came again to him" and again in Solomon's Song of Songs where he wrote, "Restore my strength with raisins and refresh me with apples!"

Sometime before 600 B.C., the Phoenicians transported grapevines to Greece and Spain. The Greeks produced wine, table grapes and raisins and grew varietals, many of which are still cultivated today. Greek wines were sought out and coveted as the best. The Greeks introduced the grape to the Romans. Both cultures literally worshipped wine and its gods, the Roman Bacchus and the Greek Dionysus. In the early part of the first century A.D., Pliny described 91 varieties of grapes and 50 kinds of wine (Hedrick, 1924).

The Roman legions brought grapevines to all parts of the Empire. Spain's climates produced a variety of grapes and raisins dried from Muscat. They were first produced along Spain's southern shores (Bonner, 1983). Eisen (1890) states that Spain exported raisins as early as 1295 A.D., and had been producing raisins for centuries before that date.

It is likely raisins and table grapes were introduced from

the eastern Mediterranean to North Africa and from there, to the Far East through Persia and India before the second century A.D. (Winkler, 1965). Islamic religious beliefs forbade the consumption of alcohol and the wine varieties contained seeds that made eating unpleasant. So varieties of grapes better suited to eating were developed and grown in Syria, Mesopotamia, Egypt, Persia and North Africa. Wine varieties virtually disappeared from Muslim areas. Developed in Persia, Sultana was the first seedless grape. When the Turks conquered Byzantium, the Greeks and Armenians, under Turkish rule, began to cultivate the Sultana as a table and raisin grape instead of cultivating wine grapes. Bonner (1983) credits the Greeks and Armenians from the Aegean with establishing the today's raisin industry, perfecting not only the cultivation of the Sultana, but also the drying process. The Sultana would change the Greek grape industry. Today Greece is no longer known for its wine but for its raisins.

Leif the Lucky was the first European to visit America according to Icelandic records and christened the new land Vinland (or Wineland) presumably for the grapes found in the new land. However, it is more likely he found cranberries. Captain John Haukins mentioned wild grapes while visiting the Spanish settlements in Florida in 1565. Sent by Raleigh in 1584, Amadas and Barlowe describe the coasts of the Carolinas of North America as being full of grapes. In 1606, Captain John Smith describes the grapes of Virginia and recommended the culture of the vine as an industry for the newly founded colony (Hedrick, 1924). While most explorers of the Atlantic seaboard mention grapes, none saw intrinsic value in these wild vines. To the Europeans, only Old World vinifera grapes were worth cultivating in the New World.

California History

The history of the grape in California really begins in Mexico. When Spain conquered the Aztec Empire and began settling the New World, missionaries followed, bringing a variety of new plants with them including grapes. Vines were planted in Mexico as early as 1520, possibly in the high plateau around Mexico City (Husmann, 1880). Vines were first planted in Baja California in 1697 by Father Juan Ugarte at Mission San Francisco Xavier (Winkler, 1965), and the vines immediately flourished in this new environment. Just 70 years after Cortez ordered plantings for the new world, the King of Spain ordered the Spanish governors of Mexico to issue edicts forbidding the planting of new vineyards in an attempt to curb surpluses (Bancroft, 1883). The Church resisted this civil decree and continued planting grapevines within the confines of the missions. The Church was therefore responsible for the growth and proliferation of grape culture in North America in the seventeenth and eighteenth centuries (Winkler, 1965).

Father Serra brought grapes from Baja California to Mission San Diego in 1769. Vines were planted at each new mission as the chain moved northward. Centrally located and ideal for grape cultivation, the vineyard at Mission San Gabriel became the "vina madre," or mother vineyard, for the missions in California (Winkler, 1965). The variety planted came to be called the Mission but appeared under many names, such as *Alicante*, *Grape of Los Angeles*, *California*, *El Paso*, and *Native Grape*. By 1775, grapes were well established in California (Bonner, 1983) and by 1800 wine was being made at missions San Diego, San Juan Capistrano, San Buenaventura, San Gabriel, Santa Barbara, and vineyards were established in San Carlos, Soledad, San Antonio, San Luis Obispo and Santa Clara (Winkler, 1965). In addition to the mission vineyards, a 100-acre vineyard was established in the area of Los Angeles in 1831. By 1848, there were a number of small vineyards at Santa Barbara, Santa Clara, and in the Sacramento Valley (Bancroft, 1883).

The Spanish had a permanent effect on the agriculture of California even though the mission period lasted only 60 years. The grapes they cultivated came to be called the Black Mission wine grape and the Muscat raisin grape, reflecting this Spanish heritage. Although seeded, the Muscat would dominate the raisin market for many years.

In 1834, a newly independent Mexican government secularized the missions. As a result, padres destroyed much of what they had developed including vineyards. The secularized lands became the basis of land grants that were sold to settlers.

After the Gold Rush of 1849, settlers made use of the agricultural practices introduced by the Missions including drying grapes in order to preserve the fruit. By one account, A.P. Smith of Sacramento introduced the Muscat of Alexandria to California and Francis Eisen introduced the Muscat Gordo Blanco, a strain of the Muscat of Alexandria, in 1861 (Winkler, 1965).

In 1890, Gustave Eisen started documenting raisin production in California. Eisen identified Colonel Agoston Haraszthy, a recent immigrant and wine expert, as being the first to introduce raisin grapes to California. Haraszthy grew Muscat seedlings in 1851 and imported the Muscat of Alexandria in 1852 and in 1861. He also imported Gordo Blanco, Sultanas from the Malaga district of Spain and white and red Corinth from Crimea for use in his San Diego County vineyard (Eisen, 1890).

California Governor Downey sent Haraszthy to Europe in 1861 to gather information on viticulture for adoption to California. Haraszthy imported 200,000 cuttings representing 300 varieties and a wealth of information about the care of different varieties for use in a winery he was building. Among the cuttings was a seedless raisin variety, however Haraszthy's interest was in wines. This oversight may have delayed development of the raisin industry in California.

Haraszthy gave his seedless cuttings to nurserymen near the northern California town of Yuba City. William Thompson propagated the seedless variety that has become the major raisin used in California. This variety became known as Thompson Seedless. (See "A Lust for the Lady de Coverly" Feature at the end of this chapter). Since it was planted in the cooler northern region of California, the potential of Thompson Seedless as a raisin grape was not recognized. Planted in the wrong climate, pruned incorrectly and allowed to sprawl on the ground, the Thompson Seedless did not do well and went unappreciated for many years. Therefore, Muscat of Alexandria, introduced by Haraszthy in 1852 remained the choice for making raisins.

The first successful raisin vineyards in California were planted by G.G. Briggs in Davisville and by R.B. Blowers in Woodland, west of Sacramento. They grew and shipped raisin grapes on a large scale. The Briggs vineyard was primarily Muscat of Alexandria while the Blowers vineyard was Gordo Blanco. The vineyards produced raisins as early as 1867, entering the market in 1873. The majority of the 6,000 boxes produced in California came from these two vineyards (Eisen, 1890). This marked the origin of the California raisin industry (Husmann, 1902). By 1882, Blowers was the largest single raisin producer in the state (Meyer, 1931), and Blowers and Biggs were operating a dehydrator by 1880 (Winkler, 1965).

Varieties, like Thompson Seedless, were routinely planted in unsuitable climates (Eisen, 1890). 1873 is considered to mark the beginning of the raisin industry in California, and is the same year raisins were being produced and sold in Northern California. Muscat of Alexandria was first planted in the San Joaquin Valley of Central California and they flourished.

By 1870, very few settlers had moved into the San Joaquin Valley as far south as the army outpost that was the seat of Fresno County. The sparse population at the army post in Fresno County consisted of the military, some gold miners and a few cattlemen and sheepherders. The settlers were unaware that this vast, warm, arid region concealed a large aquifer beneath it, but they did use the large rivers for irrigation.

In 1872, Francis T. Eisen, Gustav Eisen's brother, planted 240 acres of wine grapes six miles east of Fresno along the Kings River. Experimenting with grape varieties, he planted 25 acres to Muscat of Alexandria in the fall of 1873. Due to a very hot summer of 1876, the Muscat grapes dried into raisins on the vines. Eisen stemmed and packed the raisins and sent them to upscale grocers in San Francisco who passed them off as Peruvian imports. Eisen is credited for being the first raisin producer in Fresno County (Vandor, 1919). The events that followed are listed in **Table 2.1**.

Bernard Marks introduced the colony system to Fresno. Large tracts of land ranging from one to 10 sections (1/10 square miles) were purchased and developed with a system of irrigation canals and roadways, and subdivided into 10 to 20 farms. Long-term mortgages, water rights and the prospect of living near other settlers attracted buyers. Merchants were attracted by the availability of a variety of crops all grown in close proximity for purchase (Meyer, 1931).

In the earliest days of the raisin industry, large growers were responsible for planting, packing and marketing their own product. Demand for raisins came from the East Coast and was filled by brokers in San Francisco. The new raisin producers in Fresno filled the orders by sorting the raisins into categories of one-crown, two-crown, three-crown, four-crown grades, or top of the line Imperial clusters. Raisins were purchased from each grower with their own label. In 1886, George W. Meade of San Francisco built the first large raisin plant in Fresno, called the Raisin and Fruit Packing Company. Other packing plants paid cash for the grapes on the vine or for the raisins sold them under their own brands through eastern agents.

In 1889, growers produced a surplus of raisins. The packers decided to pay the grower a partial advance on delivery and the balance after the product had been sold. The packers charged a fixed price for packing and a 5% commission for selling the grower's crop, but did not establish a price with the grower until the crop was sold. This system left all the risk with the grower but worked well as long as it involved just a handful of packers.

Growers were still receiving more for their crop than if they sold it for cash. Packers began selling processed raisins strictly on commission, causing quality and prices to fluctuate radically. Growers were not represented in marketing raisins and found themselves under the control of the packers.

Raisin Grower Associations

Most of the growers formed the California State Raisin Growers Association (CRGA) in 1892 after the worst price decrease in history[1]. Members contracted at a minimum price of $90 per ton, but their agreement with the packers was weak and the packers quoted resale prices far below this minimum. Packers outside of the CRGA then quoted even lower prices. Panicky growers immediately withdrew from the association and rushed to sell their crops. The result was an end price to growers of about $25 per ton, instead of $90 per ton agreed upon under the CRGA. The association collapsed after one harvest and dissolved within a year. This pattern repeated many times under various scenarios, including grower-organized associations, unions and cooperatives, and grower/packer associations and companies, some organized by the growers, others by packers.

Nevertheless, the field price for raisins remained depressed from 1892 through 1912. Some packers refused to honor contracts and the growers found they had no recourse. Growers proposed another cooperative with the expectation that it would be strong enough to enforce their demands. W.R. Nutting was asked to head the committee to organize the new cooperative called the California Associated Raisin Company (CARC), because of his prior work on behalf of the industry.

Sale of capital stock in the $1,000,000 company was impaired by the poor financial condition of the raisin growers. As an alternative, the sponsors permitted the sale of the stock on the installment plan. By November 15, 1912, more than $400,000 in stocks were sold. CARC was organized under the laws of California that was later to become the Sun-Maid

Table 2.1
Prices Paid to Raisin Grower Members of the California Associated Raisin Company in 1912 and 1920 by Variety ($/ton).

Variety	1912	1920
Muscat	$69	$223
Thompson Seedless	$78	$295
Sultanas	$65	$242

raisin cooperative. The company was a grower cooperative with authority to carry on all of the functions of packer and distributor. From 1912 to 1923, CARC handled 87 to 92% of the raisins in the state (U.S. Tariff Commission, 1939).

Twenty-five trustees were elected to head the cooperative. On November 27, 1912, the articles of incorporation for the new raisin association were filed. The directors agreed on a price structure reminiscent of the one proposed by the CRGA in 1892. Growers would be paid $60 per ton for Muscat raisins if 60% of the crop was committed, and up to $70 per ton if 85% of the crop was committed.

The grower contract provided for an average minimum price that would be greater than the price paid for in the past 10 years. The first contracts were executed for the 1913, 1914 and 1915 crop years, with the option to renew the contracts for 1916 and 1917.

The new CARC began the 1913 season by purchasing raisin packing plants and contracting to handle 25,000 tons of 1912 crop surplus raisins. The cooperative contracted with 16 of the 19 major packers to pack raisins, guaranteeing the packers a $5 per ton profit. The cooperative maintained the right to sell its product directly to market.

In its first year of operation, CARC advanced growers $65 per ton on 1913 crop deliveries. By September of that year, 100,561 tons of raisins had been shipped. The new cooperative controlled the California raisin crop. In 1914, the cooperative appropriated $100,000 to launch an advertising campaign under its Bear brand. One of its more flamboyant gimmicks was to ship 1,250 tons of raisins to Chicago on a special train of 60 freight cars. Each car advertised that it was carrying California raisins grown by 6,000 growers. The campaign was successful.

The majority of grapes grown for raisins were concentrated within a 50-mile radius of the city of Fresno, now a one-crop community. Dried tree fruits, especially prunes and apricots, had a healthy export market and the merchants who exported them to Europe would later add California raisins to their inventories as the foreign users, especially bakers, began to appreciate the qualities of free flowing California seedless raisins. This emphasis on advertising resulted record sales in 1914 and by September 15 the entire 1913 crop had been sold.

War in Europe increased demand for raisin sales since Zante currant imports were curtailed. Nearly all the 1914 Sultana and Thompson Seedless crops were sold. Shipments increased from 7,183 tons to 14,784 tons. Payments to growers for the season were $70 per ton for Muscat, $65 per ton for Sultanas and nearly $80 per ton for Thompson Seedless raisins. CARC announced assets totaling $2,242,918 and paid $1,032,688 to the grower members. At the end of the 1914 crop year, the CARC announced that its growers were paid $2,275,000 more for their 1913 and 1914 crops than they would have received without a cooperative organization (U.S. Tariff Commission, 1939).

Wylie M. Giffen, a prominent Fresno rancher and grape grower succeeded H. H. Welsh as president of CARC. Giffen announced the selection of Sun Maid as the association's brand name,-"chosen because raisins are a product of our marvelous California sunshine." A new advertising program was established, directed toward popularizing CARC's new Sun-Maid brand.

In May 1915, the cooperative hired three young women to attend the San Francisco Exposition fair as ambassadors for the raisin industry. They promenaded around the fair passing out samples of Sun-Maid raisins. Each afternoon, they flew over the fair in a plane and showered the crowd with raisins. The girls were paid $15 a week for their work. Lorraine Collett was asked to pose for a watercolor wearing her own red bonnet and carrying a basket of artificial grapes. The watercolor became the Sun-Maid trademark[1].

Expansion of Thompson Seedless

Good business induced a substantial increase in raisin grape plantings, particularly of Thompson Seedless vines. Even though Thompson Seedless would prove to be a superior raisin grape variety, the conversion from Muscat of Alexandria had been slow primarily because vineyards were expensive to replant. Most growers believed that it was better to plant a grape variety that could be marketed in more than one way. The advent of prohibition destroyed the demand for Muscat of Alexandria as a wine grape and the presence of seeds in this variety reduced its value as a raisin variety (Bonner, 1983).

In 1917, the CARC announced a goal of 125,000 acres of Thompson Seedless, and offered new three-year contracts with a three-year renewal option. By February 1918, CARC claimed to have exceeded its 125,000 acre goal by 6,530 acres. This represented more than 85% of all raisin growers in the state. The cooperative's 1920 financial report showed it handled 152,500 tons of a total crop of 173,528 tons . The cooperative's growers experienced eight years of increases in prices for their crop; the price paid to the growers increased from about $70 per ton in 1912 to over $200 per ton in 1920 (**Table 2.1**).

The decade following the end of WWI was unstable in the industry with a pattern of prosperity followed by national emergencies. In 1916, a new membership drive in CARC brought in 88% of the acreage and 9,000 growers under contract. By 1918, CARC experienced legal challenges due to independent packers periodically asking the Department of Justice to investigate the legality of the company practices under the Sherman Anti-Trust Act. Withdrawals became more difficult with the 1917 contracts and complaints about the monopolistic aspects of the association were made in 1920. Following the recommendations of the Federal Trade Commission, the company reduced its percentage of control and provided more liberal terms of renewal and withdrawal (U.S. Tariff Commission, 1939).

In September of that year, a petition was filed in the case of the United States v. California Associated Raisin Company in the District Court at Los Angeles. The petition charged

[1]From a Sun-Diamond Growers of California brochure reprinting of an article by Los Angeles Times writer Dave Larsen which appeared in that publication in 1982.

CARC with monopolizing and restraining the market of raisins in violation of the Sherman Anti-Trust Act. The government contended that the price fixed by the company for the 1919 crop was excessive, and the company, controlled over 80% of the raisin crop. This allegedly constituted a monopoly that forced buyers to accept prices set by the association. CARC contested that the price was actually less than it would have been had the company not intervened, the growers would not have received the agreed price in 1918 (Colby, 1924). The company's position as a cooperative was also questioned because non-growers held stock in the company and the growers themselves had little direct involvement in formulating the policies of the association. The agribusiness community watched with interest as the outcome of this case would greatly affect the future of agricultural marketing practices.

The investigation and deliberations resulted in a compromise. The company was not forced to dissolve. Instead, a court order directed the company not to use coercion to obtain contracts and to stop certain price-fixing practices, and CARC was reorganized. The court decision also exerted influence in passing the Capper-Volstead Act which authorized the agriculture industry to organize marketing associations that uniquely met their needs (Colby, 1924).

In 1920, the Federal Trade Commission summarized the state of the industry in the twenty years prior to the formation of the CARC in 1913. Prior to the CARC's organization, growers were at the mercy of those packing and selling raisins, and they received varying and uncertain compensation for their crops. Packers seemed to insure sufficient income to operate. However, they did not always compensate the growers sufficiently to cover the cost of making the raisins. In response, growers repeatedly attempted to form cooperative organizations to bargain collectively. These cooperatives failed largely because of lack of financing and credit. The federal investigation contended that the existence of cooperatives was immaterial. The disagreement between growers and packers forced the growers to organize and protect their interests (Colby, 1924).

Sun-Maid Raisin Growers

In February 1922, CARC changed its name to Sun-Maid Raisin Growers and used the painting of Lorraine Collett as a sun maiden holding a basket of grapes. This trademark appeared on all Sun-Maid products. However, depressed raisin sales, cash advances to growers and abundant crops in 1921 and 1922 forced Sun-Maid in financial difficulty. The company was forced to reorganize again. Sun-Maid still had 87% of growers under contract, but the supply of raisins exceeded the market demand.

Fig. 2.2: John Gallo, a Biola, California raisin farmer, checks his freshly laid grape clusters on paper trays that replaced the traditional wooden trays.

The Fresno economy was dependent on raisins and community members, many of whom were facing bankruptcy themselves, wanted Sun-Maid to prosper to the extent that decisions were based more on political agendas than balancing supply and demand. This resulted in animosity between those who supported Sun-Maid and other growers and packers.

In 1922, war in Turkey destroyed Turkey's raisin production and processing facilities. Since Turkey was the largest raisin producer, this had a significant effect on the world markets. Demand for California raisins rose dramatically, doubling demand in Europe (Bonner, 1983). Increased demand encouraged planting, not only by growers, but also by entrepreneurs hoping to cash in on the increased demand for California raisins. The plantings were almost exclusively Thompson Seedless. Raisin grape vineyards expanded to 600,000 acres during the 1920s in a surge of growth not to be duplicated for another 50 years (Bonner, 1983).

Sun-Maid Growers continued in its new direction, and developed a program of advertising and promotion at a cost of $2.5 million. However, sales of the 1927 and 1928 crops did not produce enough capital to cover the promotional program after paying growers.

Sun-Maid controlled about 90% of the raisin acreage, in spite of high prices and high consumption. Outside packers thought they were finished (Bonner, 1983). Sun-Maid diverted stored raisins to other packers and payment to Sun-Maid growers was sometimes spread over several years. The financial burden caused some growers to sell part of their crop to other packers for cash. Grower contracts called for a description of the geographical boundaries of their farms with a tonnage estimate to prevent breach of contract. Nevertheless, Sun-Maid competitors were convinced that Sun-Maid was not following the Capper-Volstead Act and encouraged this breach of contract (Bonner, 1983). Soon the independent packers developed an ample supply of raisins. Due to standardized quality testing, everyone's raisins were virtually the same in quality. The independent packers did not advertise, but could still profit even when they bought Sun-Maid's surplus and sold it, along with their own, below Sun-Maid's price.

About the time the new California plantings came into production, Turkey and Greece had recovered from depressed production. England and Canada were buying more raisins from within the Commonwealth, Australia having Commonwealth preference. As a result, California raisin prices fell from $290 to $40 per ton (Bonner, 1983). The raisin industry found itself in a depression that would soon take the nation. By the beginning of 1928, Sun-Maid carried an $11 million debt (U.S. Tariff Commission, 1939). Within a year, the investors took control of the Delaware organization, which owned the packing plants contracted by Sun-Maid. They secured their loans by using the world-famous trademark on other products.

In 1929, a combined loan of $9 million was obtained from the banks of California and the Federal Farm Board for advances to the growers, and later, an additional loan of $4 million was secured from the Federal Farm Board. Nevertheless, Sun-Maid was facing bankruptcy within a year. In a move to protect its former loans, the Federal Farm Board, proposed plan to control the grape surplus and underwrote Sun-Maid's obligations (U.S. Tariff Commission, 1939). The Farm Board allocated $14 million to the raisin industry. Sun-Maid received $4 million (U.S. Tariff Commission, 1939).

Sun-Maid worked on consolidating and improving its financial base. Grower membership dropped from representing 87% of the raisin crop in 1922 to 32% (Bonner states the low as 25%). Following Sun-Maid's financial reorganization, market share rose to 41% in 1932 (U.S. Tariff Commission, 1939), and grower members paid back the banks. The organization abandoned its policy of dealing directly with retailers and began marketing its products through the customary brokerage channels the independent packers used in the United States and Canada.

WWII ended the depression for the raisin industry. Greece was occupied by Germany and was unable to export, and Australia could not safely ship to the United Kingdom. These two factors, coupled with a serious sugar shortage, opened markets for California raisins in Europe and the United States. At the beginning of the 1940s, raisin sales were approximately 225,000 tons, of which 75,000 went to Europe (Bonner, 1983). In 1943, the War Powers Act mandated growers to dry all their raisin grapes for food. California produced more than 400,000 tons of raisins. Of this tonnage, 150,000 tons were sent to the armed forces and the Allies, and 250,000, double the norm, was sold domestically in one day (Bonner, 1983). Growers again began planting Thompson Seedless so when the war ended, domestic demand decreased from 250,000 tons to about 120,000 tons, and exports fell from 150,000 tons to 10,000 tons((Bonner, 1983).

Grower Assistance Programs

From 1946 through 1948, the U.S. Department of Agriculture made huge sums of money available to the industry in the form of grower assistance programs. In 1947, the Department of Agriculture purchased more than 100,000 tons of surplus raisins. The Department warned industry leaders that the industry must take measures to correct the problem or the Department could not continue to help. By 1948, the raisin industry was in the worst imbalance of supply and demand in its history and a plan to solve the problem had not been proposed.

Again, the Department of Agriculture came to the industry's assistance, buying over 60,000 tons of the 1948 surplus raisins, with the understanding that this was the last time. The industry now put its efforts into the development of the Raisin Marketing Order, using the Marketing Act of 1937 as a model for controlling the supply of raisins. Developed by a group of industry leaders, the Federal Raisin Marketing Order went into effect on September 1, 1949 (see **Chapter 12**). The Marketing Order managed the supply and distribution of raisins in an effort to stabilize the market by determining market demand. Based on demand and production, the raisin crop was allocated into free, reserve and surplus pools.

The industry immediately took steps toward achieving economic stability, and in its first year misjudged how much to keep in reserve and how much to allocate as free tonnage for sale. As a result, the Raisin Administrative Committee (RAC), which directed the Marketing Order, sold 19,000 tons of what it considered surplus raisins for cattle feed at a price below the production costs. The 1950 crop, however, turned out to be lighter than expected and the raisins sold as surplus could have been sold at market price. The order was amended for the first time, a process it has undergone many times since.

The first chairman of the Raisin Administrative Committee was A. Setrakian, a grape grower and winery owner. "Sox," as he was known in the industry, held the position for 22 years. During his tenure, generic advertising, mandatory inspections and quality control programs were implemented. He was strong-willed and grower-oriented, perpetuating the adversarial relationship between the grower and packer. In the mid 1950s, he saved a depressed market by going overseas to sell 55,000 tons for $80 per ton from the reserve pool (Clough, 1986). He went on to try to fill all the overseas orders from the lower-priced reserve tonnage, a move that enraged packers and caused the government to step in and decree that Setrakian's methods were outside the legal parameters of the marketing order. "Sox" retired shortly before his death at age 88. After his death, the by-laws were amended to limit the term of chairman to two years.

During Setrakian's term, he received a letter from a 20 year-old grower named Ernie Bedrosian who was very unhappy with the open-price contracts for wine grapes. Bedrosian, like Setrakian, supported the Marketing Order for raisins and managing the supply of raisins and Bedrosian was a member of Allied Grape Growers, which was against a marketing order for wine grapes. Setrakian enlisted Bedrosian's help, and in 1960, Bedrosian helped Setrakian enact the first referendum for the marketing order for wine grapes (Bedrosian Interview, 1/16/1996).

In 1963, at the age of 26, Bedrosian, with Setrakian's help, was appointed to the Grape Crush Administrative Committee, which oversaw the wine grape Marketing Order for California. After criticizing Allied Grape Growers on television, accusing the marketing order for not supporting the grower, Bedrosian's Allied contract was terminated. That year the E and J Gallo Winery offered prices higher than market for wine grapes and the marketing order lost the support of the industry.

In 1963, Bedrosian was elected to the RAC. Bedrosian decided that the raisin growers not contracted by Sun-Maid Growers needed their own bargaining association because he suspected the decisions regarding the industry were dominated by Del Monte CPC, West Coast Packing, and Bonner Packing. When the Raisin Marketing Order was established, there were three options for allocating the raisin crop: reserve pool, free tonnage, and surplus. Bedrosian expressed concern about the movement of raisins from the grower to the packer and ultimately to market and believed that the surplus undermined the control of the amount of raisins made available to the packers. He wanted any association that was organized to have only free and reserve pools.

Prior to the implementation of Marketing Order, independent packers depended on Sun-Maid to offer a price before they set their price. When the Marketing Order was enacted, the packers avoided higher prices set for raisins by setting impossibly high standards for incoming raisins and discounting the field price based on defects.

In his efforts to establish a workable bargaining association, Bedrosian studied the California Canning Peach Association and adopted their concept. He believed that if he could solicit commitments for 50,000 tons of raisins, it would give him the leverage to establish a bargaining association. Setrakian arranged for Bedrosian to speak to Frank Bennett, Director of Agriculture for the State of California (today called the Secretary of Agriculture). Setrakian wanted Bedrosian to chair the new association (Bedrosian Interview, 1/16/1996).

Raisin Bargaining Associations

On October 31, 1966, Bedrosian called a meeting of 20 to 30 growers at Bruce's Lodge near Fresno. Frank Bennett gave a motivational speech describing the Peach Bargaining Association. Later, a committee of 12 spent a week at the Hotel California in Fresno, drafting the by-laws for the new Raisin Bargaining Association (RBA). Bedrosian believed that no one should become entrenched in a position of leadership or have an inordinate amount of power and the by-laws reflected this. There would be no salary for those involved in administrating the RBA, and the president would be limited to a two-year term. After the by-laws were written, the original group of 12 was expanded to 52 growers representing two from each raisin district.

The campaign to sign up RBA members included weekly articles in *The Fresno Bee*, the local newspaper. Bedrosian would give a weekly update of tonnage signed and urged growers to sign up to the extent of visiting their homes. The signed contract was held by a certified public accountant for safe keeping and the tonnage reported in the next week's *Fresno Bee*. On December 30, 1966, the RBA was incorporated, and by February 1967, the goal of 50,000 tons mark was attained and the RBA was declared operational.

The new organization immediately set about lobbying for changes in the federal marketing order. Ultimately, the surplus pool clause was removed and a preliminary tonnage was set after a price was established in the field, doing away with the open price contract. Bedrosian believed that this concept was the key element for the RBA to work. Until the price was set, packers could have only 65% of available tonnage. In the RBA's first bargaining year field price increased $55 per ton to $305 in 1967, and to $312 and $317 per ton in 1968 and 1969, respectively.

With the strength of its membership, the RBA replaced Sun-Maid in establishing field price for raisin and the competition between Sun-Maid and the RBA intensified. The RBA bargained for and negotiated higher prices throughout the 1960s. Sun-Maid was forced to offer growers a better price than the RBA in order to keep its members. There was and still is a difference in the way payments are made by the two organizations. For example, in 1984, the RBA negotiated a price of $700 per ton. By the end of the crop year, Sun-Maid announced that their members had received $790 per ton. However, RBA members received payment for free tonnage immediately upon delivery to the packer. Sun-Maid growers received a partial payment on delivery, then a series of payments over the next 15 months. Sun-Maid growers were also required to provide the cooperative with a capital fund through an annual loan of $25 a ton. This no-interest loan is kept for seven years before being returned to the growers (Clough, 1986).

Rough Times in the Raisin Industry

The 1970s and early 1980s were to be some of the most prosperous years in history of the raisin industry, due to high land values, available loans and a shortage in supply due to rain in several crop years devastating three raisin crops.

In 1973, Sun-Maid began focusing on the export market, setting a new direction for the entire industry. Throughout this period the industry expanded the foreign markets it had and opened new ones. In March 1972, the raisin vineyards experienced one of the heaviest frosts in years. The raisin harvest, expected to be 225,000 tons, came in at 110,000 tons, creating a shortage that nearly doubled the price to the grower for the next two years. In 1975, heavy rains during harvest reduced the crop to about 90,000 tons. The short crops precluded exports and in fact opened the U.S. to imports for the first time. To meet demand, domestic retail prices exceeded $2,000 per ton for the first time. Three years later, rains during harvest reduced production to 75,000 tons. Retail prices soared.

High field prices inflated vineyard value, and growers borrowed against their assets and expanded. Growers ready to retire sold their vineyards to eager newcomers for prices up to $15,000 an acre, often carrying the loan themselves. There seemed to be no end to the raisin grower's good fortune. There were rains at harvest in 1982 and the raisin crop dropped to 205,700 tons. This increased the grower a price to $1,300 per ton. In addition, many growers had purchased crop insurance and received an additional $600 a ton in insurance claims.

Prospering growers were confounded when the European Common Market's agricultural subsidies and the strong dollar gave its member nations two advantages to undercut U.S. prices for its raisins and wines. Greece sold raisins to the European community for prices so far below the U.S. that well-established foreign markets were lost. In addition, subsidies and the strong dollar allowed France, Italy and Germany to export wines to the U.S. for prices below those of domestic wines. The domestic wine market lost 25% to European imports (Clough, 1986). Thompson Seedless growers found themselves caught in a dilemma. Reduced demand for Thompson Seedless grapes for wine, in conjunction with depressed foreign sales of raisins plunged the raisin industry into the worst depression since the Great Depression. Nevertheless, overall acreage-remained unchanged.

The depressed markets had an interesting effect on the raisin market. In 1983, 48% of raisins were selling retail and 52% were selling for industrial use as ingredients in breakfast cereal, baked goods and other foods. When the price dropped, industrial users expanded the number of products containing raisins as an ingredient and increased the amount of raisins they added to their existing products. Products using raisins increased from about 18 products in 1982 to over 100 products. Typically, cereal companies report their formulations today to be from 24% to 32% raisins by dry weight. As of July 31, 1998, 31% of raisins sell retail and 68% sell for industrial use (Koligian Interview, 5/28/99). Thanks to this dramatic increase in industrial demand the market upturn came quickly and the raisin industry depression was short-lived. In 1983, the price was $599 a ton (100% basis). By 1986, the priced had climbed to $777 a ton and by 1988 it was over $900.

Raisin Promotion and Advertising

In 1986, in an effort to improve the public's perception of raisins, the California Raisin Advisory Board (CALRAB) which administrated the State Marketing Order, requested that the advertising agency Foote, Cone, Belding/Honig, to come up with a promotion for California raisins. The result was a "Dancing Raisins" promotion using "claymation" (**Figure 2.3**). The Board decided to animate the idea and test it. Nef (1998) points out that the whole concept was risky. The program was probably the first major exposure for claymation characters and did not increase domestic raisin consumption, but simply entertained using raisin figures.

Thompson Seedless use began to change in the 1980s. Wine cooler sales were steadily dropping, and new regulations were imposed on the wine industry. Prior to that time, regulations required a varietal wine be made from a minimum of 51% of the grape variety on the label. The regulations were changed to restrict the blend to 25% of varieties not appearing on the label. This reduced demand for Thompson Seedless as a foundation for varietal wines. However,

Fig. 2.3: Dancing raisins in front of the White House in Washington, D.C. as a part of the Raisin Promotion Program.

increased demand for beverages sweetened with natural juices and/or natural fruit sweeteners has increased both in the United States and in the Pacific Rim countries and has offset the reduced demand for Thompson Seedless in wines.

One purpose of CALRAB as the State Marketing Order was to develop generic advertising, such as the Dancing Raisins, to promote all brands of raisins Assessing a fee on all raisins delivered to the packer funded this program. In the mid 1980s, the State Marketing Order implemented a Credit Back Program (Koligian Interview, 5/28/99) under the premise that brand name raisins drove the market. Therefore, processors who paid to advertise and sell their raisins under a brand name were credited for the advertising expenses. When the market shifted to industrial use in the early 1990s, packers began to express dissatisfaction with the program.

Some packers who marketed nearly all their raisins in bulk for industrial use expressed concern that they received no benefit from the assessment they paid for CALRAB to promote retail raisins. Supermarket private labels were also ineligible for credit. Others who didn't have the money to qualify for credit back funds opposed the assessment. Many believed that Sun-Maid, the most widely recognized brand, was being given an unfair advantage. On the other hand, Sun-Maid did not agree it was receiving any benefits from the Dancing Raisins promotion since it already had brand recognition and its own advertising program. Sun-Maid also took issue with National Raisin Company, which prominently displayed the Dancing Raisins on its package, citing that the Dancing Raisins were meant to promote the industry at large and not a particular brand. Although the Dancing Raisins promotion was very popular, domestic raisin sales were on the decline. Unable to reach a compromise, a group of independent packers petitioned the Secretary of the California Department of Food and Agriculture to terminate the State Marketing Order for California Raisins. The order was officially terminated on July 31, 1994. Following the termination of CALRAB, the Federal Marketing Order operating under the RAC assumed responsibility for the industry's export activities.

In July 1998, an 87% vote formed the California Raisin

Marketing Board (CRMB). CRMB was considerably different from CALRAB. CALRAB had an operating budget of between $19 and $21 million received from assessments of $65 per ton from both growers and packers. Of these funds, $6 to 9 million went to advertising and approximately $4.5 million to the Credit Back Program (Koligian Interview, 5/28/99). The new board collected $4.9 million from a $15 per ton assessment of growers. There was no budget for promotion or credit back program. The new board's primary focus was outlined by Vaughn Koligian in his testimony for CRMB given on May 12, 1999, in Washington, D.C.:

> "While health and nutrition research is one key element of the Order, the ability to translate that information into effective marketing material can be found within the program as well....
>
> 1. There will be food processing research directed toward increasing the use and consumer acceptance of raisins in manufactured food products.
> 2. Market research will also be conducted to improve knowledge about new and existing markets for raisins, to improve sales, distribution, consumption and overall usage of raisins.

One section of the Order is devoted entirely to communications and the importance of educating the general public, both domestically and abroad, about the availability, uses and healthful properties of raisins. There will also be communications activities directed toward non-consumer buyers of raisins and raisin products. The new board's funds will be used for promoting raisins and educating the public about this ancient food whose ease of use as well as nutritional and health benefits fit today's life-style as easily as it fit that of the ancient Hebrews, Greeks and Romans."

Acknowledgments

I would like to thank the following industry people for allowing me to interview them for the purposes of this book: Ernie Bedrosian, President, National Raisin Company; Leo Chooljian, President, Chooljian Bros. Packing Co., Inc.; Barry Kriebel, President, Sun-Maid Growers of California; Vaughn Koligian, CEO, Raisin Bargaining Association; Clyde Nef, retired manager, RAC; Jerald Rebensdorf, President, Fresno Raisin Cooperative Growers, Inc.; Richard Van Diest, USDA.

Photo Credits

Figure 1: Biola Packing Company
Figure 2: Biola Packing Company
Figure 3: California Raisin Advisory Board

References

Bancroft, H. H. 1883. History of Mexico. Vol.III. The History Company, San Francisco, CA.

Bonner, Charles W. 1983. A rocky road. The pilgrimage of the grape. Pioneer Publishing, Fresno, CA.

Clough, Charles W. 1986. Fresno County in the 20th century. Panorama West Books, Fresno, CA.

Colby, Charles Carlyle. 1924. The California raisin industry: a study in geographic interpretation. Assn. of American Geographers, Albany, N.Y.

Elliott, Wallace W. 1919. History of Fresno County with biographic sketch of the leading men and women of the county.... Wallace W. Elliott, publ., Los Angeles, CA.

Eisen, Gustav. 1890. The raisin industry. H. S. Crocker, San Francisco, CA.

Hedrick, U. P. 1924. Manual of American grape-growing. MacMillan, N.Y.

Howard, Fred K. 1922. History of the Sun-Maid raisin growers. F. K. Howard, Fresno, CA.

Husmann, George C. 1902. Grape, raisin and wine production in the United States, Yearbook, Dept. of Agriculture. Washington, D.C.

Levy, Louis C., ed. 1928. History of the co-operative raisin industry of California. Fresno, CA.

Meyer, Edith Catharine. 12/1931. The development of the raisin industry in Fresno County. MS Thesis, University of California, Berkeley.

Nef, Clyde E. 1998. The fruits of their labors.... Malcolm Media Press, Clovis, CA.

Nehr, Stanley. 12/1967. Competition in the world raisin markets. Foreign Agr. Serv. U.S. Dept. of Agr. Washington, D.C.

Stuller, Jay and Glen Martin. 1989. Through the grapevine. Wynwood Press, N.Y., N.Y.

U.S. Dept. of Agriculture. 1975. Marketing California raisins. Ag. Mktg. Serv. Mktg. Bul. No. 58, Washington, D.C.

U.S. Tarriff Commission. 1939. Grapes, raisins and wines report #134. Washington, D.C.

Vandor, Paul. 1919. History of Fresno County California. History Record Co., Los Angeles, CA.

Walker, Ben R. 1941. The Fresno County blue book. Arthur H. Cawston, Mngr. Ed. and Publ., Fresno, CA.

Winkler, A. J., J. A. Cook, W. M. Kliewer and L. A. Lider. 1974. General viticulture. Univ. Calif. Press.

A Lust for the Lady de Coverly

Allan Shields

The Grape that can with Logic absolute
The Two-and-Seventy jarring Sects confute.

[Stanza 57, Omar Khayyam (Transl. Fitzgerald)]

The Origin of the Thompson Seedless Vine in California

The longer the remove from the actual events concerning the origin of the Thompson Seedless grape in California, the more numerous become the doubtful accounts, for any misstatement of fact early on gets repeated, and even amplified, through the retelling. The paucity of reliable, written documents on the subject presents a formidable obstacle. Thus, any effort to rediscover the exact facts about William Thompson's saga faces two essential problems: first, to correct any mythical account, and second, to replace the familiar and traditional, but fictitious, accounts with the accurate ones. Myths die hard, even about a bunch of grapes, because people tend to hang onto long held beliefs instead of changing them to be true. Even more difficult an obstacle is a printed account claiming to be a true account. We are disposed to accept the printed word at face value, even when shown it to be in error. Many of the accounts about William Thompsons' great vine have fostered incorrect information.

In broad strokes, the familiar accounts are true. William Thompson immigrated from Yorkshire, England with his family in 1851, eventually arriving in Marysville, California, on August 25, 1863. In 1872, he received three cultivars of the grape Lady de Coverly (a synonym for Sultanina) from the firm Ellwanger and Barry in Rochester, New York. Of the three cultivars Thompson grafted onto his own rootstock, only one survived a flood. This vine produced a prodigious amount of fruit in the fall of 1875, leading to a display at the Marysville Fair in the fall of 1875. To identify the exhibit, a Fair clerk labeled the grapes, "Thompson's Seedless," only as a temporary name during the Fair. In a meeting of the Sutter County Horticutlural Society on August 16, 1888, J. P. Onstott made a motion to name Thompson's grape variety the "Thompson's Seedless;" the motion passed unanimously despite an earlier objection by Professor Frederic Theodore Bioletti of the College of Agriculture, University of California, Berkeley. One source claims they voted to call it simply "Thompson." From 1875 to 1876, Thompson freely gave cultivars from his already locally famous vine to fellow farmers, including J. P. Onstott, the Harter Brothers, and the Stabler family (S. J. and Harry P. Stabler, especially), among others not recorded. Thompson's son, George, tried to dissuade his father from giving cultivars away, urging him to seek a patent on what he believed to be a new variety, hoping to reap dividends on the sale of propagated vines. From Thompson's original vine, thousands of acres of California vineyards were planted with descendant cultivars between 1876 and 1910. In stark outline, this is the true story, often told and retold. Unfortunately, there is much more to

Chronology of the Thompson Seedless Vine

1816
William Thompson is born on September 26 in Wistow, Yorkshire, England.

1818
Ann Marie Whiteley (Thompson) is born September 12 in York, Yorkshire.

1839
George Thompson, son of William and Ann Marie, is born September 4 in Selby, Yorkshire. Two other sons, William and Thomas, were born later in Wistow.

1847
Gustav Eisen is born in Stockholm, Sweden.

1850
Francis Eisen becomes the first vineyardist in Fresno County, California.

the story. The mystery, devil, beauty, and the truth lie in the expanded details.

In an age of jet plane travel, it requires an effort of imagination to return to the mid-1880s, when William Thompson and his young family sailed away from England headed for an unknown future in America.

William Thompson was born in Wistow, Yorkshire, England, on September 26, 1816, and his wife, Ann Marie Whiteley (Thompson) was born on September 12, 1818 in York, Yorkshire. George, their first son, was born on September 4, 1839, in Selby, Yorkshire. Two other sons, William and Thomas, were born in Wistow, Yorkshire. All five family members left Wistow in 1851, crossing the Atlantic Ocean to settle in Macoupin County, Illinois, now called Carlinville. (See "The Bond's Genealogy" at http://homepages.rootsweb.com/~radbud/gedpages/fam01747.htm, for complete genealogical record.) After farming tobacco in Illinois for twelve years, William and Ann sold their farm in 1863, bought two wagons, seven horses, and a cow, then traveled across the continent with others by way of the Platte River and Salt Lake City, arriving in Marysville on August 25, 1863.

From 1863, William Thompson, with the considerable help of George, by then age 24, farmed diversely—pigs, bees, horses, cows, grain, and grapes—while continuously expanding their acreage. The family records fail to specify how many or which grapes they had planted, but it is known that the Muscat Gordo Blanco and Muscat Alexandria were popular varieties in the area.

One document, "Thompson History" by Barbara Vorous, which has some ring of authority nevertheless, as some documents do not, states:

"In 1872, William received a catalogue from Ellwanger and Barry of Rochester, New York who had 6,000 grape cuttings for sale. He ordered three grape cuttings under the name Lady DeCoverly at one dollar each. When they arrived, William successfully grafted them onto three Los Angeles grapevines. Then he ran into trouble. The Sacramento River overflowed its banks and washed away two of the grapevines. The remaining vine proved disappointing because it failed to produce grapes.

"William had pruned it the same as he did his Los Angeles grapes cutting the canes back to three buds and providing no trellis for the vines to climb on. Disgusted with its failure to bare fruit he abandoned his vine, neither pruning nor cultivating it during the following season. The vine, left on its own, continued to grow and climbed a nearby tree. When William next checked on the vine, it had 56 lbs of grapes. The grapes were oval, yellow seedless, with thin skin, good but not strong flavor, and low in acidity, bunches large or very large, and the vine an enormous bearer.

"From this vine the seedless raisin grape propagated into California." (Vourous, 1989)

Later we shall see that there are a number of details in this account that require amendment. Especially questionable is the assurance that Ellwanger and Barry identified the three cuttings as the Lady de Coverly in one of their catalogues, or that the number of cuttings for sale numbered 6,000. Also, the precise number of pounds the vine produced is doubted in other sources, though this is relatively unimportant. The value of this document is that it spells out one doubtful, traditional account of an historical event of some importance.

William Thompson's family history fails to record many details about his years in Yorkshire. We do learn that he was a farmer, and that, in his earlier years, he worked for a time in the glasshouses (greenhouses in the United States) of a local nobleman who grew the vine, the Lady de Coverly. It is said about him that he never forgot the flavor and characteristics of the grape, though he did not remember how to properly prune the vine in 1872-73. His family recalled that he often recalled his experiences with that particular grape and that he recalled the name. In California, he found no grape (between 1863 and 1875) to match the qualities of his favorite. He could not have known in 1873 or 1874 that the vine he ignored was the Lady de Coverly by taste, for it bore no fruit during those two years. It remains to discuss the evidence from documents whether or not he could have known the vine was the Lady de Coverly in another way. There is widespread agreement among documents of various dates that he did mistakenly prune the vine severely in

"A Lust for the Lady de Coverly" first appeared in the Wayward Tendrils Quarterly, Vol.12, No. 1, January 2002 and Vol. 12, No. 2, April 2002.

1851
The William Thompson family leaves Wistow to settle for 12 years in Macoupin County, Illinois, to farm tobacco.

1863
William Thompson family arrives in Marysville, California, on August 25. From 1863-1875, the family farms with diverse crops and stock.

1872
William Thompson receives 15 different varieties of grapes he ordered from Ellwanger & Barry nursery, finds three cultivars E&B have added to his order without charge. Thompson grafts the three unknown cultivars onto his own rootstock. Only one vine survives.

1875
The single Thompson Seedless vine produces about 50 pounds of excellent grapes, which are displayed in the Marysville Fair as "Thompson's Seedless."

1875-76
William Thompson freely gives cultivars of his 'wonder grape' to friends, including John Paxton Onstott, the Harters, and the Stablers—over the objections of George Thompson, who believes the grape to be a new variety.

Fig. 1: George Thompson in his field from Peter Delay's "History of Yuba and Sutter Counties, California," 1924, Historic Record Company, Los Angeles, CA.

1872 and 1873. In the first pruning period, winter of 1872-73, he removed all of the essential bearing wood, thus producing a barren vine in the fall of 1873. For the year of 1874, he ignored the vine, so that in the fall of 1875, with the previous year's rank cane growth (bearing wood), the vine, now vigorously growing and "trellised" in a nearby tree, was discovered to have produced a hefty 50 to 56 lbs of luscious grapes. Two sources state that the amount was an impossible 500 lbs.

Looking back from 2001, we now know the vine to have been a Sultanina, which is synonymous with the Lady de Coverly and the Thompson Seedless. We know this from ampelographical data given in many recent sources. We know this from expert testimony of horticulturists and nurserymen who were contemporaries of Thompson, such as the Harter Brothers, Stablers, J. P. Onstott, and Gustav Eisen and his father, Francis. In the fourth edition of his book, *California Fruits and How to Grow Them,* (1909) Edward J. Wickson, Dean of the College of Agriculture, University of California, Berkeley, following the legendary Dean Eugene Woldemar Hilgard (1833-1916), identifies the relevant varieties:

"*Sultana;* syn. *Seedless Sultana*—'Vine vigorous, upright; leaves large, five-lobed, with rather large sinuses, light colored and coarsely toothed; bunches large, long-cylindrical, with heavy shoulders or wings, well filled when not cultured, but not compacted; berries small, round, firm and crisp, golden-yellow and without seeds.' In California the variety is apt to have some seeds. It has more acid, and therefore more piquancy of flavor, than Thompson's Seedless, but the latter has recently far outstripped in popularity among growers.

"*Thompson's Seedless*—Named by Sutter County Horticulture Society, after W. Thompson, Sr., of Yuba City, who procured the cuttings in 1878, from Ellwanger & Barry of Rochester, New York. It was by them described as 'a grape from Constantinople, named Lady Decoverly.'(Eisen, 1890) When it fruited in Sutter County, it was seen to be superior to the Sultana and has been propagated largely. It was first widely distributed by J. P. Onstott of Yuba City, and others, and is now to be found in all parts of the State [in 1909]. The variety is described by Dr. Eisen as follows: 'Oval; greenish-yellow; as large as a Sultana; seedless with a thin skin; good but not strong flavor, and without that acid which characterizes the Sultana grape and raisin; bunches large or very large; vine an enormous bearer.' Mr. Bioletti considers the variety identical with the Sultanina of Asia Minor, and gives this description: 'Vine very vigorous and with large trunk and very long canes; leaves glabrous on both sides, dark yellowish-green above and light below, generally three-lobed with shallow sinuses, teeth short and obtuse, bunch large, conico-cylindrical, well filled, on herbaceous peduncles; berries under medium, ellipsoidal, crisp, of neutral flavor, with moderately thick skin of a fine golden yellow color'" (Bioletti, 1918).

Some questions in this important quotation will be discussed later. For now, there is a serious question to raise, whether William Thompson and/ or George Thompson knew at the time of planting that the vine was the Lady de Coverly. If, indeed, they did correctly identify the vine *prima facie*, then why did they believe it needed the new name, Thompson's Seedless? This legitimate question leads us into documents that may or may not settle this question and others.

Stalking the Lady de Coverly

According to numerous newspaper and magazine stories, as well as family documents collected over the years, William and George were examining the Spring 1872 catalogue of Ellwanger and Barry nursery of Rochester, New York, in which 6,000 grape cuttings were advertised for $1 each. Just how many varieties or which varieties are represented in the 6,000 is not recorded. It is said that Thompson placed an order for 15 different varieties of

1882
J. P. Onstott establishes nurseries in Fresno and Los Angeles, California, specializing in the Thompson Seedless rooted stock.

1888
J. P. Onstott, in a meeting of the Sutter County Horticultural Society, in Yuba City, makes the motion to name Thompson's grape the "Thompson's Seedless." The motion passes unanimously.

1890
Gustav Eisen's book, *The Raisin Industry*, is published.

1892
George Thompson and William Thompson try to get information from Ellwanger and Barry about the source of their vine, without success.
Ann Whiteley Thompson dies June 12 in Sutter County, California.

1898
William Thompson dies February 25 at home near Butte, California.

grapes. Subsequently, he received a shipment of cultivars, presumably all fifteen varieties. It isn't clear how many of each. It is at this point in the saga where the data become obscure. The traditional accounts maintain that there were in the shipment three cuttings of the Lady de Coverly (or unclassified cuttings, by some accounts), and these three were grafted onto his own rootstock, probably his "Los Angeles" vines. According to George Thompson's own testimony, given in an affidavit published as a letter to the editor many years later in 1911, two of the grafts failed; only one survived. George Thompson never mentions the Lady de Coverly. The apocryphal story is that the Sacramento River flooded in 1872-73, carrying two of the vines away. Other accounts maintain the two simply "perished." George also testifies in the 1911 letter that 20 years later Ellwanger and Barry were contacted by William Thompson and himself to find out the source of their great vine, but they received no reply.

Recent diligent research in the Special Collections Library at the University of Rochester, the locale of all archives for Ellwanger and Barry, by Melissa Mead, Special Projects Librarian, discovered that the Lady de Coverly cultivars are never named in Ellwanger and Barry Catalogues from 1871-1880. The name, "Sultanina," also does not appear in any of the Ellwanger and Barry Catalogues for the relevant years. "Sultana" does appear, but with a clear description distinguishing it as the different grape from Asia Minor. How can this apparent historical anomaly be explained plausibly? One account to be quoted states that a note from Ellwanger and Barry that came with their order requested William Thompson to try their three cuttings and then to let them know the results:

"In 1873, a Sutter County farmer received an order of nursery stock from Ellwanger and Barry, Nurserymen of Rochester, New York. Included in the shipment were three European grape vines. The farmer was asked to plant them and report his success or failure. Two of the vines died, but the third grew. The grapes from the vine had no seeds and local growers soon realized they would make good raisins. The Sutter County Horticultural Society named the grape 'Thompson Seedless' after the farmer, William Thompson.

"While the Thompson Seedless is no longer produced commercially in Sutter County, more acres are planted with this variety than with any other raisin grape…." ("Yuba City's Agricultural Heritage")

This account suggests that the cuttings were not included in William Thompson's order for vines, but added gratis by the supplier, who hoped to enter the California market with a new variety, new to cultivation in the United States, and, of course, in the California environment. If the three cultivars were tagged (named) separately from the other fifteen varieties then clearly William Thompson and George could, at that point, have known they were the Lady de Coverly variety. Moreover, they did, in fact, keep the three identifiably separate from the others when planted, as subsequent events prove. A reputable nursery of the stature of Ellwanger and Barry would be extra careful to tag "plant material" in order for the buyer to retain control of his plantings. It is reasonable to believe that Ellwanger and Barry did, in fact, tag the three cultivars to retain their identity

Setting this obvious explanation aside for the moment, it is at least plausible to believe that William and George may not have known the identity of the cuttings when they grafted them onto rootstock, so that not until 1875, when the sole surviving vine fruited could William identify the vine as the Lady de Coverly ampelographically by the taste and characteristics he remembered from his experiences in Yorkshire years before.

A plausible link in the Lady de Coverly variety's arrival in California is the real possibility that Patrick Barry, the Irishman in the firm, Ellwanger and Barry, may have known about the variety from his viticulture studies in the nurseries in England. As he was a devoted experimentalist in viticulture, it makes sense to believe that it was he, and not Ellwanger, who obtained the Lady de Coverly from a source in England, hoping to establish yet another *Vitis vinifera* in the United States. Though this is, admittedly, a speculation, it is an informed one under the circumstances. This possibility augments other explanations; it doesn't displace them.

Then too, the source might easily have been William Robert Prince and his family who extended back three generations of botanical experimenters, notably in grapes. Ulysses Prentiss Hedrick has extremely high regard for Prince's botanical work:

"Three generations of Princes tried to grow European grapes in America. William Robert Prince devoted his life

1911
George Thompson publishes an affidavit regarding the "true origin" of the Thompson Seedless grape in the Daily Appeal, Yuba City.

1915
George Thompson and Sarah Ann Burgett Thompson become Sutter County spokesmen for the Thompson Seedless grape and Sutter County at the Panama-Pacific International Exposition in San Francisco.

1928
Sarah Ann Burgett Thompson dies September 29 in Tudor, Sutter County, California.

1934
September 17, George Thompson dies in Tudor, Sutter County, California.

1940
Gustav Eisen dies.

1980
September 15, a memorial plaque to William Thompson and the Thompson Seedless grape in California is dedicated as an Historical Landmark # 929 on California State Highway #20, located 2-1/2 miles west of Sutter.

to grape culture. He tried all of the varieties of *Vitis vinifera* to be obtained in the several countries in Europe where grapes were grown, but after fifty years of work with foreign sorts, he gave the rest of his life to improving and distributing varieties of native grapes. Because of his nursery, his book, and his writings, he must be ranked with Adlum and Longworth as one of the three geniuses of American grape growing. His *A Treatise on the Vine* was the first good work to appear on viticulture in America." (Hedrick, 1950)

Fig. 2: William Jr. and William Thompson Sr. in their field, circa late 1880s.

Unfortunately, Prince's book contains no mention of the Lady de Coverly or its synonyms, such as Sultanina, or the many other names used in many countries. Mystery unsolved, so far.

Thus far, there are these explanations about how William Thompson knew his famous vine was originally the Lady de Coverly, that is, a variety which already had a given name: 1) The cultivars were tagged and named the Lady de Coverly; 2) William Thompson recognized the fruit and vine characteristics (the ampelography) in 1875; and 3) An Ellwanger and Barry catalogue named the grape among the 6,000 cultivars they had for sale. The third alternative must be set aside, for the simple reason that the Ellwanger and Barry catalogues do not in fact contain the name, Lady de Coverly. It is possible that both of the other interpretations are true. The cultivars were tagged and William Thompson did recognize the vine in the 1875 vintage.

Since these conclusions contradict the testimony of George Thompson and conflict with the long-standing, oft-quoted traditional story of the vine's origination, further questions must be raised and answered.

If the 1872 catalogue of Ellwanger and Barry did not contain the grape name, Lady de Coverly, why is George Thompson described as claiming that it did? George Thompson does not himself make such a claim in the records. Why did George Thompson and his father drop the name Lady de Coverly so quickly in 1875, if they did, agreeing with the Marysville clerk to name the exhibit, "Thompson's Seedless" *pro tempore*? Why did it require thirteen years (in 1888) for the Sutter County Horticulture Society to vote to name the grape "Thompson's Seedless"? Why did Professor Frederic Bioletti refrain from extending his professional advice to George Thompson about establishing any official name for the vine, urging him to seek approval through the California Legislature offices? Why did George Thompson, and maybe William as well, try to get more information from Ellwanger and Barry 20 years later, in 1892, about the name and source of their vine? Why did J. P. Onstott, whose considerable business interests were involved, step forward to be the one to urge the Sutter County Horticulture Society to vote unanimously to name the grape "Thompson's Seedless" as its "official" name, the one that has since become so well-established, and never to be replaced, at least in California?

For a time early in the research, it seemed possible to answer all of these questions with some finality. Now, nearing the end of the study, the answers appear to be less clear and certain.

From George Thompson's statement and articles about him, it was he, and not William, who had ambitions for the vine's propagation from the first discovery. It was George who urged his public-spirited, Baptist, teetotaler father to cease giving away cultivars from the vine, to retain control of their distribution, even to seek a patent for the vine as a new variety. Clearly, George did believe the Thompson Seedless was a new variety, a species, a hybrid, and thus eligible for a patent. It was no such thing, of course, as even his contemporary ampelographers could have told him. Apparently it was George who approached Professor Frederic Theodore Bioletti (1865-1939), the young viticulturist in the College of Agriculture, University of California, Berkeley, where Eugene Woldemar Hilgard was the austere dean, seeking to confirm the novelty of the Thompson Seedless vine as a variety. In the two-page document called, "Thompson History," these passages are evidence that George was intent on his objective:

"William wrote to the nursery firm 20 years afterwards (1892) to find out where his three cuttings were grown and where they came from, as their booklet did not give this information in 1872.

"Professor Bioletti of the University of California said they were mixed with fifteen other varieties and that the legislature ought to name this grape. The Thompsons claimed the right to the name, for out of all the cuttings, only one grew." (Vorous, 1989)

Additionally, George Thompson himself paraphrases Professor Bioletti in his important letter-to-the-editor in 1911.

"Bioletti of the University of California said they were hoodoo [doubtless a colloquialism from the nineteenth century for 'all messed up'] with 15 other varieties. He said the legislature ought to have the grape. We have the right to the name for out of 6,000 cuttings only one grew. My father gave the cuttings away to several families." (Bird, 1961)

What is important to note about this imprecise statement is that George Thompson did take considerable pains to consult a professor of viticulture in the College of Agriculture about the possible justification for a claim for a patent on the vine.

Professor Bioletti's reply was recorded, though not verbatim. He stated that such an important, essentially scientific claim, had to be decided by the California State Legislature through the official process by the California Board of State Viticultural Commissioners and that their Thompsons' Lady de Coverly could not be separated from the other fifteen varieties short of the official process. By implication, Professor Bioletti may also have been suggesting that the Sutter County Horticultural Society was not legally empowered to name a new grape variety in California simply on a motion by J. P. Onstott, and approved, even unanimously, by a voice vote. Writing in 1909, Dean Wickson states about the Thompson Seedless, "...Mr. Bioletti believes the variety identical with the Sultanina of Asia Minor and gives this description..." The 1909 edition is the fourth. Presumably, the first edition would contain the same information even earlier.

In her thesis for the degree Master of Arts in the History Department, University of California, Berkeley, in 1931, Edith Catharine Meyer states categorically, "The most important seedless varieties grown are the Sultana (f.n. The Sultana grape is extensively grown near Smyrna in Asia Minor and was first brought to California by Colonel Haraszthy in 1861) and the Sultanina which has been erroneously called Thompson Seedless." Edith Meyer is correct: The Thompson Seedless grape was named erroneously and by fiat of an unauthorized source. Even prior to its arrival in California in 1872, the grape was widely recognized in the rest of the world as synonymous with the Sultanina, and not the near relative, Sultana. As the Lady de Coverly, it was known in South Africa, where it was re-named Sultana, from where (as Lady de Coverly) it was carried by ship to Australia and where it is yet known as Sultana. By tracing the lineage, one can find more than 20 names for Sultanina, usually listed as "synonyms." Strangely, no standard references on wines and grapes published to the date of writing—encyclopedias, ampelographies, wine dictionaries, etc.—that have been consulted ever include the name Lady de Coverly as a synonym; neither is the name listed nor defined. Correspondence between Professor Vincent E. Petrucci, retired viticulturist at California State University, Fresno, and Peter Clingeleffer in Australia, clearly shows the pattern of distribution. In "Dried Fruit in Australia," under the heading, "Sultana Types," we read: "The Sultana vines now in the Sunraysia and Riverland regions came from material that can be traced to Sultana vines held in glasshouses in England and known as Lady de Coverley's grape. These vines provided the material for the vineyards established in the Cape of Good Hope, near Cape Town in South Africa. These vines in turn provided cuttings that were brought by sailing ship to Australia. In 1867, Sultana buds from the Cape of Good Hope were successfully grafted onto the root systems of eight other grape vines to establish a propagation source for the Sultana variety." (Clingeleffer, 1999)

This document establishes the source for the Lady de Coverly in "glasshouses" (greenhouses in the United States) in England in the 19th century.

Risking redundancy, it is helpful to add the following confirming summary of the naming problem written by H. W. Wrightson in 1925:

"Following is an interesting technical description of the Sultanina [source not identified]—Synonyms: Thompson, Thompson's Seedless (in California), Lady de Coverly (English hothouses), Sultanieh, Oval-fruited Kishmish (Turkey, Palestine). This variety is grown in collections or in small quantities as a table grape throughout the Mediterranean region. It is grown largely in the Levant, more particularly in the warmer parts of Asia Minor, as a raisin grape. It appears to be widely distributed in Asia as far east as Persia and probably beyond. From it are made the genuine Sultana raisins of Smyrna.

"It was brought to California in 1872 by Wm. Thompson, Sr., of Sutter County, who obtained it from Ellwanger and Barry of Rochester, New York, under the name of Lady de Coverly, a name by which it is known in English hothouses. It was distributed here under the name of Thompson's Seedless to distinguish it from the Seedless Sultana, a grape grown more sparingly in the same Asiatic regions. Its name of Sultanina, by which it is known in most countries, or Sultanieh as it is sometimes spelled, a name derived from the town of Soultaneih, in Persia.

"Mr. Thompson deserves great credit for having introduced this valuable grape into California, but it seems hardly necessary or desirable to change the euphonious and appropriate name by which it has been known in most of the regions where it has been grown probably for hundreds of years..."

Virtually all of the questions raised previously can be resolved, if it be assumed that, after about 1880, when Harter Brothers, J. P. Onstott, the Stablers, and others in Sutter County and vicinity were actively propagating the vine, George Thompson was still continuing a campaign for the Thompson Seedless patent rights, futile though that objective was on any ampelographical basis. Informally, by popular acceptance, "illegally," "erroneously," calling the vine Thompson Seedless, instead of Sultanina, or even Sultana (which it eventually was and is called by some in the San Joaquin Valley), George Thompson became the proud, committed spokesman for his father's "discovery," and their good name. After 1897, when William Thompson died, George became recognized as the representative of his father's legacy. By 1915, George and his wife, Sarah Ann, were made official spokesmen for the Sutter County Exhibit in San Francisco during the Panama-Pacific International Exposition, where George repeatedly recounted, orally, the story of the Thompson Seedless vine and its importance in California viticulture. Did hyperbole and embellishment, spawned and enhanced by the public performances for days, enter into his spiel, his account?

In an illustrated article, "Thompson Seedless Grape Had Start in Sutter County: Chance Cutting Almost Discarded, Produced State Raisin Industry," by Margaret Kimerer, in *The Appeal Democrat,* 1961, the caption for the two pictures reads: "The scene of origin of the Thompson Seedless grape, that gave rise to California's gigantic raisin industry and almost turned Sutter County into a vineyard, is shown above, the old home of the late Mr. and Mrs. George Thompson. It was Mr. and Mrs. Thompson who found unclassified cuttings in a shipment of grape stock and to which they gave their name, who are the virtual "parents" of the state's famous raisin grape."

This is the first recorded instance found where George Thompson purportedly admits that the three cuttings were "unclassified," despite other suggestions that they were "tagged," and even "named." The author of the article, Margaret Kimerer, may have based her data on personal interviews with George Thompson, though she does not say so. She did go to the trouble to interview W. Ray Chandler, a close friend of J. P. Onstott's son Jacob Onstott, who had his own first-hand version of the events leading to the propagation of the Thompson Seedless vine in Sutter County. In his version, William Thompson gave J. P. Onstott cuttings from his "non-bearing" vine, and Onstott developed his own vineyard and nursery business from those cuttings. Indeed, after subsequent trials, when Onstott found the trick with the vine was to prune long bearing wood, he made a point of returning to the Thompson place to tell William about his success (Bird, 1979). This trail is a dim one. However, the stuff of controversy, as indeed it became. The point of Chandler's account is this: When J. P. Onstott received his cuttings, William Thompson believed his vine to be barren of fruit, and William even told Onstott it was, for Onstott only wanted a decorative cover for an arbor at his home.

The most interesting observation in the caption, for our purposes, is the description of the three cultivars as "unclassified" cultivars, which simply appeared in the larger shipment of grape stock. Now unverifiable as fact, it does confirm our hypothesis that William Thompson and George did not actually place an order for the three cultivars, and were thus unable to know the vine with certainty until the first fruiting, at the earliest, when William especially could have recognized the Lady de Coverly.

In the *Pacific Rural Press* for September 30, 1893, the lead article, "Thompson's Seedless Grape," states, "This variety was brought to this State from Ellwanger & Barry's nurseries in Rochester, New York, about 1872 and was by them said to be from Constantinople and named Lady Decoverly. No [reference] work on European grapes accessible here has any grape by that name, and the Eastern importers can give no further information about it…"

It is of importance to know that the "editor" of this journal, which was popular among California farmers, was Dean Edward J. Wickson, College of Agriculture, University of California, Berkeley. His many books were directed to the farmers, though some of the technical ones were of greater interest to scientists. It can be inferred from the quotation that Wickson actually did attempt to ascertain the correct name for the variety. His major work, which went through 10 editions, *California Fruits and How to Grow Them,* was first published in 1897, seven years after the work on raisins by Gustav Eisen (1847-1940).

As we have seen, there is no evidence in the University of Rochester archives to show that Ellwanger & Barry actually listed any vine with the name, Lady de Coverly. The passage quoted above from the *Pacific Rural Press*, was taken, in 1893, verbatim from Gustav Eisen's book, *The Raisin Industry* (1890). Indeed, the passage goes on to quote Eisen's ampelographical data of the grape.

Unfortunately, we are left with a fundamental question: Where did Gustav Eisen find his information about the Lady de Coverly?

We know something about Eisen's voracious curiosity and his expert management of his elder brother's vineyards in Fresno. In his book, he makes it plain that he actually spoke at length with many farmers, some in Sutter County and the vicinity of Marysville and Yuba City. William Thompson is not so identified as having been interviewed. It stretches credulity to think that he wasn't made totally aware of Thompson's special grape and its bizarre history. Gustav Eisen, a master vineyardist and viticulturist, must have been fascinated to learn what he could about the facts. He was also a scientist—maybe a scientist first.

The Eisen vineyards in Fresno were a family affair. Francis Eisen (1826-1895) was born in Stockholm, Sweden, educated in Bremen, immigrating to America early in the Gold Rush period in 1850. He is said to be the very first vineyardist in Fresno County, but only after becoming a lucky striker in the Comstock Mines, which made his San Francisco business expansion and his Fresno vineyards and winery possible. The Eisen 650 acres purchased the year of the start of the Thompson Seedless saga, in 1872, lay very near South Clovis Avenue, adjacent to the railroad and a major irrigation canal.

August Eisen, his son, handled Francis' business interests after Francis' death, but it was Francis' younger brother, Gustav, who was the manager of the vineyards and the extensive experimental viticulture plot where, it is said, Gustav tended vines and recorded detailed data to help advance the burgeoning agricultural businesses, including their own.

Gustav Eisen, was a near-polymath, who earned the degree Doctor of Philosophy from the truly ancient, academic citadel, the Swedish University, Uppsala. He arrived in California in 1871. Gustav's "groundbreaking " book on the raisin industry contains many historical gems, yet fails to reveal, in a tantalizing passage, where and when he became acquainted with the Lady de Coverly. This is the passage quoted by others, notably Wickson, both in his book and the article in the *Pacific Ru-*

Fig. 3: Historical monument marking the site of the first propagation of the Thompson Seedless vine in California in 1872. The "California State Historic Landmark #929" is located 2-1/2 miles west of Sutter on State Highway 20.

ral Press:

"*Thompson Seedless*—This variety has been growing in California for many years, but has only lately come into notice. It was imported from Rochester, New York, from the establishment of Ellwanger & Barry, about 1872, and was by them described as a grape from Constantinople under the name of Lady Decoverly. Thompson Seedless is the name given this grape by the local growers around Yuba City, and not the original name. I am inclined to believe that this grape is related to, but not identical with, the oblong seedless grape which is grown around Damascus in Asia Minor and there dried into a raisin of very good quality..."

Earlier, Eisen writes the understatement of the 19th century: "...The Muscatel grape is planted to some extent [in Yuba County], but the favorite grape is the Thompson Seedless, a new variety of great promise." (Eisin, 1890). Actually it was not a new variety, though new to the region.

Obviously, Gustav was well acquainted with the Thompson Seedless, though the Eisens' operation was mainly devoted to wine grapes and their winery, from which they "grew" excellent wines favored in Europe as well as the United States. It cannot be proved by documentation how Gustav came to identify the Lady de Coverly as synonymous with the Thompson Seedless, but surely, he must have discussed the identities with Prof. Frederic Bioletti, and even with William and George Thompson personally in Sutter County during his horticultural travels. With an experimental vineyard of 300 varieties, exclusive of their producing vineyards, it seems implausible that by, say, 1885, he would not have placed orders himself with Ellwanger & Barry, if not other nurseries in the eastern United States. His *magnum opus, The Raisin Industry,* was published in 1890, and so must have been in preparation for eight years or more. If Gustav Eisen did in fact order from eastern nurseries, which had traveling agents taking orders, it may be that he came onto the Lady de Coverly independently and concurrently with the Thompsons. More likely, Eisen probably ordered the Lady de Coverly from either J. P. Onstott, or the Stabler nurseries, both being ready to fill orders for rooted cultivars by the early 1880s. In the back of Eisen's book is a full-page advertisement by the B. G. Stabler nursery of Yuba City, offering Thompson Seedless "rooted vines." In very tiny print at the bottom of the advertisement is a carefully worded statement purporting to be an endorsement by Gustav, which reads:

"In a communication to *California, a Journal of Rural Industry*, May No., 1890, entitled, 'With the Fruit Growers in Sutter County,' Professor Eisen thus refers to Mr. Stabler, and his work: 'Mr. B. G. Stabler makes a specialty of dried peaches and seedless raisins and has succeeded well with both. The principal raisin-grape of this vicinity is the little-known seedless grape, Lady Decoverly, here known as the Thompson Seedless, he being the first to grow it. Years ago, about 1872, this gentleman saw advertised in an eastern catalogue a seedless grape, said to come from Constantinople, and was called the Lady Decoverly. It proved to be very different from the common Sultana, being of yellow color, and of oblong shape. It is certainly strange that this singular variety of grape should have existed here so many years, and failed to attract general attention. It is an enormous bearer, heavier even than the Sultana, and ripens in early August. It makes very choice raisins for cooking purposes...'"

This text mimics his statements in the body of his book.

On the other hand, Gustav had been a world traveler. It is likely that he met the Lady de Coverly in England, as Patrick Barry and William Thompson both did. Gustav may have carried cultivars across the Atlantic. If Agoston Harazsthy could do that, so could he. It would not have been the first time some nobleman's lady was carried across shire borders.

Getting to the Root of the Vine

From 1890 on, the Thompson Seedless, gradually at first and then by an explosion, became the single most popular grape planted in California and probably in the world in the 20th century. In her study in 1931, Edith Catharine Meyer found that 95% of raisin vines planted in the San Joaquin Valley, with an equally high percentage of these made up of the Thompson Seedless grape. She goes on to say, "We have no positive proof as to who produced the first commercial raisins in the state, but in 1863 raisins were exhibited at the [California] State fair by Dr. J. Strentzel. The only grape that had

Fig. 4: V. E. Petrucci, Professor Emeritis, California State University, Fresno, and Thomas Pinney, Professor Emeritis, Pomona College, in front of the Thompson Seedless vine along Barstow Avenue outside of the Viticulture and Enology Research Center at California State University, Fresno.

been dried successfully up to that time was the White Muscat of Alexandria."

What she seems not to have known is that Dr. J. Strentzel was the father-in-law of Naturalist John Muir, no mean horticulturist himself. Muir inherited Strentzel's fruit growing business, managing it successfully by sales in the Bay Area.

By 1872, as we have seen, Francis Eisen had planted a pioneering vineyard in Fresno, and soon afterward began experimenting with raisin varieties, such as Muscat of Alexandria, Muscat Gordo Blanco, and Malaga, experimenting for the climate of the San Joaquin Valley, and discovering in the process that the Fresno area was ideal for growing raisin grapes. Only slowly at first did the Thompson Seedless (Sultanina) become popular, due in no small measure to the business efforts of J. P. Onstott and Stabler. It was the quality of being seedless that brought a kind of planting frenzy of the grape, but it's versatility soon became apparent, for it could be used as a shippable table grape, juice, canned fruit, varietal wine or wine supplement, for brandies, etc. Acreage of the Thompson Seedless burgeoned from roughly 1900 to 1915.

In an article in 1985, Carol Withington states that J. P. Onstott did more than propagate and grow the Thompson Seedless in Yuba-Sutter Counties. "By 1882, Onstott had also established nurseries in Fresno and Los Angeles to propagate the grape. He supplied growers in the San Joaquin Valley and in Southern California with Thompson Seedless roots." Ernest Elmer Sowell states it was about 1890.

J. P. Onstott was not alone in the promising business of propagating the Thompson Seedless. In Yuba and Sutter counties, the Harter Brothers and the Stabler family planted well over 1,000 acres before 1892, but because of greater profits from stone fruit and other crops, and especially due to the grape scourge of phylloxera, acreage of the Thompson Seedless in the Yuba and Sutter counties dwindled rapidly.

Unstable Times

Unfortunately in 2001, a towering surfeit of the Thompson Seedless crop has brought about a sharp decline in the price of raisins; bumper crops on too great an acreage have greatly exceeded demand. Some growers believe, with evidence, that a free market for imported grapes from South American countries, such as Chile, has exacerbated the glut—and even displaced the market in the United States for California grapes. The success of the Thompson Seedless industry in the United States is once more facing a serious challenge in the marketplace. Many small farmers are finding it impossible to remain on their land, when costs of equipment, equipment parts and repairs, fertilizers, labor, water, electricity, have all risen to that proverbial point of "no return" (or no returns) on the investment. Unlike diversified farming, vineyardists cannot simply yank out 100-year-old vines (or younger, for that matter), and start raising almonds or cotton or some other crop needing years to develop. In Fresno county in the summer of 2001, a lot of acreage of Thompson Seedless is being left on the vines, going unpicked, languishing for buyers, wasted, acreage listed for sale in increasing numbers, signaling family tragedies, or the fruit being sold to the huge wineries for salvage prices.

Doris and Walter Halemeier are victims of cancelled contracts for their Thompson Seedless grapes in 2001, a cancellation for which there is no remedy. In their case, it is no immediate disaster. Because their security is the result of a long family history of farming success. They are not hurt as much as some of their neighbors; they are, however, unhappy with the clouded future for the Halemeier Vineyards on South Armstrong Avenue in Fresno County.

Doris is a 4th generation farmer in California, originally from the Sacramento River area in Clarksburg, 25 miles south of Sacramento. Her family arrived in California in 1848. For 40 years in Fresno County, she was a valued substitute teacher in the schools, an active member of several women's organizations, and, with Walter, a member of the Emanuel Lutheran Church. When they cleared a part of their vineyard to build a new home, it was Doris who drove the tractor to pull out vines, while Walter manhandled the unwilling chains.

Walter was born in 1918 and was raised on their present place in southeast Fresno County. His grandfather, August, arrived in the Fresno area in 1886, emigrating from Wallenbruck, Germany. In 1912, August, Jr., was married to Sophia Albrecht of San Jose, California, and in 1912, Walter's father, August, planted 30 acres of Thompson Seedless vines which are still producing. Over the years, on expanded acreage, the family planted other varieties:

Carignane, Malaga, Muscat of Alexandria, Grenache, Malvoisie, and Sultana. Encroachment on their ranch by developers has proven to be an uncontrollable force they have been unable to meet successfully. By selling acreage sections of their land, sales forced by developers' contiguous encroachments, they have become financially secure, managing to hold onto enough acreage to prevent a total takeover. The cost in lost heritage for their descendants cannot, of course, be folded into the loss. That is simply a given.

Doris and Walter raised two daughters, Christine Halemeier Raymond and Elizabeth Halemeier Hudson. Elizabeth and her husband farm in the Sanger, California, area nearby, making her a 5th generation California farmer.

For two seasons, their Thompson Seedless crop was bought under contract by a firm in Oregon, whose business it is to make sweet juices used in various other products, an arrangement Doris and Walter welcomed, for they were relieved entirely of the labor of picking and shipping. Unfortunately, the beneficial contract was not renewed for 2001, the reason given being that the firm was able to buy Chilean "white grapes," pears and apples for less money. As a result, the Halemeier Vineyards are placed in a severely disadvantaged financial position this year. What can be done with 50 acres, 400-plus tons of high quality, fast-ripening grapes when no one wants to buy them?

Just down the road a ways from the Halemeiers, Walter Cucuk raises eight varieties of wine grapes, such as Sirah, Zinfandel, Cabernet Sauvignon, Ruby Cabernet, etc., plus 90 acres of Thompson Seedless. Walter can trace his Yugoslavian family roots back 400 years. His father, Vido, arrived in Fresno County in 1913 at the age of 13, helping to plant some of the original vines of Thompson Seedless. Walter was born in 1932 "right over there beyond those trees" near his present home and never found a good reason to leave. Like the Halemeiers, Walter is watching the slow demise of local farming, a degradation he believes is nationwide. This year, his Thompson Seedless grapes were sold at a new low price at the huge local winery: $75/ton. In 2000, the price was only $125, down from 1999's price of $225. "Just drive along any road here and look at all the Thompsons. We have succeeded in growing ourselves out of business. Only the large corporations—agribusinesses—can succeed now."

Walter also showed us new grape varieties, which today are able to produce enormous crops per vine, as a result of horticultural experimentation. Such breeding success only hastens the eventual failure of the small farmer. With his neighbors, Walter and his school teacher wife, realize the commercial value of their property for development is so great they could never afford to continue farming. Selling the land becomes a hard-headed, business decision they must eventually face, however reluctantly.

Summary

Through most of his adult life, George Thompson was intent on preserving the name, Thompson Seedless, for his father's vine. From 1873 to 1910, he continued his personal campaign to preserve the name, and in the face of some strong criticism in the national and local press. By 1915, at the age of 76, he continued his campaign as spokesman for the Thompson Seedless and for Sutter County in San Francisco at the Panama-Pacific International Exposition. In 1888 when J. P. Onstott, his friendly competitor in the propagation of the vine, offered the motion before the Sutter County Horticultural Society to give their imprimatur to the name, Thompson's Seedless, William and George Thompson felt honored, though George deferred to his father as the senior "discoverer." In about 1910, a farmer or nurseryman in Tulare County sought to be given credit for engineering the vine, prompting George to defend their title, and doing so successfully. By 1915, the name and the vine were firmly established in California, and soon to be recognized throughout the world as the Thompson Seedless (in California), synonymous with many other names for the same vine and grape.

To this day, there is the serious question of whether the name, Thompson Seedless, has been legally and formally recognized in California, where varietal names have long been the responsibility of the California Department of Food and Agriculture throughout its official agency.

In retrospect, it appears that William Thompson was an accidental hero for being the source for the first Sultanina vine to be recognized in California and from which subsequent cultivars stem. No one was more surprised about the development of the vine than William Thompson, even seemingly unconcerned, at least compared with his son. It would even seem that he would have been pleased just to see it happen, for it was he who happily shared his Midas-tainted vine with the Harters, Stablers, and, especially, J. P. Onstott.

Undocumented Conclusions Reached by the Author

• William Thompson did not order three cuttings of the Lady de Coverly vine; he did receive three from Ellwanger & Barry.

• The three cuttings were tagged and named by Ellwanger & Barry, and they did request information on how the vine flourished in California. (George Thompson could never report his findings to Ellwanger & Barry using the pseudonym, Thompson Seedless because it would have undermined his personal campaign to establish the name Thompson Seedless.)

• George Thompson, from the start in 1872 suppressed the knowledge about the Lady de Coverly, especially after William Thompson recognized the fruit in 1875. George grew to believe his claim that the three cultivars were originally "unclassified."

• At the Marysville Fair in 1875, when George hesitated to use the known name of the vine, the Clerk simply called it "Thompson's Seedless," for identification. This was a historic moment gone awry, for had George named it the Lady de Coverly to the Fair clerk, that name might have become the one used worldwide for the California vine, at least. Instead, Thompson Seedless became just another synonym for

Sultanina.

• George Thompson invented the fiction that the three cuttings were unclassified, in his later years, to justify the Thompson Seedless name.

• It is doubtful that George Thompson ever actually wrote to Ellwanger and Barry regarding the source of the grape, for he would have had to make reference to the Lady de Coverly, revealing his misguided campaign.

• William Thompson is an accidental hero who blundered into "discovering" the Thompson Seedless only because the vine itself thrust itself bodily onto him. He certainly did virtually nothing to "engineer" the vine. "History" fell into his lap all unbidden.

• George Thompson was absolutely denied his patented new variety, because it was not a new variety.

• John Paxton Onstott deserves all the credit for planting, propagating, and distributing the Thompson Seedless vine. In a sense, John Paxton Onstott is the unsung hero in the saga of the vine, not William Thompson.

• There is classic irony in the fact that William Thompson was an innocent bystander whose name became famous, while George Thompson, who truly lusted for recognition through the vine, is no longer identified with the grape, except in a narrow circle, such as his descendants.

Will DNA Testing Settle the Naming of Vines?

In 2001, there is really no scientific problem about the synonymity between Thompson Seedless and Sultanina. That issue has been settled decades ago, but in 1875, for both horticulturists and farming empirics (to use Aristotle's term) like William and George Thompson and J. P. Onstott, men who were only starting to become proficient vineyardists, let alone proficient ampelographers (a term that was likely foreign to them), the identity of the Lady de Coverly, Thompson Seedless, and Sultanina, not to neglect the close relative, Sultana, was a puzzle. Amateur ampelography, and vested business interests, bred many barnyard and kitchen table disputes in the late 1800s, disputes that can now be settled by more objective genetic studies.

Carole Meredith, in a brief article with a long title, "North American Geneticists Untangle the Vine Variety Web," argues that advances in DNA profiling of grapevines have already shown the eventual successes of identification of varieties and species that will survive any possible argument from cumbersome ampelographical data. Studies already completed have promised more clarification of varieties and their sources to come. Research in various countries continues.

"These studies can be expected to converge eventually to produce a family tree of sorts that will include most of the major varieties. We can look forward to an increasingly clear picture of the ancient migrations and couplings that gave us the classic grapes we so appreciate today" (Cass, 2000).

Acknowledgements

Without the constant, generous help of Jay Bond, this work would have been impossible to accomplish. Jay is the "second great-grandson" of William Thompson, who serves as a recorder and collector of the Thompson family documents and pictures. His web site on genealogy is cited in the bibliography. Another Thompson descendant, Barbara Vorous, the Thompson family historian, was instrumental in sending critical documents needed to clarify important details.

It was Julie Stark, Director of the Community Memorial Museum of Sutter County, who put me in touch with Jay Bond. Julie also checked her records in answer to my questions during the process of writing.

The key questions about William Thompson's contacts with Ellwanger & Barry by direct examinations of the original catalogues by Melissa Mead, Special Projects Librarian, University of Rochester Library. Her help at a distance (e-mail) was the key to unlock the Thompson Seedless puzzle.

Other librarians rendered professional assistance: At California State University, Fresno, Alev Akman (Vincent E. Petrucci Library, Viticulture and Enology Research Center); Jan Bird (Madden Library, Interlibrary Loans); Jean Coffee (Sanoian Special Collections, Madden Library); Christy Hicks (Madden Library, Wine Collection). At Fresno Free Library, California History Section, Ray Silvia, Librarian, was especially kind and helpful concerning one important document. Tom Pinney, Emeritus Professor of English Literature, Pomona, and prolific author on wine history, helped me in a pinch, as he has done before.

Professor Vincent E. Petrucci followed the study with particular interest. It was he who put me in touch with sources in Australia, where the Lady de Coverly is still a princess.

Doris and Walter Halemeier were gracious hosts for a pleasant morning among their 90-year-old Thompson Seedless vines in Fresno County.

Walter Cucuk, vineyardist, shared his rich experience with Thompson Seedless, and provided delicious bunches of other grapes he is tending.

Jack Sowell, 86-year-old 30-year veteran of the U. S. Marine Corps, a nephew of Ernest Elmer Sowell, who gave timely help and materials on the genealogy and history of the Sowell family. *Semper fi!*

Bernice Shields, as always, gave helpful reflections on the manuscript and library assistance, even as we sailed past our sixtieth wedding anniversary, an unevent we celebrated by yet another library visit.

Photo Credits

Figures 1, 2 and 4: Jay Bond
Figure 3 Ronald Unzelman

References

"A Lady Scorned," unsigned article about the Lady de Coverly, *Cornerstone 1968,* The Jostal Company, Publishers, Sacramento, California.

Adams, Leon David, *Grapes and Grape Vines of California*, Harcourt, Brace, Javanovich, Publishers, New York and London, 1981.

Agricultural Crop and Livestock Report, 2000, Fresno Department of Agriculture, Fresno, California.

Bespaloff, Alexis, *New Frank*

Schoonmaker Encyclopedia of Wine, William Morrow and Company, Inc., New York, 1988.

Bioletti, Frederic Theodore, "The Seedless Raisin Grapes," Bulletin no. 298, September 1918. Agriculture Experiment Station, College of Agriculture, University of California, Berkeley. University of California Press, Berkeley. Pp. 75-86.

Bird, Jessica, "The Thompson Seedless," by Carol Withington (Compiled from articles published in *The Appeal Democrat* in 1961) *Sutter County Historical Society New Bulletin,* Vol. XIX, No. 1, January 1979, Yuba City, California. Pp. 11-20.

Bond, Jay, 2nd great-grandson of William Thompson, may be reached at suttergen@iname.com He is a basic source for the Thompson family archival materials.

California Grape Acreage 2000, California Agricultural Statistics Service, Sacramento, California, June 2001 nass-ca@nass.usda.gov.

Cass, Bruce, *The Oxford Companion to the Wines of North America,* Oxford University Press, Oxford, 2000.

Clingeleffer, Peter, "Dried Vine Fruit in Australia," fax pp. To Prof. Vince E. Petrucci, January 20, 1999, p. 2. The author is Ross Skinner.

Community Memorial Museum of Sutter County, Julie Stark, Director museum@sylx.com.

Delay, Peter, *A History of Yuba and Sutter Counties, California,* Historic Record Company, Los Angeles, California, 1924, pp. 371-372.

Eisen, Gustav, "The Fruit Growers of Sutter County," *California, A Journal of Rural Industry,* May No., 1890.

Eisen, Gustav, *The Raisin Industry: A Practical Treatise on the Raisin Grapes, Their History, Culture, and Curing.* H. S. Crocker Company, San Francisco, 1890.

Ellwanger and Barry Papers 1818-1963. Sixty-seven boxes, 266+ volumes. Special Collections Library, University of Rochester, Rochester, New York. http://www.lib.rochester.edu/rbk/ell&bar.htm.

Galet, Pierre, *A Practical Ampelography: Grapevine Identification.* Translated and adapted by Lucie T. Morton, Comstock Publishing Associates, Cornell University Press, Ithaca and London, 1979.

Harters' history. Harter Brothers http://harters.com/History/history.htm.

Hedrick, Ulysses Prentiss, *A History of Horticulture in America to 1860.* Oxford University Press, New York, 1950.

Heintz, William F. "Thompson Seedless in California," *Wines and Vines,* September, 1980, p. 82.

Hilgard, Eugene Woldemar, *Report of the Agricultural Experiment Station of the University of California,* 1887-1888; 1888-1889;1894-1895.

"Historic Farm Near Sutter Is Purchased [in 1928]", source unknown, typescript, date probably shortly after the San Francisco Panama-Pacific International Exposition in 1915, based on internal evidence.

Hyams, Edward, Editor, *Vineyards in England,* Faber and Faber, Limited, London, 1953.

Johnson, Hugh, *Hugh Johnson's Modern Encyclopedia of Wine,* 4th Edition, Simon and Schuster, 1998.

Karp, David, "Where Have All the Great Grapes Gone?" *Los Angeles Times,* Food Section, August 15, 2001.

Kimerer, Margaret, "Thompson Seedless Grape Had Start In Sutter: Chance Cutting, Almost Discarded, Produced State's Raisin Industry." No source, 2 pp. Published after 1915.

Laney, Honora A., "The Raisin of the raisin: Green Thumb with Grape," *Daily Independent Herald,* Yuba City, California, special for National Realtors Week, 1973.

Mead, Melissa—See University of Rochester.

Meyer, Edith Catharine, *Development of the California Raisin Industry in Fresno County,* History thesis for the M.A., University of California, Berkeley, 1931.

Nef, Clyde E., *The Fruits of Their Labors: A History of the California Raisin Industry,* First Edition, Malcolm Media Press, Clovis, California, 1998.

Olmo, Harold P., *Plant Genetics and New Grape Varieties,* An Interview Conducted by Ruth Teiser, The Bancroft Library, University of California, Berkeley. California Wine Industry Oral History Project, 1976.

Peninou, Ernest E. and Gail G. Unzelman, Compilers, *The California Wine Association and Its Member Wineman: 1894-1920,* Nomis Press, Santa Rosa, California, 2000.

Perold, A. I., *A Treatise on Viticulture,* Macmillan and Company, Limited, London, 1927.

Prince, William Robert, *Treatise on the Vine,* T. & J. Swords, etc., New York, 1830.

Raisin Administrative Committee: 2000/2001 Analysis Report, Fresno, California, October, 2000.

Robinson, Jancis, *Oxford Companion to Wine,* 2nd Edition, Oxford, Oxford University Press, 1999.

Robinson, Jancis, *The Oxford Companion to the Wines of North America,* Oxford, New York, Oxford University Press, 2000.

Shields, Allan, "'The Professor's' Singular Vine," *The Wayward Tendrils Quarterly.* Vol.12, No.1, January 2002;

and Vol.12, no.2, April 2002Profusely illustrated version copy in the Vincent E. Petrucci Library, Viticulture and Enology Research Center, California State University, Fresno, July 3, 2000. Traces the various names for the Thompson Seedless grape.

Sowell, Ernest Elmer, *John Paxton Onstott (1841-1914): Pioneer Developer of California's Thompson Seedless Grape and Raisin Industry.* [Ms. of 70 pp. Typescript, California State Library, California History Section], Pacific Grove, California, by Author, Illustrated, 1960. (Available on microfilm from the California Section, California State Library, CA.)

Sowell, Ernest Elmer, *Purple Gold: The Birth of California's Thompson Seedless Grape and Raisin Industry, with a Brief History of Grapes, Raisins and Wines.* [Ms. of 450 pp. Typscript, California State Library, California History Section], Pacific Grove, California, the Author, 2 volumes, Illustrated, 1960. (Available on microfilm from the California Section, California State Library, Sacramento, CA.)

Standerfer, Steve, "Grape Industry Started In Y-S," *Appeal Democrat,* (Marysville-Yuba City, California), Saturday, September 30, 1978.

Stark, Julie, See Community Memorial Museum of Sutter County.

Stephan, Harry J., Editor, *Wines and Brandies of the Cape of Good Hope: The Definitive Guide to the South African Wine Industry,* Stephan Phillips (Pty) Ltd., 1997.

Sullivan, Charles L., *A Companion to California Wine,* University of California Press, Berkeley, California, 1998.

Thompson Collection. http://homepages.rootsweb.com/~photos/geovin.htm.

Thompson, George, "Certificate of Birth," [photocopy].

Thompson, George, "The True Origin of the Thompson Seedless Grape," Typescript copy signed affidavit, n.d. in Sutter City, California, 1 p.

Thompson, George, "Untitled Account of the Thompson family's life in England and their journey across the plains to California from Illinois." Transcribed by Barbara Vorous from the original holograph written in pencil, n.d. 2 pp.

"Thompson History," by Barbara Vorous, 1 p., May 1989.

"Thompson's Seedless Grape," *Pacific Rural Press,* Vol. XLVI, No. 14, Saturday, September 30, 1893 [Unsigned article] The Dewey Publishing Company, San Francisco. [*Pacific Rural Press,* established 1870, was edited by "Prof. E. J. Wickson," Dean of the College of Agriculture, University of California, Berkeley, who published works about agricultural topics for public consumption over many years. This article, based on Gustav Eisen's book, *The Raisin Industry,* is probably authored by Wickson.]

Thompson, William; Genealogy:
http:// www.homepages.rootsweb.com/~radbud/gedpages/fam01747.htm
Thompson, William Boyce (1869-1930) A wealthy magnate and horticulturist, erroneously identified sometimes with the William Thompson family. No relation.
http://www.bartleby.com/61/59/TO175950.htm

University of Rochester, Rochester, New York, Special Collections Library, Melissa Mead, Special Projects Librarian. 716-275-9293.
mmead@rcl.lib.rochester.edu

Viala, P. and V. Vermovel, *Ampélographie,* Tome II, Paris, 1901.

Voorhees, Louise Viola Thompson, "Daughter of Seedless Grape Grower Founder Went Barefoot to West Butte School," as told by Lu Voorhies (sic), Thursday, May 5, 1960, *Independent-Herald*, Yuba City, California. Louise Voorhees is the granddaughter of William Thompson, daughter of George Thompson.

Voorhees, Louise Viola Thompson (Lulu), "Grandparents in England," n.d., 1 p.

Vorous, Barbara, "Thompson History," Typescript, 2 pp., n.d.; source: Jay Bond. Barbara Vorous is the great-granddaughter of George Thompson.

Wickson, Edward J., *California Fruits and How to Grow Them,* Fourth Edition, The Kruckeberg Press, Publisher, Los Angeles, 1909. {Wickson's book went through 10 editions.]

Withington, Carol, "Yuba City Wanderings: Thompson Seedless Grape Feature of 1875 State Fair," *Leisure Time Magazine,* July, 1985, p. 2.

Wrightson, H. W., "The Sultanina (Thompson Seedless)," *California's Most Important Table Grape Varieties,* California Grape Grower, San Francisco, 1925, pp.13-14.

"Yuba City's Agricultural Heritage," 2 pp. http://www.syix.com/yuba-city/ycag.htm

Plate 3.1: Cluster Shapes - a. Conical, b. Conical Shouldered, c. heavily Cylindrical Shouldered, d. Cylindrical, e. Globular, f. Globular Shouldered, g. Cylindrical Winged.

As stated earlier, the grape varieties used to make raisins are relatively few. Cluster shape (**Plate 3.1**) and berry shape (**Plate 3.2** specimens for **Plates 3.1** and 3.2 provided by Craig Winn II) are the most obvious fruit characteristics. Color plates of each variety show these characteristics as well as vegetative characteristics which include shoot tip, the first leaf and the mature leaf. These characteristics are shown in color plates in this chapter, and complement the following description of ampelographic characteristics.

Thompson Seedless

Thompson Seedless (**Plate 3.3**) is by far the dominant variety grown in California and is the grape of choice for making raisins throughout the raisin-producing countries of the world.

Synonyms (According to Pierre Galet): Sultanine Blanche, Sultanina Kechmish jaune, Kis-Mis Alb (in Rumania), Kismis Belii Ovalnti (in the former USSR countries), Bealo Bez Seme (in Bulgaria), Cekirdeksiz (in Turkey), Feherszultan (in Hungary), Sultana (in Australia and South Africa) and Oval Kishmish.

Origin: Anatolia, Asia Minor, now the Asiatic area of modern Turkey (Perold, 1927)

Description

Shoot tip (**Plate 3.3.1**) - cobwebby and green.

First leaves (**Plate 3.3.1**)- glabrous green with a faint tinge of bronze.

Mature leaf (**Plate 3.3.2**) - medium to large, glabrous top and bottom, three-lobed with shallow sinuses and petiolar sinus closed to overlapping.

Cluster (**Plate 3.3.2**) - large, cylindrical, heavily shouldered and loose to well-filled, sometimes compact. Berries are medium small size, ellipsoidal elongated, greenish white to yellow and seedless (**Plate 3.3.3**). Pleasing sweet, slightly acid to neutral flavor.

Attributes

As mentioned earlier, the Thompson Seedless variety is very versatile and plays a significant role in all five commercial classes of grapes. All of the golden seedless, dipped seedless and oleate seedless raisins are made from Thompson Seedless. It ripens in mid-August to the first week in September in the San Joaquin Valley of California. Thompson Seedless characteristically makes excellent quality raisins. Since the majority of Thompson Seedless are sundried on paper trays in the vineyard, the fruit normally requires about three weeks to dry. Because of its rather late ripening and the time it takes to dry, it is subject to rain damage in years when the first autumn rains come early.

Table 3.4
Level of Airstream Sorter Grade with Respect to Drying Method.

	Harvest (°Brix)						
	16	17	18	19	20	21	22
Drying Method	Airstream Grade (A & B)						
NTS	9.0a	34.4a	64.7a	62.8a	62.1a	63.3a	84.7b
DIPPED	10.8a	25.7a	71.0b	74.1b	80.3c	64.6a	62.2a
SOT	29.9b	53.7b	79.7c	59.3a	74.3b	88.8b	91.3c
DOV	38.3c	61.4b	71.1b	76.9b	84.1d	89.9b	91.5c
Significance of F- Value	1%	1%	1%	1%	1%	1%	

Different letters indicate mean separation at 5% level. Duncan's New Multiple Range Test
Source: Clary, C.D. AND V. E. Petrucci, California State University, Fresno, 1981.

Plate 3.2: Berry Shapes - a. Round, b. Oblate, c. Oval, d. Ovoid, e. Ovid Elongated, f. Obovoid, g. Obovoid truncated, h. Ellipsoidal, i. Cylindrical, j. Falcoid.

Growth Characteristics

Thompson Seedless is very vigorous and because the basal buds on the canes are mostly sterile; cane pruning is required.

Raisin Characteristics

Natural sun-dried - 310,060 tons, 281,873t. Dried-on-the-ground (DOG) using paper trays (**Plate 3.3.4** and **5**), Thompson Seedless raisins have a brown to purplish color with varying degrees of wrinkling dependent upon maturity at harvest, and patchy bloom as a result of handling during harvest (**Plate 3.3.6** and **3.3.7**).

Dried-on-the-vine - 5,000 tons, 4545 t. The dried-on-the-vine (DOV) raisin (**Plate 3.3.8, 9, 10**) is more elongated, has a lighter brown to purple color with thin wrinkles which parallel the berry's length (capstem end to berry tip) (**Plate 3.3.11**). The bloom is completely intact. The taste, however, is the same as that of natural sun-dried.

Golden seedless - 19,258 tons, 17,507 t. The golden seedless (**Plate 3.3.23, 24**) has a distinct golden color as a result of its exposure to sulfur dioxide prior to dehydration. All other characteristics are the same as for the natural sun-dried except for texture and taste. During the dipping process the skin is checked (finely cracked) to facilitate drying, allowing sugars to seep out. The sugar then adheres to the outer surface of the raisin, causing stickiness. Because of the dipping process, golden seedless has a more tender texture which may not be a significant factor since most goldens are used in baked or cooked products. The flavor, however, is definitely influenced by the use of sulfur dioxide during the bleaching process. Most-end users require 2000 PPM of Sulfur Dioxide (SO_2) to maintain the gold color. It should also be pointed out again that the sulfur taste dissipates rather quickly when used in cooking or baking.

Dipped seedless - (12,424 tons, 11,295 t) The dipped seedless (**Plate 3.3.25** and **26**) has a brownish red translucent color. The waxy bloom has mostly been removed during dipping prior to dehydration. During the dipping process the skin is checked (see **Chapter 8**) to facilitate drying, allowing sugars to seep out. The sugar then adheres to the outer surface of the raisin, causing stickiness. The texture is more tender than that of the natural sun-dried raisin and the flavor is sweet with less oxidative character than that of the sun-dried raisin.

Oleate seedless - 1500 tons, 1364 t.. These raisins are sprayed with an oleate emulsion similar to the process used in making sultanas in Australia (see **Chapter 7**). They are either sprayed and dried-on-the-vine (DOV oleate) (**Plate 3.3.13, 14, 15**) or SOT (**Plate 3.3.17, 18, 19, 20**). These raisins have an amber to dark yellow color, similar to Australian sultana. The flavor is more or less neutral, but sweet and mild with a characteristic all its own. The DOV oleate and SOT oleate raisins have a significantly higher USDA airstream sorter grade of B or Better than either the natural sun-dried or the dipped Thompson Seedless (**Table 3.4**). Some have described the oleate seedless as having a "buttery" aftertaste which could be related to the fatty acids contained in the spray solution. The oleate raisins are cold dipped into an oil emulsion solution (methyl oleate and potassium carbonate) which does not crack the skin but alters the waxy cuticle on the grape berry surface to facilitate the drying process (Kriedeman and Possingham, 1967). The oleate raisin is less sticky, therefore more free-flowing, than either the dipped seedless or golden seedless.

Fiesta

Plate 3.4. *17,500 tons, 15,910 t.*

Fiesta is a relatively new raisin variety, which was introduced in 1973 by the USDA/ARS Horticulture Station in Fresno, California, and selected by plant breeders John H. Weinberger and Frank N. Harmon (1974).

Origin: The genealogy of the Fiesta is (Muscat of Alexandria X Thompson Seedless) X (Tafafihi Ahmr X Red Malaga).

Plate 3.3.4: Natural dried on the ground (DOG), full row view. Note brownish purple colored raisins

Plate 3.3.16: Tractor spraying grapes for production of sprayed on the tray (SOT) raisins over Vaughn terrace.

Plate 3.3.17: Sprayed on the tray, full row view. Note amber colored raisins.

Description

Shoot tip (**Plate 3.4.1**) - cobwebby and reddish.

Young leaves (**Plate 3.4.1**) - shiny green with a bronze tint.

Mature leaves (**Plate 3.4.2**) - medium to large, glabrous top and bottom, three to five lobes with very shallow sinuses and a darker green color than Thompson Seedless.

Cluster (**Plate 3.4.2**) - medium to large, cylindrical, moderately shouldered, with a density more loose than Thompson Seedless. Berries are medium size, oval, white to yellowish gold color and seedless (**Plate 3.4.3**). The taste is pleasant and sweet and when sun-dried has the characteristic oxidized flavor.

Attributes

The Fiesta ripens 12 to 14 days ahead of the Thompson Seedless (USDA, 1997). Therefore, it is a good candidate for making natural DOV raisins (**Plate 3.4.4, 5, 6,7**). Compared to Thompson Seedless, Fiesta raisins are slightly larger, plumper and meatier, have a more tender skin with finer wrinkles and a darker color which make them more attractive than Thompson Seedless raisins (**Plate 3.4.8**). Under comparable conditions of crop load and vine vigor, Fiesta vines have consistently produced higher quality raisins than Thompson Seedless.

Fiesta grape vines must be cane pruned for maximum yield. San Joaquin Valley farmer Earl Rocca averages three tons of raisins per acre using a two wire/cross arm vertical trellis system while valley farmer Lee Simpson produces as much as five to six tons of raisins per acre using a unique overhead trellis system (see **Chapter 5**). When dried on the vine and sprayed with the methyl oleate-potassium carbonate emulsion spray (**Plate 3.4.9, 10, 11, 12**), the raisin color was

Plate 3.4.9: Fiesta DOV showing canopy separation.

much darker than either the Thompson Seedless or DOVine varieties. In some instances, the capstems are not completely removed during processing resulting in a tiny fragment referred to as a "nub," but it is hardly detectable when consumed, either out of hand or in baked products.

Black Corinth

Plate 3.5. *4,726 tons, 4296 t.*

Origin: Greece. It is said to be a seedless mutation of a Greek variety called Liatiklo (Galet, 1979).

Synonyms (According to Pierre Galet): Corinthe Noir, Zante Currant Grape (England), Corino Nero, Passera, Uva Pasa (Italy), Stephidampelos (Greece).

Description

Shoot tip (**Plate 3.5.1**)- woolly and reddish.

First leaves (**Plate 3.5.1**)- cobwebby and yellowish with a bronze tint.

Mature leaf (**Plate 3.5.2**)- medium large and roundish, five lobed with rather shallow sinuses. Varies from a dark

Thompson Seedl

3.3.1 Shoot tip growth and first leaves.

Top

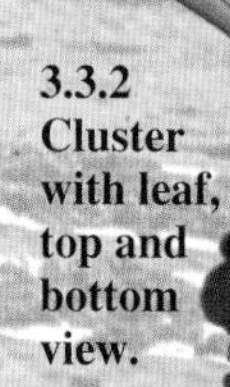

3.3.2 Cluster with leaf, top and bottom view.

3.3.9 Dried on the vine, raisin cluster.

3.3.10 Dried on the vine, dish sample.

3.3.11 Dried on the vine single raisin (right) and DOG raisin (left).

3.3.13 Dried on the vine cluster treated with oleate.

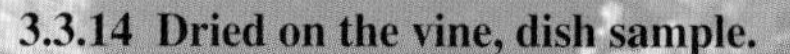

3.3.14 Dried on the vine, dish sample.

3.3.15 Dried on the vine, single raisin.

3.3.5 Natural dried on the ground, single paper tray.

3.3.19 Sprayed on the tray, dish sample.

3.3.20 Sprayed on the tray, single raisin.

3.3.18 Sprayed on the tray, raisin cluster.

3.3.12 Dried on the vine Hiyama sprayer with return system.

S

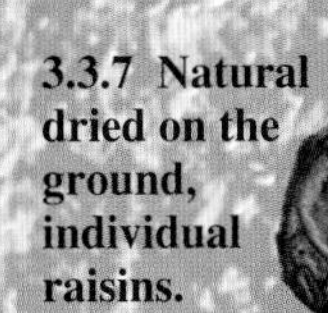

3.3.7 Natural dried on the ground, individual raisins.

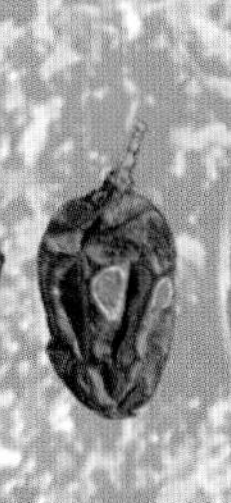

3.3.6 Natural dried on the ground, dish sample.

3.3.3 Full berry and longitudinal section.

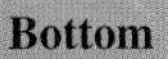

Bottom

3.3.25 Dipped seedless raisins, dish sample.

3.3.26 Dipped seedless raisins, single raisin.

3.3.23 Golden seedless raisins, dish sample.

3.3.24 Golden seedless raisins, single raisin.

3.3.22 Trays of grapes stacked on dehydrator cars.

3.3.21 Dehydrator at California State University, Fresno.

3.3.8 Dried on the vine (DOV), canopy separation (severed canes).

Fiesta

3.4.1 Shoot tip growth with first leaves.

3.4.2 Cluster with leaf top and bottom views.

3.4.3 Berry with longitudinal section (note small seed trace).

3.4. Dr on gro raisin, sin rai

3.4.5 Dried on the vine, raisin cluster.

3.4.6 Dried on the vine, raisin cluster, dish sample.

3.4.7 Dried on the vine, single raisin.

3.4.8 Dried on the ground raisin, dish sample.

3.4.10 Dried on the vine cluster treated with oleate.

3.4.11 Dried on the vine, dish sample, oleate.

3.4.12 Dried on the vine, single raisin, oleate.

Black Corinth

3.5.1 Shoot tip growth with first leaf.

3.5.2 Cluster with leaf top and bottom view.

3.5.3 Berry and longitudinal section with seed trace.

3.5.4 Natural dried on the ground raisin, dish sample.

3.5.5 Natural dried on the ground raisin, single raisin.

3.5.6 Dried on the vine raisin cluster.

3.5.7 Dried on the vine raisins, dish sample.

3.5.8 Dried on the vine, single raisin.

green, thick, almost glabrous on upper surface, to slightly woolly on the underside. Petiolar sinus wide V to an almost closed, narrow V.

Cluster (**Plate 3.5.2**)- Well-filled cluster, usually straggly, but compact when commercially treated with a growth regulator. Small to medium sized, cylindrical, always winged, sometimes double. Berries are round to oblate, quite small (5 mm) and delicate with tender skin, reddish black and seedless (**Plate 3.5.3**). May have an occasional hard nonviable seed in berries much larger than normal. Flavor is neutral to vinous (tart).

Attributes

Black Corinth variety is best known in California and most parts of the world as Zante Currant. It is early maturing (about August 10 in the San Joaquin Valley), very vigorous and quite productive when ample berry set is accomplished by applying growth regulators at the time when three-fourths of the flowers' calyptra have fallen. It matures about eight days ahead of the Fiesta and is best adapted to drying in the field on the ground (**Plate 3.5.4, 5**). It is readily adapted to DOV raisin production utilizing the cane cutting technique (**Plate 3.5.6,7,8**). Earl Rocca, a Fresno area raisin grower, recom-

3.6.1 Shoot tip growth with first leaf.

3.6.2 Cluster with leaf top and bottom view.

3.6.3 Berry and longitudinal section showing large rudimentary seed coat.

3.6.4 Natural dried on the ground raisin, dish sample.

3.6.5 Natural dried on the ground raisin, single raisin.

3.6.6 Natural dried on the vine raisin cluster.

3.6.7 Natural dried on the vine raisins, dish sample.

3.6.8 Natural dried on the vine, single raisin.

mends cutting the canes around August 10 to 15, allowing approximately 21 days for drying on the vine. The raisins are then harvested mechanically directly into receiving bins (1,000 lb capacity). Because of its thin skin and small berry size, this variety dries the fastest of all and is seldom damaged by the fall rains that occur in the San Joaquin Valley. The raisins are very small, black, very tender and have a tart spicy flavor. Zante Currant raisins are used almost exclusively in baked products.

Black Monukka

Plate 3.6. *600 tons, 546 t.*

Origin: May have originated in India (Hogg, 1884 (IN:) Perold, 1927)

Description

Shoot tip (**Plate 3.6.1**) - woolly and red.

First leaves (**Plate 3.6.1**) - shiny and reddish bronze.

Mature leaf (**Plate 3.6.2**) - medium large, glabrous both sides, five lobed, petioles slightly red. Main veins occasionally reddish at their base.

Cluster (**Plate 3.6.2**) - large to very large, long conical and shouldered, loose to well-filled, berries medium size, ovoid and elongated (**Plate 3.6.3**). Color is uneven and reddish

Sultana
3.7.4 Natural dried on the ground raisin, dish sample.
3.7.3 Berry and longitudinal section with slight seed trace.
3.7.1 Shoot tip growth with first leaf.
3.7.5 Natural dried on the ground raisin, single raisin.
Top
Bottom
3.7.2 Cluster with leaf top and bottom view.
3.7.8 Natural dried on the vine, single raisin.
3.7.7 Natural dried on the vine raisins, dish sample.
3.7.6 Natural dried on the vine raisin cluster.
3.7.10 Oleate treated dried on the vine raisins, dish sample.
3.7.11 Oleate treated dried on the vine, single raisin.
3.7.9 Oleate treated dried on the vine cluster.

black with a silvery bloom. Soft tasteless rudimentary seed traces (**Plate 3.6.3**). Rich, sweet, vinous flavor. Berries shatter readily from rachis.

Attributes

Black Monukka ripens slightly later than Thompson Seedless and is very vigorous, with yields of two and a half tons of raisins per acre when cane pruned. It has a slightly thicker skin than Thompson Seedless, hence requires a longer drying time of three to four weeks. Acreage continues to decline because of the popularity of new, improved raisin varieties. The raisins are larger than Thompson Seedless and are generally plump and meaty (**Plate 3.6.4, 5**). They are black in color and nutty due to its relatively hard, rudimentary seed coat and has a fig-like flavor. It's possible only in the very early districts to make DOV raisins which are large, dark, finely wrinkled and meaty (**Plate 3.6.6, 7, 8**). Some gourmet food stores provide a small outlet, mainly due to its size and attractive black color. Because of berry shatter, more than average field losses occur which decreases potential yield.

Sultana

Plate 3.7. *250 tons, 227 t.*

Origin: Unknown, but most likely the Caspian Sea area.

Synonyms: Round Seedless (Winkler, 1974) Sultana, a known *Vitis vinifera* variety, is not to be confused with the Thompson Seedless or the Australian Sultana. For further explanation, please see "Some Definitions" in **Chapter 1**.

Description

Shoot tip (**Plate 3.7.1**)- woolly green.

Young leaves (**Plate 3.7.1**)- green and shiny.

Mature leaves (**Plate 3.7.2**)- medium large, five lobed, glabrous on both sides. Petiolar sinus shallow and narrow U shape.

Cluster (**Plate 3.7.2**) -large and very heavily shouldered, well-filled to compact. Berries small, round to oblate, greenish with a tendency to sun blush (**Plate 3.7.3**). Some berries will have a partially hard seed (**Plate 3.7.3**).

Attributes

This once popular variety has less than 200 bearing acres in California (CASS, 1998). It has been largely replaced by Thompson Seedless and cannot compete with the new varieties as a raisin grape (**Plate 3.7.4, 5**). Its raisins are light reddish purple with deep wrinkles and possess a rather tart flavor

3.9.6 Natural dried on the vine showing canopy separation and severed canes.

when naturally sun-dried on the vine (**Plate 3.7.6, 7, 8**). As DOV oleate sprayed, its quality improves but not enough to be competitive (**Plate 3.7.9, 10, 11**).

Muscat of Alexandria

Plate 3.8. *200 tons, 182 t.*

Origin: North Africa

Synonyms: Muscat Romain, Panse Musquee, Zibibbo (Italy), Muscatel Romano (Spain), Iskenderiye Misketi (Turkey), White Hanepoot (South Africa), Gordo (Australia).

Description

Shoot tip (**Plate 3.8.1**)- Downy, white with some redness.

First Leaves (**Plate 3.8.1**)- Cobwebby on top, with some bronzing.

Mature leaf (**Plate 3.8.2**)- Medium sized, round, three to

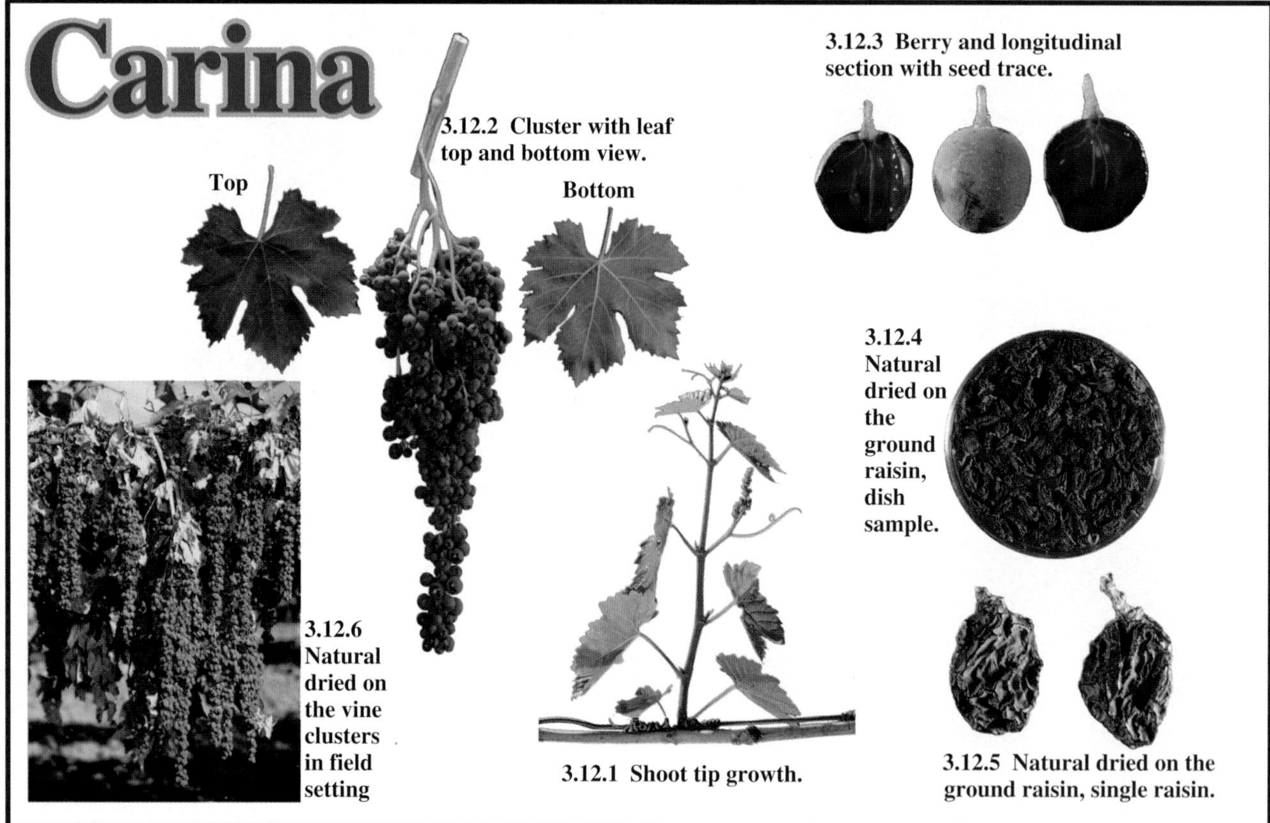

five lobed, petiolar sinus narrow lyre shape, glabrous on top, light to medium tomentum on underside.

Cluster (**Plate 3.8.2**)- Medium large, conical. Cluster density varies from straggly to loose to compact. Berries are seeded, white, large to very large, and obovoid with a strong muscat flavor (**Plate 3.8.3**).

Attributes

Muscat of Alexandria is a multiple use variety and today is used mostly for wine, raisins and some table use. This variety was once a leading raisin grape, second only to Thompson Seedless. Currently, about 1% of Muscat of Alexandria is used for raisins. The rest is almost exclusively used for wine. This variety makes a delicious, meaty, highly flavored and tasty raisin but has fallen out of favor as a raisin primarily because of its seeds (**Plate 3.8.3**). Because of its large size and tough, thick skin (**Plate 3.8.3**) it requires the longest drying time of any variety. It will soon be replaced by new USDA muscat seedless types such as the Summer Muscat released in 1999, which is seedless and makes a raisin

with a strong and pleasing muscat flavor (Ramming and Tarailo, 1999).

DOVine

Plate 3.9. *No official reported production tonnage for DOVine available at the time of publication.*

Origin: Released by Ramming and Tarailo, USDA in 1995. It is the result of hybridization of two seedless genotypes, P79-101 X Fresno Seedless.

Description

Shoot tip (**Plate 3.9.1**) - cobwebby and reddish, first tendrils red.

First leaves (**Plate 3.9.1**) - shiny with bronze tint.

Mature leaf (**Plate 3.9.2**) - medium to large, dark green five lobed with shallow upper sinus, petiolar sinus lyre shape. Main veins red at base.

Cluster (**Plate 3.9.2**)- medium to large, conical shouldered, well-filled to slightly compact. Berries are medium sized, slightly larger than Thompson Seedless, oval and greenish-white (**Plate 3.9.3**). The flavor is neutral and sweet.

Attributes

DOVine is an early-season, white seedless grape which matures about seven days before Fiesta and 14 to 17 days before Thompson Seedless, about the second week of August in the San Joaquin Valley. Sugar development reaches 22% in early August allowing the canes to be cut and the fruit to sundry on the vine naturally (**Plate 3.9. 6, 7, 8, 9**). The DOV raisins consistently grade higher (95% B or Better than when dried on trays). The vine is very vigorous and requires cane pruning for maximum production with yields similar to that of Thompson Seedless, but less than that of Fiesta. The DOV oleate sprayed raisins have a higher percent of yellow to gold color raisins (**Plate 3.9.10, 11, 12**) than either Thompson Seedless or Fiesta. DOVine may also be dried on the ground (DOG) (**Plate 3.9.4, 5**).

Ruby Seedless

Plate 3.10. *2000 tons, 1818 t.*

Origin: Dr. H.P. Olmo. Released by the University of California in 1968, it is a cross of Emperor X Provano 75.

Description

Shoot tip (**Plate 3.10.1**) - red with straggly hairs.

First leaves (**Plate 3.10.1**) - red tint, some straggly hairs on upper surface.

Mature leaf (**Plate 3.10.2**) - medium to large three to five lobes, petiolar sinus closed to overlapping, slight tomentose on main veins on bottom side of leaf. Basal portion of main veins have a reddish tint.

Cluster (**Plate 3.10.2**) - very large, conical, shouldered and well-filled, berry skin reddish-black to dark red. Berries medium size, oval with firm flesh (**Plate 3.10.3**). Sweet neutral flavor.

Attributes

Ruby Seedless is a mid- to late-season table grape but makes a rather nice raisin when the sugar content is high. Ruby Seedless raisins are larger than Thompson Seedless raisins (**Plate 3.10.4, 5**). Ruby Seedless raisins are very dark in color and are made primarily from table grape strippings (fruit left on the vine after normal table grape harvest). Because it matures two to three weeks after Thompson Seedless, it is not recommended for use as a natural sun-dried raisin. However, it makes a good raisin if commercially dehydrated. It is not suitable for DOV raisin production.

Flame Seedless

Plate 3.11. *3000 tons, 2727 t.*

Origin: A USDA/ARS selection made by J. H. Weinberger and F. N. Harmon in 1973. It is the result of the cross (Cardinal X Thompson Seedless) X [(Red Malaga X Tafafih Ahmr) X (Muscat of Alexandria X Thompson Seedless)].

Description

Shoot tip (**Plate 3.11.1**) - cobwebby red, red tendrils.

First leaves (**Plate 3.11.1**) - very reddish and shiny, some straggly hairs on veins on underside.

Mature leaf (**Plate 3.11.2**) - medium to large, five lobes, petiolar sinus wide V, basal portion of main veins, both top and bottom, are red.

Cluster (**Plate 3.11.2**) - medium size, conical, slightly shouldered, loose to well-filled, berries well spaced, small, round, firm, crisp and bright red (**Plate 3.11.3**).

Attributes

The Flame Seedless is an early ripening table grape and attains high sugar. It makes an excellent raisin that is very dark in color, plump, meaty and pliable (**Plate 3.11.4, 5**). The raisins are made from field strippings left on the vines after the normal table grape harvest. The raisins are very large as a result of the girdling and gibberellin treatment for table grapes. It does, however, make a good raisin in its natural state, primarily because of its loose cluster, early ripening and high sugar. It would also be a good candidate for natural sundried DOV raisins (**Plate 3.11.6, 7, 8**). It can have a large seed trace.

Carina

Plate 3.12. *Not commercially available in the USA.*

Origin: Australia. CSIRO (1975) release. Cross of Shiraz X Sultana.

Description

Shoot tip (**Plate 3.12.1**) - cobwebby and green.

Young leaves (**Plate 3.12.1**)- green, shiny, with a few straggly hairs.

Mature leaf (**Plate 3.12.2**)- medium, mid-green, glabrous, three to five lobed, petiolar sinus wide V.

Cluster (**Plate 3.12.2**)- long, narrow cylindrical, many shouldered and loose. Berries small, round, black and seedless (**Plate 3.12.3**). Flavor sweet and tangy.

Attributes

"Carina is a vigorous vine and produces fruit whose ripening berries, unlike Zante Currant, are tolerant to damage by rain even though it ripens later. In Australia, it produced raisins of higher grade than Zante Currant. Yields are satisfactory enough for commercial purposes, however, its flowering habit is problematical, particularly if the weather during flowering is unfavorable for pollination. It appears that the standard GA-CCC growth regulator spray such as is used in Zante Currants is effective in obtaining a commercial berry set in Carina. It is doubtful that Carina would be a replacement for the Zante Currant in California because it is later ripening. In Australia it is frequently grown on tall trellis systems, such as the Shaw swing arm trellis, dried on the vine and mechanically harvested." (*Journal of the Australian Institute of Ag Science,* Dec. 1975). **Plates 3.12.4** and **5** show Carina raisins dried-on-the-ground (DOG) and **Plate 3.12.6** shows Carina clusters drying on the vine in a vineyard setting.

Summer Muscat

Plate 3.13. *No official reported production tonnage for Summer Muscat available at the time of publication.*

Origin: USDA/ARS release, February 1999 and selected by Ramming and Tarailo. It is the result of a cross from A4-162 X P100-111 is a seedling of C15-133 = (Calmeria X Blackrose) X Autumn Seedless made in 1980. A4-162 is a muscat flavored seedling of B37-45 =(Blackrose X P64-18) = (Muscat of Alexandria X Sultanina) X Flame Seedless. P100-111 is a seedling of C15-133 = (Calmeria X Blackrose) X Autumn Seedless.

Description

Shoot tip (**Plate 3.13.1a) -** cobwebby green with slight tinge of red. Sometimes shows faciation, (double tip) (**Plate**

3.13.1b).

First leaves (**Plate 3.13.1a**) - cobwebby upper surface, glabrous below.

Mature leaf (**Plate 3.13.2**) - medium to large, three to five lobed, petiolar sinus narrow slit to overlapping, glabrous top and bottom.

Cluster (**Plate 3.13.2**) - medium to small (0.25 to 0.5 lbs in weight) but long in length, conical with shoulders. Clusters are loosely-filled. Berries are seedless, medium sized (1.4 to 1.9 g) and are oval in shape (**Plate 3.13.3**), sweet in taste with strong muscat flavor that remains when the fruit is dried.

Attributes

Summer Muscat is an early-season, white seedless grape that matures with Fiesta about the third week of August in the San Joaquin Valley. Sugar development reaches 22% by the second to third week of August, allowing canes to be cut (**Plate 3.13.4**) and fruit to dry on the vine (**Plate 3.13.5, 6**). The raisin quality based on percent B or Better grade is similar to Fiesta and has averaged 85% to 100%. The skin is medium thick, medium tough, adheres to the flesh and is similar to Fiesta (**Plate 3.13.7, 8**). The berries contain one to two aborted seeds, which are small and rarely detectable. (**Plate 3.13.3**) It is a very productive variety averaging two clusters per shoot (USDA/ARS, 1999). Summer Muscat can be dried on the ground (**Plate 3.13.9, 10**).

Two new raisin varieties the Diamond Muscat and the Selma Pete were recently released, but not in time to be described in their entirety for inclusion in this Treatise.

Diamond Muscat

Plate 3.14. *No official reported production tonnage for Diamond Muscat available at the time of publication.*

Origin: USDA/ARS release, February 28, 2000, selected by Ramming and Tarailo. It is the result of a cross from A13-2 and B2-11 made in 1985. For a complete genealogy, refer to **Appendix C**.

Description and Attributes

Diamond Muscat is an early-ripening white seedless Muscat raisin grape suitable for cutting canes and drying on the vine. It ripens about the same time as Fiesta during the second and third week of August in Fresno with a sugar content of 22 °Brix. It is very productive averaging 2.1 clusters per shoot. The clusters are medium to large averaging 0.63 lbs. They are conical with shoulders, well filled to slightly loose which is a good attribute for DOV. The berries are medium size (1.8 to 2.2 g) and are oval in shape. The Muscat flavor is pronounced and remains so in the raisins. The raisins are of high quality with a B or better grade averaging 85 to 100%.

Diamond Muscat is vigorous in growth, hence a "T" trellis is recommended to spread the fruiting canes to allow for more air circulation when drying on cut canes, DOV.

Diamond Muscat is very productive, averaging 2.1 clusters per shoot compared to 1-1/2 clusters per shoot for Fiesta.

Source: USDA/ARS. 2000. Notice to fruit growers and nurserymen on the release of Diamond Muscat raisin grape.

Selma Pete

Plate 3.15. *No official reported production tonnage for Selma Pete available at the time of publication.*

Origin: USDA/ARS release, May 16, 2001, selected by Ramming and Tarailo. It is the result from a cross of C66-144 and DOVine made in 1988. The seedling was developed using embryo rescue method. For a complete genealogy, refer to **Appendix C**.

Description and Attributes

Selma Pete is an early-ripening, white seedless raisin grape adapted for mechanical harvesting by cutting the canes and producing DOV raisins; or by picking and drying the fruit on trays. The fruit is of a neutral flavor and does not contain any Muscat flavors. It has many attributes and has a great future as a raisin variety. It consistently ripens with a sugar content of 21 °Brix or higher during the first week in August, or one week before DOVine, two weeks before Fiesta and three or more weeks before Thompson Seedless. Because of its early ripening, caution must be taken when drying the fruit on trays because of the potential for caramelization in the hot August sun. It is recommended that Selma Pete only be dried for raisins since the fresh fruit can develop an astringent flavor. It is very productive, equal to Fiesta and more productive than DOVine and Thompson Seedless. The clusters are medium size, conical and well filled. The berries are medium in sized averaging 2.05 g fresh weight. Raisin quality, when dried on trays, averages 89% B or Better. However, when dried on the vine, Selma Pete has consistently averaged 95 to 100% B or Better raisin quality.

It is moderately vigorous with lateral growth similar to Thompson Seedless. A "T" trellis is recommended for spreading the canes to facilitate air circulation for DOV production. Cane pruning is recommended leaving four to five canes.

Production is a good as Fiesta and better than Thompson Seedless and DOVine.

Source: USDA/ARS. 2001. Notice to fruit growers and nurserymen on the release of Selma Pete raisin grape.

Summary

The Thompson Seedless grape variety will continue to be the dominant variety producing California's natural sun-dried raisins, followed by Fiesta and Black Corinth (Zante Currant). It is predicted that Black Monukka will occupy a very small share of the California market and will be solely dependent on demand from gourmet food outlets. The Muscat of Alexandria will be phased out and replaced by the newly introduced seedless variety, Summer Muscat, which has a pronounced muscat flavor, matures early and is suited for DOV raisin production. There are additional seedless muscat type varieties under development. The newly released DOVine variety will continue to increase in planted acreage, but will not surpass Thompson Seedless for several years. The Sul-

3.13.1a Shoot with single tip and first leaf.

3.13.3 Berry with longitudinal section with seed trace.

3.13.1b Shoot with double tip and first leaf.

3.13.2 Cluster. (a) Leaf, top view. (b) Leaf, bottom view.

3.13.6 Natural dried on the vine raisin cluster.

3.13.9 Dried on the ground raisins, dish sample.

3.13.7 Natural dried on the vine raisins, dish sample.

3.13.10 Dried on the ground raisins, single raisins.

3.13.8 Natural dried on the vine, single raisin.

3.13.4 Natural dried on the vine showing placement of severed cane.

3.13.5 Natural dried on the vine raisin clusters in field setting.

3.14.1: Grape cluster.

3.14.2: Winged grape cluster.

3.14.3: Natural DOV cluster.

3.15.1: Grape cluster.

3.15.2: Natural DOV cluster.

CHAPTER 4.

Methods of Developing Improved Raisin Cultivars

David W. Ramming, Ph.D.

Current Breeding Programs

United States

The grape breeding program at the Horticultural Crops Research Laboratory, United States Department of Agriculture, Agricultural Research Service (USDA/ARS) in Fresno, California, was started in 1923 under the direction of Elmer Snyder, when the first grape hybrids were made. One of the main objectives at that time was the development of seedless grapes (Snyder, 1937). From 1955 to 1973, under the direction of John H. Weinberger (Ramming and Fear, 1993), seedless selections were made from hybrid table grape seedling families and then tested for their value by drying them into raisins. John Weinberger (Weinberger and Loomis, 1973) released the first raisin cultivar, Fiesta, in 1973. It resulted from crossing two unnamed white selections, of which the seedless pollen parent is a sibling of Flame Seedless. The parentage of Fiesta includes Calmeria, Maraville, Tafafihi Ahmr, Muscat of Alexandria, Cardinal, and Sultanina. Fiesta was released to provide an early-ripening raisin grape which ripens seven to 10 days ahead of Sultanina to escape early fall rains. It is also adapted to mechanical harvest by cutting the fruiting canes and allowing the fruit to dry on the vine without applying chemical drying aids. As of 1998, 4,983 acres (2,017 hectares) of Fiesta were planted (Calif. Agr. Statistics Service, 1999) and many are being used for mechanical harvest as dry on the vine raisins.

Under the current direction of David W. Ramming, the selection of raisin grapes from table grape families continued. Embryo rescue tissue culture for the development of seedless grapes was started in 1982 when the first plant was grown from Sultanina aborted seeds (Emershad and Ramming, 1982). The development of embryo rescue procedures for seedless grapes allows for the direct hybridization of seedless grapes (Emershad and Ramming, 1984; Emershad et al., 1989; Emershad and Ramming, 1994a). This research was supported by the California Raisin Advisory Board (CALRAB) until 1993 when CALRAB was disbanded. In 1992 raisin selections were made from families made specifically for table grapes and for raisin grapes. Selections were made for white seedless berries the size of Sultanina with small aborted seeds. The fruit was dried and then evaluated for its raisin characteristics. From 1,141 seedlings of seedless x seedless table grape crosses, 186 (18%) initial selections were made. After drying, only 26 (2%) of the original selections were kept as potential raisin types. This compared to 121 (30%) initial selections made from 405 seedlings of seedless x seedless raisin crosses. The percent selections saved as potential raisins was greater, amounting to 52 (12%) selections. This showed the value of making hybrids between parents chosen specifically for the purpose of raisin cultivar development. Selecting from seeded x seedless raisin crosses was even less efficient than selecting within seedless x seedless table grape populations. From 1218 seedlings, 50 (4%) were initially selected and only nine (0.7%) were saved after the fruit was dried and evaluated as raisins.

See Appendix C for an extensive list of seedless raisin cultivars, their origins, sources of reference, attributes and parentage.

The DOVine raisin grape was developed and released in 1995. It was the result of hybridization of two seedless genotypes, P79-101 x Fresno Seedless, made in 1983 (USDA/ARS, 1995). DOVine produces a raisin grape ripening seven days ahead of Fiesta and is suitable for mechanical harvest by

Table 4.1

The Number and Percent Embryos and Plants Recovered From Hybrids Between Seedless Genotypes in the USDA/ARS Fresno Breeding Program, 1983-1998.

Year	# of Ovules Cultured	# of Embryos	Percent Emb/Ov	No. Plants	Percent Plt/Emb	% Plants from Ovules
1983[zy]	4,075	647	15.9	50	7.7	1.68
1984[zy]	8,859	3,203	36.2	725	22.6	7.9
1985[zy]	16,942	3,781	22.3	1,602	42.4	9.0
1986[zy]	7,484	1,124	15.0	341	30.3	4.5
1987[zy]	10,670	2,737	25.7	276	10.1	2.5
1988[zy]	31,435	4,545	14.5	2,466	54.3	7.8
1989[y]	49,706	10,900	21.9	4,932	45.2	9.9
1990[y]	3,900	705	18.1	370	52.5	9.5
1991[w]	28,721	5,825	20.3	2,371	40.7	8.3
1992[w]	30,547	3,224	10.6	1,643	51.0	5.4
1993[w]	30,409	5,326	17.5	3,046	57.2	10.0
1994[w]	43,718	5,183	11.8	2,223	42.9	5.1
1995[v]	14,546	3,527	24.2	1,490	42.2	10.2
1996[v]	38,385	7,190	18.7	4,309	59.9	11.2
1997[vu]	7,141	2,425	33.9	1,010	41.7	14.1
1998[vu]	15,778	4,564	28.9	1,826	40.0	11.6
Total	**342,316**	**60,796**	**17.8**	**28,680**	**47.2**	**8.4**

[z]Years embryos cultured on filter paper, 20-25 per vessel.
[y]Years liquid ovule and embryo culture medium used.
[w]Years liquid ovule culture medium and agar solidified embryo culture medium used.
[v]Years agar solidified ovule and embryo culture medium used.
[u]Years not all ovules cut open or not all plants planted.

cutting canes for drying on the vine without applying chemical drying aids. Summer Muscat (B1-88) was released in 1999 to provide a seedless muscat-flavored raisin that could be dried on the vine for mechanical harvest (USDA/ARS, 1999). Diamond Muscat (C96-54) was released in 2000. It's muscat flavor is more fruity than Summer Muscat and is earlier-ripening (USDA/ARS, 2000). Grape hybridizations are now being made specifically for the development of raisin grapes, however, at a much-reduced level after CALRAB was disbanded. Development of raisin genotypes suitable for mechanical harvest is currently one of the major objectives, along with powdery mildew (*Unincula necator* (Schw.) Burr) resistance (see **Chapter 11**).

The observation that somatic embryogenesis was occurring during embryo rescue procedures led to the development of a plant regeneration system that is suitable for genetic transformation (Emershad and Ramming, 1994a). This procedure has allowed for the incorporation of foreign genes into grapes and the regeneration of plants containing these genes (Scorza et al., 1996; 1997). This opens new possibilities for the improvement of raisin grapes.

Australia

The Commonwealth and Scientific Industrial Research Organization (CSIRO) grape breeding program at Merbein, Victoria, Australia, started in 1964 and included crosses made specifically for raisins. Allan Antcliff led the program from 1964 to 1982. Peter Clingeleffer has been the program leader from 1982 to the present. The original objectives were to develop seedless raisin types with rain tolerance, early-ripening, increased fruitfulness of basal buds to allow spur pruning, disease resistance, and nematode tolerance in direct producers to avoid use of rootstocks (Clingeleffer, 1985; 1998). Embryo culture methods (Barlass et al., 1988) to recover plants from seedless x seedless hybrids are being used to make the development of seedless raisin cultivars more efficient. Carina, a Zante Currant type, was released in 1975 (Antcliff, 1975). Carina resulted from the cross Shiraz x Sultanina and is stenospermocarpic seedless, compared to Zante Currant, which is parthenocarpic seedless. Carina has the potential for large crops because of its large cluster size. The clusters are loose, making it tolerant to rot and the berries tolerant to rain. Because the flowers are female, the fruit set can be light and the size small without pollination. The application of gibberellic acid and Cycocel (chlormequat) growth regulators appear to promote good yield of even-sized fruit (Leamon, 1982). Merbein Seedless, released in 1981, resulted from the cross of Planta Pedralba (syn. Farana) x Sultanina (Antcliff, 1981). It produces a Sultanina type raisin with larger crops but generally ripens later. Sunmuscat has been jointly released by CSIRO and USDA to provide a seedless muscat flavored raisin cultivar for Australia (USDA/ARS and CSIRO, 1999). Sunmuscat, a sibling of Autumn Seedless table grape, resulted from the cross Calmeria x P64-18 =(Muscat of Alexandria x Sultanina).

Current objectives include early-ripening, low basal bud fruitfulness to facilitate trellis drying, high productivity, high sugar levels, small- and large-berried types, light-colored types for treatment with drying emulsion, rain tolerance, and disease resistance. Embryo rescue techniques and genetic transformation studies are being used to facilitate development of new cultivars (Clingeleffer and Scott, 1995). Sultanina has been transformed with a gene that silences or down regulates polyphenol oxidase activity and hopefully will produce naturally light-colored raisins (Thomas et al., 1998). This would allow raisins with a higher grade to be produced. Financial support is received from the raisin industry.

South Africa

A raisin breeding program at NIVROOP, Stellenbosch, South Africa, has been in place for some time. No cultivars have been named and currently because of low commercial interest, few raisin hybrids are being produced. Their objectives were the development of seedless genotypes suitable as replacements for Black Monukka, Sultanina, Muscat of Alexandria, or Zante Currant. Especially important are high yielding types that ripen early with consistent yields and resistance to rain and decay. Embryo culture is being used for the development of seedless grape genotypes (Burger and Trautmann, 1997).

Objectives for Breeding Raisins

The development of seedless fruit is one of the major objectives of raisin cultivar breeding programs. Seedlessness can be parthenocarpic or stenospermocarpic. Parthenocarpic seedless results from pollination, but no fertilization occurs (Pearson, 1932). In this case the ovules (aborted seed) are so small they are barely distinguishable. The berries are also very small, usually less than 0.025 oz (0.5 gm) in weight. Zante Currant (Black Corinth) is an example of parthenocarpic seedlessness. Stenospermocarpic seedlessness results from pollination and fertilization but the ovule (seed) aborts sometime during its development (Stout, 1936). The time of abortion and amount of seed coat development, scarification and endosperm development are factors that determine the final size of the aborted seed. Even though the aborted ovule may be classified as seedless, it may not be acceptable to the consumer because of its size or evidence when eaten. The aborted seed must be small enough so it is not noticeable when eating the fruit. Embryo development occurs in stenospermocarpic grapes but is incomplete. Abortion can occur but viable embryos at fruit maturity have been observed (Stout, 1936) and grown into plants (Zhang and Ramming, 1994). Embryos can be rescued from ovules of stenospermocarpic grapes by embryo culture techniques and grown into plants on artificial sterile nutrient media (Emershad and Ramming, 1984; Spield-Roy et al., 1985; Emershad et al., 1989; Ramming, 1990). An average of 17.8% embryos per ovules cultured has been obtained since 1984 from seedless x seedless crosses (**Table 1**). An average of 8.4% plants per ovules cultured has been the average recovery rate. The recovery rate varies from year to year due to climatic influences on seed/embryo development and abortion, and parents chosen. Generally, genotypes with smaller aborted seeds

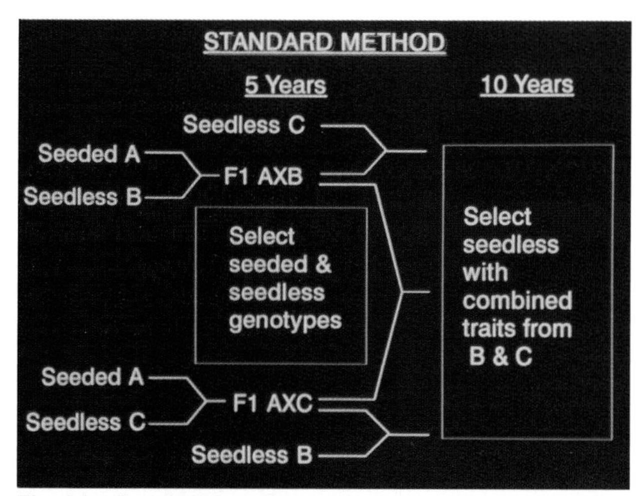

Fig. 4.1a: Standard breeding method using a seeded parent as the female for seedling production.

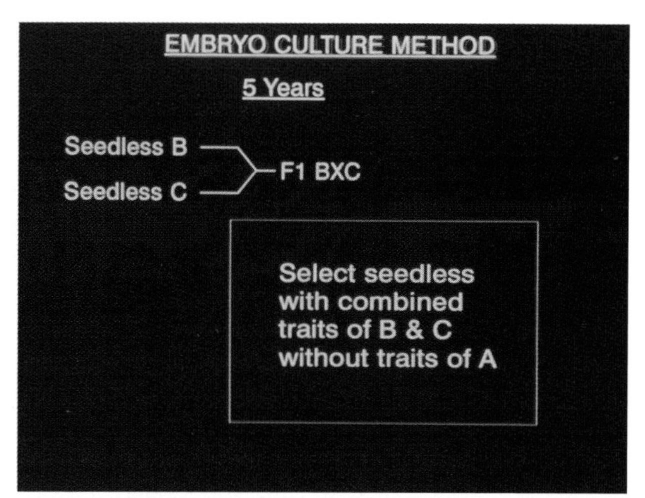

Fig. 4.1b: Embryo culture method for recovery of seedlings when using a seedless parent as the female.

have fewer viable embryos. Improvements in culture media and rescue techniques have improved the recovery rate the last several years. The direct hybridization of seedless x seedless eliminates the necessity of a seeded female as a genetic bridge and reduces the generations needed from two to one (**Figures 4.1a** and **b**). The generation time averages four to five years. The number of seedless individuals obtained is also increased from 0-45 to 55-100% making the breeding of seedless grapes more efficient (Ramming et al., 1990; Ramming, 1997).

Another important trait is early-ripening fruit. Early fall rains in California and Australia can damage or destroy the crop before it is dry (**Figure 4.2**). Cultivars with early-ripening fruit can be picked earlier and therefore increase the chances of the fruit drying before fall rains start. Even in years where rainfall comes at an average time, a shortage of labor may result in late harvest that can be damaged by fall rains. Due to this shortage and labor regulations, the need for mechanized harvesting is evident (May, 1995; Grubich, 1998). Mechanical harvesting of raisin grapes has been increasing in Australia and now accounts for 10% of the crop (Clingeleffer, personal communication). It is also increasing in California with a number of different trellising systems and machine harvesters being used (see **Chapter 5**). The disadvantage of most cultivars currently used for mechanical harvest is their late maturity. This necessitates the need to cut canes to start fruit drying, as well as the application of chemical drying aids or the use of special trellises as is the case for Sultana in Australia and Thompson Seedless in California, respectively. The early-ripening traits of Fiesta and DOVine allow them to dry adequately by cutting canes without the application of chemical drying aids to induce fruit drying (**Figure 4.3**). Cutting canes to induce fruit drying has been reported to cause an average of 10% reduction in crop in Australia (Scholefield et al., 1977). When over 50% defoliation resulted from cutting canes, crop was reduced the following year (Macrae, 1985). Some of this decrease was overcome by using a wider trellis. Cutting canes and trellis drying actually increased production when vines were grown on Ramsey rootstock. Vines grown on Ramsey are usually vigorous, resulting in poor fruit set and reduced yield. Cutting canes allows more light penetration for better flower bud development and cane maturity resulting in increased production. Cultivars that are even earlier ripening than Fiesta and DOVine might allow fruit drying and mechanical harvesting without cutting canes (**Figure 4.4**).

Berry size is generally an easy characteristic to evaluate for when selecting raisins. However, berry size can be affected by the crop size and therefore must be evaluated over several years. The size of the fruit must fall within commercially accepted classes such as currants, midgets, Sultanina (Thompson Seedless) size, Monukka size, or large size (Muscat of Alexandria) to be acceptable. Since the majority of raisins produced in the world are from Sultanina clones, it is the most desirable size.

Production is one of the most important traits needed for profitable returns from the grower's viewpoint. The components that make up production are number of clusters, number of flowers and berries set per cluster, berry size and sugar development. Excess crop, if the plant cannot mature it, however, reduces or delays sugar accumulation, resulting in poor quality fruit. Therefore, combined yield and fruit quality traits need to go hand in hand. For mechanical harvest, genotypes with low basal bud fruitfulness are desir-

Fig. 4.2: Early fall raisin damage that destroyed raisins drying in the field on paper trays.

able. This reduces the amount of fruit left in the head of the vine behind the cut canes. This fruit does not dry and, if excessive, could increase moisture of mechanical harvested raisins above acceptable levels, necessitating supplemental drying.

Muscat-flavored raisins enjoy a specialty market niche for use in baking, eating fresh or candy coating. The majority of muscat-flavored raisins are made from Muscat of Alexandria. The major objection to it is its seeded fruit. When deseeded mechanically, the integrity of the fruit is damaged and a sticky, difficult to manage product is produced. Seedless muscat flavored types are needed to overcome this problem. Several seedless cultivars with muscat flavor exist but have not become popular. They are Sultana Muscato, Thomuscat, and Seedless Muscat. USDA and CSIRO jointly released Sunmuscat as a seedless muscat raisin for Australia. The fruit is larger than Sultanina and it ripens late, necessitating the use of chemical drying aids. In California, Sunmuscat's flavor is very mild, probably a result of hot night temperatures inhibiting the muscat flavor development. USDA released Summer Muscat in 1999 as an early-ripening seedless, muscat raisin that has potential for mechanical harvest. Its fruit is similar in size to Sultanina and ripens about a week earlier. Its loose cluster makes it suitable for drying on the vine (DOV) mechanical harvest raisin production. Diamond Muscat, released in 2000, ripens earlier than Summer Muscat, has a fruity muscat flavor and is also suitable for DOV.

Light-colored or golden raisins are produced in Greece, Turkey, Australia, Iran, Afghanistan, and to a limited extent in California. Their quality/grade is based on the lightness of the color. These products are made by dipping the fruit in methyl oleate and potassium carbonate or exposed to sulfur dioxide gas. Since a portion of the population is asthmatic and sensitive to sulfites, it is desirable to have raisins with natural light-colored fruit. Phenolic compounds are the chemicals that cause the raisins to turn brown when oxidized during drying. Raisin genotypes with low levels of phenolic compounds will produce naturally light colored raisins without chemical additives. Bruce's sport (Antcliff and Webster, 1962) and Sultana Marble are types with low phenolic compounds (Antcliff and Webster, 1962) and produce light-colored raisins (Barrett et al., 1969). The polyphenol oxidase gene, which produces the enzyme to cause phenolic oxidization and browning, has been identified and sequenced from potato (Hunt et al., 1993). The gene has been introduced into potato in the antisense construct and has lowered browning in field grown potatoes (Bachem et al., 1994). Currently, genetic transformation efforts in Australia using antisense constructs of the polyphenol oxidase gene to obtain grapes with inhibited production of polyphenol oxidase is underway. It is hoped this will lead to the natural production of a light colored Sultana after drying (Thomas et al., 1998).

Disease resistance is needed for all raisin production areas. However, some diseases are more prevalent than others. Powdery mildew (*Unincula necator*) is the major fungal disease on grapes in California. Resistance to it and downy mildew (*Plasmopora viticola* Berl. and Toni.) for other grape production regions would greatly reduce grower production costs. Botrytis bunch rot (*Botrytis cinerea* Pers.) is also a major fungal disease of the fruit (see **Chapter 11**). Genotypes with resistance to berry cracking and rain damage will also reduce the incidence of mold in the raisins and reduce crop loss. Fruit with thicker skin and flexible berries will generally have the most resistance. Various species with disease resistance exist in grape germplasm.

Fig. 4.3: DOVine raisins drying on the vine after the canes were cut.

Fig. 4.4: Early-ripening grapes drying naturally without cutting canes.

Most wild species usually have low fruit quality, which must be improved to meet raisin standards when they are used as a source of disease resistance in breeding programs. Olmo (1971) suggested the use of *V. rotundifolia* Michx. as a source of disease and pest resistance as others have done in the past. Wang et al. (1995) reported 31 clones of 10 native *Vitis* species in China that were highly resistant to powdery mildew. Roy and Ramming (1990) reviewed the literature on powdery mildew resistance of grape germplasm and found a number of cultivars reported to be resistant. The results varied among different programs due to differing environmental conditions and differing methods used to identify resistance. Field observations have been used to determine resistance by allowing natural infection (Doster and Schnathorist, 1985a; Eibach, 1994a). This takes a long time and infection may or may not occur. Inoculation of plants in the greenhouse (Doster and Schnathorist, 1985b) or of leaf disks in the laboratory (Staudt, 1997) with powdery mildew can detect resistance/susceptibility sooner, but only evaluates the leaf not the fruit. In vitro dual cultures of powdery mildew or downy mildew with grapes have also been used to identify resistance (Klempka et al., 1984; Barlass et al., 1986). For all these tests, observations must be correlated to ratings in unsprayed field trials for validation. The production of natural chemicals such as the phytoalexin resveratrol in the grape plant has been linked to resistance (Pool et al., 1981; Dercks and Creasy, 1989; Langcake, 1981) along with mechanical barriers (Eibach, 1994b). If the mode of resistance is known, the mechanism for resistance can be searched for instead of relying on natural field infections, which are time consuming and variable. Molecular markers that are linked to disease resistance also have the potential to aid selection of resistant genotypes. They would allow for selection at a very early stage of plant development.

Breeding Methods

Clonal Selection

Clonal selection as a method of cultivar improvement has been practiced since observances of differences in plants started. Clonal selection has been used extensively for the improvement of wine cultivars since the early work of Sartorius (1926, 1928). Clonal selection has been used in both Australia (Woodham and Alexander, 1966; Antcliff, 1973; Antcliff and Hawson, 1974) and the United States (Christensen and Bianchi, 1994) to identify superior clones of Sultanina. As a result of these programs, Sultanina types with improved yield and berry size have been selected for table grapes (TS H5) or raisin production (TS 2A) and planted as the preferred clonal material. When clonal selection was initially practiced, differences were difficult to attribute to genetic effects because of unknown virus status of the different selections. Just selecting plants without viruses had a positive effect on yield. A number of natural mutations have been selected in Sultanina that have resulted in dramatic morphological changes. Clones with different fruit shapes, sizes, and colors have been identified (Barrett et al., 1969). Most mutations were a normal yellow-green color except the red-fruited clone, Red Sultana. Pale colored fruit clones were Kennedy and Bruce's sport. Berry size ranged from small in B27-11, H23, and Bruce's sport to large in Denham, H25, Kennedy and H5. Gigas Thompson Seedless and N22-24 are tetraploids with very large berries.

Conventional Breeding by Sexual Hybridization

Methods to make controlled pollinations for sexual hybrids in grapes have been summarized by Snyder (1937), Einset and Pratt (1975), and Reisch and Pratt (1996). The breeding programs in existence today use sexual hybridization to combine traits from two different parents. Simply put, it consists of emasculating hermaphroditic flowers (removing anthers from flowers with male and female parts), covering them with a paper bag until they are receptive, and then pollinating them with pollen from the selected male parent. The pollen can be either fresh or from storage. Female flowers do not need to be emasculated as their pollen is nonfunctional and their reflex anthers identify them. However, if female flowered genotypes are hybridized with heterozygous hermaphroditic genotypes (those carrying a recessive gene for reflex anthers), only 50% of the offspring will have hermaphroditic flowers. Since female flowered genotypes are unacceptable for commercial production, their use in the breeding program is inefficient unless homozygous hermaphroditic males are used as pollen parents. This results in 100% hermaphroditic seedlings. Staminate (male) flowers produce pollen and nonfunctional ovaries. They can only be used as pollen parents, except in some cases where the application of cytokinins stimulates ovary development and seed production (Negi and Olmo, 1971; Moore, 1970). Grape pollen is carried by the wind. Therefore, the flowers must be protected from contamination by unwanted pollen. Pollen can be stored but viability decreases quite rapidly. Olmo (1942) reported successful pollen storage for four years at 10.4 °F (-12 °C) and 28% relative humidity. Good fruit set was obtained from pollen with as low as 6% germination. By hitting clusters with 50% of the flowers open (from clusters that have been bagged) against a glass plate, the pollen can be collected with a razor blade. If large quantities of pollen are needed, the clusters are collected just as several flowers are opening. The unopened flowers are rubbed off the rachis into a shallow pan and dried under a low wattage incandescent light bulb set 18 inches (45.72 cm) above the flowers. Within 24 to 48 hours the flowers are dry and can be stored in containers in a freezer. To extract the pollen, the unopened container is hit against the hand and the pollen collected off the side of the container with an artist's paintbrush and applied to the emasculated flowers. Pollen germination can be tested on medium containing 20% sucrose and 1% agar in deep well slides covered with cover slips. Germination can be counted after four hours at 78.8 °F (26 °C).

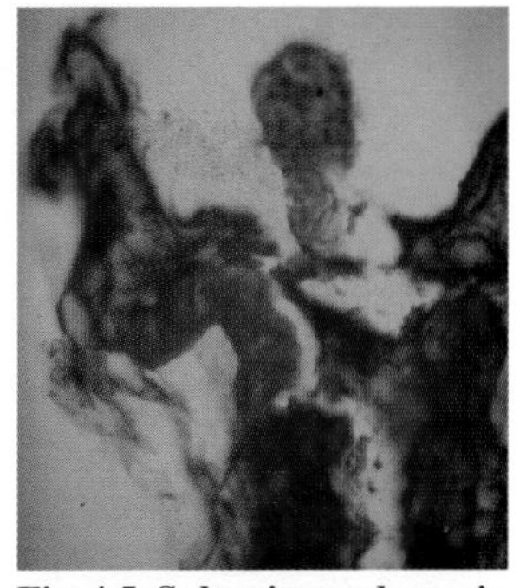

Fig. 4.5: Sultanina embryo six weeks after bloom with approximately 50 cells.

When the fruit is ripe, it is collected, the seeds extracted, and put in plastic bags with damp medium. Rives (1965) reported nearly as good germination from seeds collected two weeks after veraison and stratified at 50 °F (10 °C) for three weeks as those harvested a month later. The seeds) are generally given a cold treatment at 33.8 °F (1 °C) for three to four months, or until it is time to plant them. Sand is a good germination medium, and the seed will germinate in two to four weeks at 75.2 to 84.2 °F (24 to 29 °C). When seedlings are large enough, they are transplanted to the field for fruiting and evaluation. In some cases, the seedlings can be pre-selected in the greenhouse for desirable traits before planting to the field.

Embryo Rescue of Early-Ripening, Seedless Grapes

In the past, seedless genotypes were only useful as pollen parents as no viable seed was produced. Stout (1936) described embryo development in stenospermocarpic seedless grapes. He found that the embryo is fertilized, starts to develop but then stops. With the advent of tissue culture techniques and the rescue of embryos that abort due to incomplete development (Smith et al., 1969; Stewart and Hsu, 1977 and 1978), the rescue of embryos from seedless grapes was made possible (Cain et al., 1983; Emershad and Ramming, 1982 and 1984; Emershad et al., 1989). The plant recovery rate from seedless embryo culture is affected by cultivar, ripening date, aborted seed size and culture date (Pommer et al., 1995). The procedure involves making the sexual hybrid as usual and collecting the fruit six to eight weeks after bloom. At this time, embryos in Sultanina ranged from four to 50 cells in size (**Figure 4.5**) (Emershad and

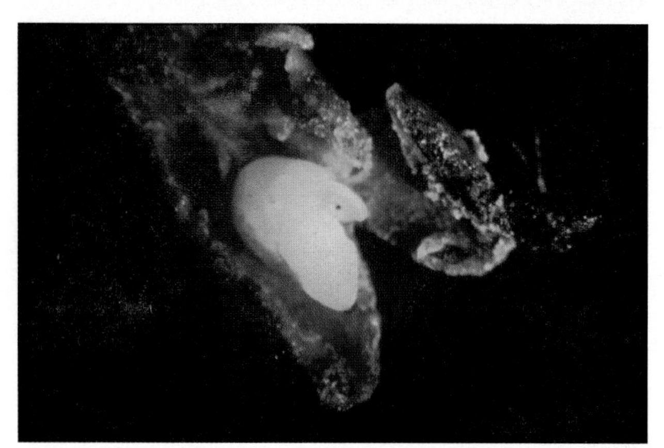

Fig. 4.6: Enlarged embryo after two months of invitro culture.

Ramming, 1984). The ovule is extracted under sterile conditions and cultured on Emershad and Ramming medium (Emershad et al., 1989) for two months at room temperature in ambient light. Then the enlarged embryos (**Figure 4.6**) are extracted and germinated on Woody Plant medium (Lloyd and McCown, 1981) under light and 84.2 ºF (29 ºC) (**Figure 4.7** and **4.8**). The seedlings are acclimated to soil in a double cup system (**Figure 4.9**). When the seedlings are large enough, they are planted in the field (**Figure 4.10**).

The embryo rescue procedure allows the direct hybridization of seedless x seedless genotypes (Ramming, 1990). The advantages provided are the elimination of the seeded parent as a genetic bridge. Therefore, instead of requiring two generations to combine traits from two seedless genotypes (**Figure 5.1a**) and numerous seedlings from which to select desirable genotypes, only one generation is needed (**Figure 5.1b**). The efficiency of the program is also increased because 50 to 100% seedless offspring can be obtained in a family instead of the 0 to 45% when seeded are hybridized with seedless (Ramming et al., 1990; Ramming, 1997). Embryo rescue has also been used to introgress seedless with disease resistance from *V. vinifera* and *V. rotundifolia*, respectively (Goldy et al., 1988).

Genetic Transformation

Genetic transformation of grapes provides the possibility of adding specific genes to existing cultivars while keeping all the characteristic traits of the original cultivar. Ideally this will allow the addition of traits such as disease or insect resistance to existing commercial cultivars without adding other detrimental genes, which would be the case if wild germplasm containing a resistant gene were sexually hybridized with the commercial cultivar. However, it might be useful to introduce a gene into any grape germplasm if it did not exist there before. The germplasm could then be sexually hybridized with commercial cultivars, from which seedlings could be selected for potential new resistant cultivars. To apply genetic transformation in a breeding program, several critical steps are necessary. The first step is the development of an efficient transformation, selection, and plant regeneration system that will provide 50 to 200 plants, each from a different transformation event for a specific gene. It is necessary to be able to identify, sequence and have available useful

Fig. 4.7: Enlarged embryo germination on woody plant medium.

genes for transformation. The development of promoters that will induce gene expression in specific plant tissues will also advance the usefulness of gene transformations. Finally, evaluation of the transformed plants in the field is necessary to determine the best gene expression and if plant performance has been affected. Embryogenic cultures have been the most successful tissue of use in transformation experiments. Reisch and Pratt (1996) have reviewed the progress made in developing somatic embryogenic cultures. Success has been obtained from various grape cultivars and species and from different organs. The response achieved, however, varies greatly depending on the species, cultivar, organ and environment the tissue was grown in. Zygotic tissue (young embryos) has been the easiest tissue from which to achieve somatic embryogenesis in many plants, including grapes. Emershad and Ramming (1994a) reported finding multiple embryos in ovules cultured from seedless grapes (**Figure 4.11**). When subcultured, these embryos continue to multiply (**Figure 4.12**) indicating this is a good proliferation method. Embryos from seedless grapes (Emershad and Ramming, 1994b) and seeded grapes (Stamp and Meredith, 1988) were also induced to produce multiple embryos, and these could be germinated and grown into plants. Durham et al. (1989) showed by isozyme analysis that these embryos were generally the same, and this has been

Fig. 4.8: Plant development after one to two months on woody plant medium.

Fig. 4.9: Double cup system used to acclimate plants.

Fig. 4.10: Seedlings being planted in the field for evaluation.

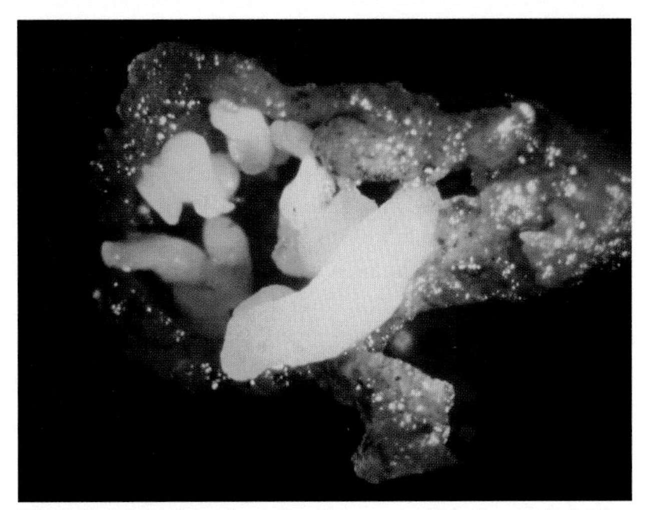

Fig. 4.11: Multiple embryo development within the ovule during invitro ovule culture.

Fig. 4.12: Proliferation of embryos during subsequent subcultures.

confirmed in additional tests (Ramming, personal communication). Margosan et al. (1994) showed that these embryos developed from a single cell on the epidermal layer of the zygotic embryo (**Figure 4.13** and **4.14**). This is the ideal situation for transformation by *Agrobacteria*, which infects the plants outside layer first. Embryos transformed with a gene can be selected on medium containing kanamycin (**Figure 4.6f**) and the transformed embryos verified by ß-glucuronidase (GUS) assay (**Figure 4.15** and **4.16**). The development of somatic embryogenic tissue from zygotic embryos is easy and provides a simple way to introduce new genes into grape germplasm. However, if one wants to insert a new gene into a specific cultivar, embryogenesis must come from somatic tissue of that cultivar.

The first transformed grape vines were produced by Mullens et al. (1990) with *Agrobacterium tumefaciens* (E. F. Smith & Towsend.) Conn. They introduced the neomycin phosphotransferase II (NPTII) gene for selection with the antibiotic, kanamycin (KAN) and the ß-glucuronidase (GUS) gene as a marker that stains blue (Jefferson, 1987). Somatic embryos were produced from anthers of *V. rupestris* Scheele for transformation and regeneration into plants. Others also used these selection and marker genes to transform *V. vinifera* L. CV Koshusanjaku (Nakano et al., 1994), *V. rupestris* (Martinelli and Mandolino, 1994), *V. vinifera* zygotic embryos (Scorza et al., 1995), *V. vinifera* CV Sultanina (Franks et al., 1998) and *V.* sp. CV Chancellor (Kikkert et al., 1996). Later, other genes to control disease were also introduced into grapes. They include the coat protein gene of grape fan leaf nepovirus for resistance to grape fan leaf virus introduced into rootstocks 41B and SO4 and *V. vinifera* CV Chardonnay (Mauro et al., 1995) and into rootstocks *V. rupestris* du Lot and 110 Richter (Krastanova et al., 1995). They also include the coat protein gene of grapevine chrome mosaic nepovirus for resistance introduced into rootstock 110 Richter (Le Gall et al., 1994). They also include the coat protein gene of tomato ringspot virus for resistance introduced into Thompson Seedless (Scorza et al., 1996); and the lytic peptide Shiva-1 gene for potential resistance to bacterial diseases introduced into Thompson Seedless (Scorza et al., 1996). Perl et al. (1996) transformed *Vitis vinifera* CV Superior Seedless with resistance to hygromycin or basta using engineered *Agrobacterium tumefaciens* and the antioxidants polyvinylpolypyrrolidone and dithiothreitol during the cocultivation period. To date, the only gene introduced into grape for raisin improvement was the antisense constructs of a polyphenol oxidase gene into Sultanina by Thomas et al. (1998) for the inhibition of browning caused by phenol oxidation. The purpose of this experiment is to pro-

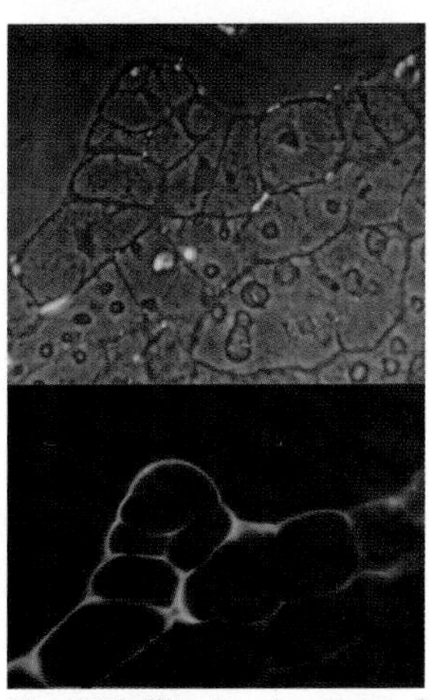

Fig. 4.13: The development of an embryo during proliferation from the epidermal layer of the original embryo (light and florescent microscopy).

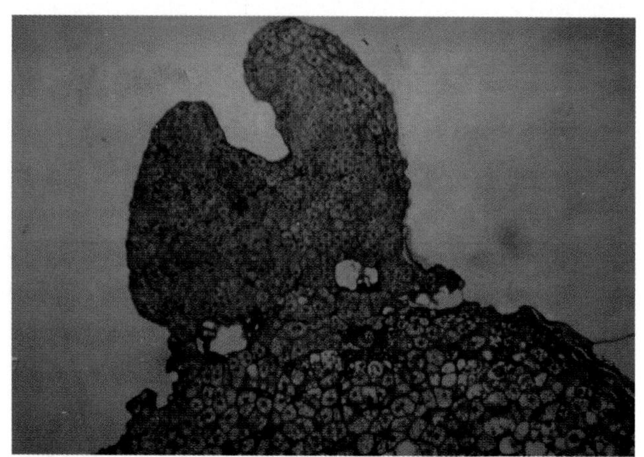

Fig. 4.14: Enlarged embryo showing cotyledon development that proliferated from epidermal layer of the original embryo.

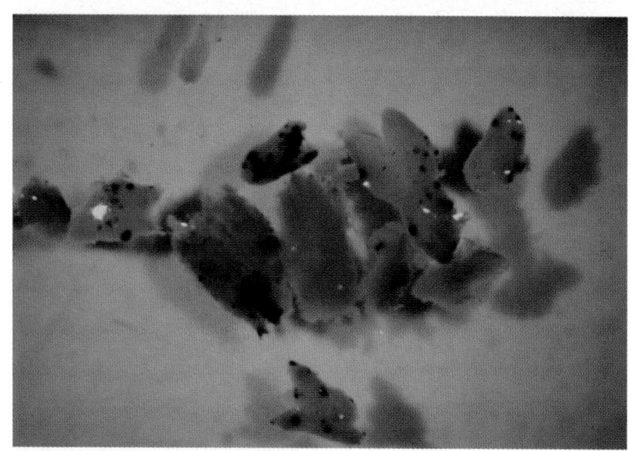

Fig. 4.15: Embryos showing positive for ß-glucuronidase transformed where areas have occurred.

Fig. 4.16: Selection of transformed embryos during germination on a medium containing kanamycin.

duce golden raisins without the use of sulfur dioxide or methyl oleate and potassium carbonate. Somatic embryos were produced by anther culture and transformed with engineered *Agrobacterium tumefaciens* (Franks et al., 1998).

Production of transformed grape vines will be an additional tool to be used in grape breeding programs for the improvement of grape germplasm and development of new cultivars. However, transformed grapes will have to be field tested to determine if they are producing the desired effects

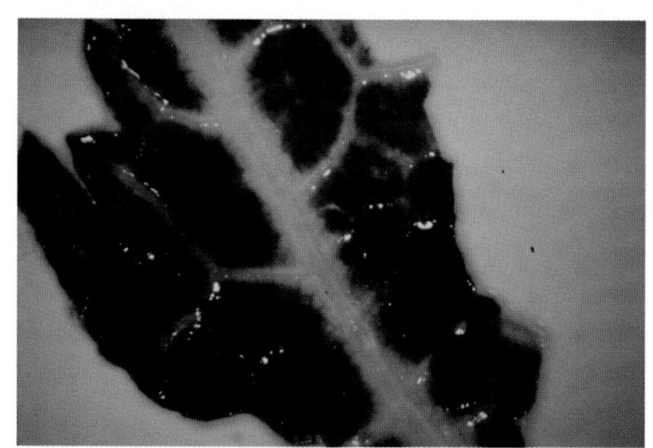

Fig. 4.17: Leaf of transformed plant showing positive for ß-glucuronidase, verifying transformation.

without deleterious effects on production or other traits. When trying to transform a cultivar, a number of transformation events need to be produced so selection can be made among the transformed individuals to find the best one that produces the most desirable results. Genetic transformation will be most useful when a gene is needed that does not exist in grape germplasm but exists in other organisms. In other cases, genetic transformation will be beneficial when trying to introduce only one gene into a specific cultivar. When sexual hybridization is used, many genes are added along with the desired trait that can change the major characteristics of the cultivar.

Molecular Markers

Molecular markers are beneficial to breeding programs to provide a quicker method to identify desirable genotypes without having to test the phenotype. This means that selection of individuals that contain desirable genes can be made without having to grow the plant in the greenhouse or field where the plant would normally produce the trait. Time is saved, allowing the production and screening of large numbers of individuals without having to grow them in the field. A number of methods have been used to develop molecular markers. Among them are restriction fragment length polymorphism (RFLP), random amplified polymorphic DNA (RAPD), sequenced characterized amplified regions (SCARs), amplified fragment length polymorphism (AFLPS), and simple sequence repeats (SSR); and their application in plant breeding has been reviewed by Staub et al. (1996). Initially, these methods were used for finger printing different cultivars, enabling them to be positively identified (Bowers et al., 1993; Thomas et al., 1994; Xu et al., 1995; Lamboy and Alpha, 1998; Lin and Walker, 1998; Ye et al., 1998). More recently, they have been used to identify linkages with desirable traits. Lodhi et al. (1995) developed a molecular map from a single interspecific hybrid grape population, Cayuga White x Aurore. Over 200 RAPD markers plus 16 RFLP and isozyme markers were used to develop the maps and their linkage groups. Development of such maps are important because they contain a series of markers that can be linked to genes for viticultural traits to aid in selection. However, for these markers to work they have to be tightly linked to the desired trait, or they must be the actual DNA sequence for the gene controlling the desired trait. Molecular markers for seedlessness have been reported (Striem et al., 1996; Lanhogue et al., 1998). Since inheritance of seedlessness is not simply inherited and molecular markers not tightly linked with seedlessness, seven molecular markers were needed to identify 72% of the seeded individuals (Striem et al., 1996). Striem et al. (1994) showed significant correlation of muscat flavor or berry color with four molecular markers each. A. Walker (personal communication) has developed an AFLP marker linked to rootknot nematode resistance and at the same time has developed a molecular map for the rootstock parents used. Markers for powdery mildew resistance, *Botrytis* rot resistance, and flower sex have been located for grapes in B. Reisch's laboratory (Reisch, 1998). Additional markers have been identified for a single dominant gene for powdery mildew resistance from *V. rotundifolia* (Pauquet et al., 1998).

Bulk segregant analysis is a method used to identify markers linked to specific traits. Two bulked DNA samples from a segregating population are made. Each bulked group is different only for the desired trait and screened to identify primers that are different between the two groups (Michelmore et al., 1991). The closest linked primers would be the best for selecting the desired trait. Once markers have been established, those flanking the desired gene on both sides are identified. The genetic code for the desired gene between the two markers can then be sequenced. Another method to identify genes is to determine the enzyme protein that is responsible for the trait in question and then construct the genetic code to match the amino acid sequence that makes up the enzyme. The proof for correct sequence identification of a gene is to determine if the gene sequence will produce the expected product in a test organism such as tobacco. This was done for the resveratrol gene, which was sequenced from grape, cloned, and then introduced into tobacco. Resveratrol was produced and the plant shown to resist *Botrytis* infection or growth (Hain et al., 1993).

Molecular markers are most beneficial in identifying simply inherited traits although methods are being developed for quantative trait loci (QLT). The number of QLT markers for traits in tomatoes, for example, was reviewed by Staub et al. (1996). The number of markers used to identify specific quantative traits ranged from three to 11 and explained 18 to 72% of the phenotypic variability. As of 1996, no cultivar selected by molecular markers has been released publicly. Currently, the cost of using molecular markers can out weigh their effectiveness when compared to selection based on the plant appearance. With laboratory costs decreasing, this tool should have a place in breeding programs in the future.

Genetic Control of Desired Traits

De Lattin (1957) has compiled a list of 53 grape genes. Many of the genes important for raisin production such as yield, berry size, sugar production, and early-ripening are quantatively inherited. Although heritability of yield is low (Firoozabady and Olmo, 1987), traits that contribute to yield such as cluster number and weight could be selected for to increase yield potential. Heritability was also generally high for berry weight and cluster compactness. It is important to understand genetic correlation between traits to know if they can be obtained together by breeding. Fanizza (1979) found no correlation between yield, sugar content, acidity, and pH, leading to the conclusion that cultivars with high yield could be combined with high sugar content, low acidity, and low pH. Other fruit traits, such as color, are controlled by two pair of genes with epistatic effect, where black is dominant to red and both are dominant to white (Barritt and Einset, 1969). According to Wagner (1967), five complementary dominant genes control Muscat flavor.

Ledbetter and Ramming (1989) and Reisch and Pratt (1996) have reviewed theories for inheritance of seedlessness. Many different theories for seedless inheritance probably result from the different methods used to define seedlessness. More recently, Ledbetter and Burgos(1994) have proposed a three complementary dominant gene theory for seedlessness. Bouquet and Danglot (1996) subsequently proposed three complementary recessive genes regulated by a dominant gene for control of seedlessness. Genetic control according to this theory is quantitative in nature. Families have now been reported from seedless x seedless crosses that produced all seedless individuals. DOVine raisin grape has given many families that contain 100% seedless offspring. When hybridized with an early-ripening seeded grape, 85% seedless offspring were recovered and 99.8% seedless offspring were recovered when hybridized with Redglobe (Ramming, 1997).

Reisch and Pratt (1996) have reviewed the inheritance of resistance to powdery mildew, downy mildew, *Botrytis*, Pierce's disease, phylloxera and nematodes. The control of these traits is usually multigeneic and complex. Sources of resistance to these diseases and pests have been identified in wild species. The production of stilbene phytoalexins such as resveratrol have been associated with downy mildew and *Botrytis* resistance (Langcake, 1981; Langcake and McCarthy, 1979).

Understanding the inheritance of traits needed for improving raisins aids the selection of the most appropriate breeding methods. At the same time, it allows for the selection of parents that will provide the best populations from which selection of desirable raisin genotypes can be made.

Photo Credits

Figures 4.1a and b, 4.2, 4.3, 4.4 and 4.10: David W. Ramming

Figures 4.5, 4.6, 4.7, 4.8, 4.9, 4.11 and 4.12: Richard L. Emershad

Figures 4.13, 4.14: Dennis Margosan

Figures 4.15, 4.16 and 4.17: Ralph Scorza

References

Antcliff, A. J. 1973. Evidence for a genetic difference in

berry weight between Sultana vines. Vitis 12:16-22.

Antcliff, A. J. 1975. Four new grape varieties released for testing. J. Austral. Inst. Agri. Sci.41(4):262-265.

Antcliff, A. J. 1981. Merbein Seedless: a new white seedless grape for drying. J. Austral. Ins. Agri. Sci. 47(3):167-168.

Antcliff, A. J. and H. Hawson. 1974. The Australian Sultana clones: Rapid adoption of improved planting material. J. Austral. Inst. Agri. Sci. 40:109-113.

Antcliff, A. J. and W. J. Webster. 1962. Bruce's sport - a mutant of the Sultana. Austral. J. Exp. Agri. and Animal Husbandry 2:91-100.

Bachem, C. W. B., G.-J. Speckmann, P. C. B. van der Lende, F. T. M. Verheggen, M. D. Hunt, J. C. Steffens, and M. Zabeau. 1994. Antisense expression of polyphenol oxidase genes inhibits enzymatic browning in potato tubers. Bio/Tech. 12:1101-1105.

Barlass, M., R. M. Miller and A. J. Antcliff. 1986. Development of methods for screening grapevines for resistance to infection by downy mildew. I: Dual culture in vitro. Am. J. Enol. Vitic. 37:61-66.

Barlass, M., D. W. Ramming and H. P. Davis. 1988. In-ovulo embryo culture: a breeding technique to rescue seedless x seedless table grape crosses. The Australian Grapegrower and Wine Maker 259:123-125.

Barrett, H. C., G. H. Kerridge and A. J. Antcliff. 1969. The drying characteristics of several Sultana clones. Food Technology in Australia 21:516-517.

Barritt, B. H. and J. Einset. 1969. The inheritance of three major fruit colors in grapes. J. Am. Soc. Hort. Sci. 94:87-89.

Bouquet, A. and Y. Danglot. 1996. Inheritance of seedlessness in grapevine (*Vitis vinifera* L.). Vitis 35:35-42.

Bowers, J. E., E. B. Bandman and C. P. Meredith. 1993. DNA fingerprint characterization of some wine grape cultivars. Am. J. Enol. Vitic. 44:266-274

Burger, P. and I. A. Trautmann. 1997. Breeding seedless grapes in South Africa by means of embryo culture. South African Inter. Table Grape Symposium, Cape Town, Abstr poster 30.

California Acreage Statistics Service. 1999. California grape acreage. P.O. Box 1258, Sacramento, CA 95812.

Cain, D. W., R. L. Emershad and R.E. Tarailo. 1983. In-ovulo embryo culture and seedling development of seeded and seedless grapes (*Vitis vinifera* L.). Vitis 22:9-14.

Christensen, L. P. and M. L. Bianchi. 1994. Comparisons of Thompson Seedless clones for raisin production. Am. J. Enol. Vitic. 45:150-154.

Clingeleffer, P. R. 1985. Breeding table grape varieties. Austral. Grapegrower. Winemaker 256:117-119.

Clingeleffer, P. R. 1998. Breeding table grape and raisin varieties. In: Omura, M.; Hayashi, T., and Scott, N.S. eds. Breeding and biotechnology for fruit trees: Proceedings of the 2nd Japan-Australia Inter. Workshop; Tatura, Merbein, Adelaide, Australia. Tsukuba, Japan: National Institute of Fruit Tree Science: 46-49.

Clingeleffer, P. R. and N. S. Scott. 1995. Molecular techniques and grapevine breeding. In: Goussard, R. G., Archer, E., Saayman, D., Tromp, A. and Van Wyk, C. J. eds. "Proceedings of the first SASEV Inter. Congress", 8-10 November 1995, Cape Town, South Africa, pp. 63-66. (Dennesig: South African Society for Enology and Viticulture).

De Lattin, G. 1957. On the genetics of grapes. Present results of factor analysis in the genus *Vitis* (in German). Vitis 1:1-18.

Dercks, W. and L. L. Creasy. 1989. The significance of stilbene phytoalexins in the *Plasmopara viticola* - grapevine interaction. Physiological and Molecular Plant Pathology. 34:189-202.

Doster, M. A. and W. C. Schnathorst. 1985a. Comparative susceptibility of various grapevine cultivars to the powdery mildew fungus *Uncinula necator*. Am. J. Enol. Vitic. 36:101-104.

Doster, M. A. and W. C. Schnathorst. 1985b. Effects of leaf maturity and cultivar resistance on development of the powdery mildew fungus on grapevines. Phytopathology 75:318-321.

Durham, R. E., G. A. Moore, D. J. Gray and J. A. Mortensen. 1989. The use of leaf GPI and IDH isozymes to examine the origin of polyembryony in cultured ovules of seedless grape. Plant Cell Rpts. 7:669-672.

Eibach, R. 1994a. Investigations about the genetic resources of grapes with regard to resistance characteristics to powdery mildew (*Oidium tuckeri*). Vitis 33:143-150.

Eibach, R. 1994b. Defense mechanisms of the grapevine to fungus diseases. Amer. Vineyard. Jan. p8-10.

Einset, J. and C. Pratt. 1975. Grapes. p. 130-153. In: J. Janick, and J.N. Moore eds. Advances in Fruit Breeding.

Purdue Univ. Press, West Lafayette, IN.

Emershad, R. L. and D. W. Ramming. 1982. In-ovulo embryo culture of *Vitis vinifera* L., c.v. Thompson Seedless. HortScience 17(4):576. Abstr.

Emershad, R. L. and D. W. Ramming. 1984. *In-ovulo* embryo culture of *Vitis vinifera* L. cv. 'Thompson Seedless'. Amer. J. Bot. 71:873-877.

Emershad, R. L. and Ramming, D. W. 1994a. Somatic embryogenesis and plant development from immature zygotic embryos of seedless grapes (*Vitis vinifera* L.). Plant Cell Reports 14:6-12.

Emershad, R. L. and Ramming, D. W. 1994b. Effects of buffers and pH on *Vitis* somatic embryo proliferation and enlargement. Inter. Symp. Table Grape Production. Special issue Amer. Soc. Enol. and Vit. pg 219-222.

Emershad, R. L., D. W. Ramming and M. D. Serpe. 1989. In-ovulo embryo development and plant formation from stenospermic genotypes of *Vitis vinifera*. Amer. J. Bot. 76(3):397-402.

Fanizza, G. 1979. Genotypic and phenotypic variation and correlations in wine-grape varieties *Vitis vinifera* grown in Apulia (Italy). Genet. Agr. 33:37-44.

Frioozabady, E. and H. P. Olmo. 1987. Heritability and correlation studies of certain quantitative traits in table grapes, *Vitis* spp. Vitis 26:132-146.

Franks, T., D. G. He and M. Thomas. 1998. Regeneration of transgenic *Vitis vinifera* L. Sultana plants: genotypic and phenotypic analysis. Molecular Breeding 4:321-333.

Goldy, R., R. Emershad, D. Ramming and J. X. Chaparro. 1988. Embryo culture as a means of introgressing seedlessness from *Vitis vinifera* L. to *V. rotundifolia* Michx. HortScience 23(5): 886-889.

Grubich, L. 1998. Mechanical laborers. Fruit Grower. Nov. p12A, B & D.

Hain, R., H. J. Reif, E. Krause, R. Langebartels, H. Kindl, B. Vornam, W. Wiese, E. Schmelzer, P. H. Schreier, R. H. Stocker and K. Stenzel. 1993. Disease resistance results from foreign phytoalexin expression in a novel plant. Nature (London) 361:153-156.

Hunt, M. D., N. T. Eannetta, H. Yu, S. M. Neroman and J. C. Steffens. 1993. cDNA cloning and expression of potato polyphenol oxidase. Plant Mol. Biol. 21:59-68.

Jefferson, R. A. 1987. Assaying chimeric genes in plants: The GUS gene fusion system. Plant Mol. Biol. Rpt. 5:387-405.

Kikkert, J. R., D. Hebert-Soule, P. G. Wallace, M. J. Striem and B. I. Reisch. 1996. Transgenic plantlets of 'Chancellor' grapevine (*Vitis* sp.) from biolistic transformation of embryogenic cell suspensions. Plant Cell Rpt. 15:311-316.

Klempka, K. C., C. P. Meredith and M. A. Sall. 1984. Dual culture of grape powdery mildews (*Uncinula necator* Burr.) on its host (*Vitis vinifera* L.). Am. J. Enol. Vitic. 35:170-174.

Krastanova, S., M. Perrin, P. Barbier, G. Demangeat, P. Cornuet, N. Bardonnet, L. Otten, L. Pinck and B. Walter. 1995. Transformation of grapevine rootstocks with the coat protein gene of grapevine fanleaf nepovirus. Plant Cell Rpt. 14:550-554.

Lahogue, F., P. This and A. Bouquet. 1998. Identification of a codominant scar marker linked to the seedlessness character in grapevine. Theoretical and Applied Genetics 97:950-959.

Lamboy, W. F. and C. G. Alpha. 1998. Using simple sequence repeats (SSRs) for DNA fingerprinting germplasm accessions of grape (*Vitis* L.) species. J. Amer. Soc. Hort. Sci. 123:182-188.

Langcake, P. 1981. Disease resistance of *Vitis* spp. and the production of the stress metabolites resveratrol, g-viniferin, a-viniferin and pterostilbene. Physiol. Plant Path. 18:213-226.

Langcake, P. and W. V. McCarthy. 1979. The relationship of resveratrol production to infection of grapevine leaves by *Botrytis cinera*. Vitis 18:244-253.

Le Gall, O., L. Torregrosa, Y. Danglot, T. Candresse and A. Bouquet. 1994. Agrobacterium-mediated genetic transformation of grapevine somatic embryos and regeneration of transgenic plants expressing the coat protein of grapevine chrome mosaic nepovirus (GCMV). Plant Sci. 102:161-170.

Leamon, K. 1982. Sprays for setting currants and Carina. Agnote No. 2005/82. Dept. of Agri., Government of Victoria.

Ledbetter, C. A. and L. Burgos. 1994. Inheritance of stenospermocarpic seedlessness in *Vitis vinifera* L. J. Hered. 85:157-160.

Ledbetter, C. A. and D. W. Ramming. 1989. Seedlessness in Grapes. Plant Breeding Reviews. Vol. 11:159-184.

Lin, H. and M. A. Walker. 1998. Identifying grape rootstocks with simple sequence repeat (SSR) DNA markers. Am. J. Enol. Vitic. 49:403-407.

Lloyd, G. and B. H. McCown. 1981. Commercially feasible micropropagation of mountain laurel, *Kalmia latifolia*, by use of shoot tip culture. Proc. Inter. Plant Prop. Soc. 30:421-427.

Lodhi, M. A., M. J. Daly, G.-N. Ye, N. F. Weeden and B.I.Reisch. 1995. A molecular marker based linkage map of *Vitis*. Genome 38:786-794.

Macrae, I. 1985. Vine management for drying on the vine. p 11-19. In Proceedings "Trellis Drying and Mechanical Harvesting of Grapes". Sunraysia College of T.A.F.E., Mildura, Australia 12-12-85. Dried Fruits Research Committee Seminar.

Margosan, D. A., Emershad, R. L. and Ramming, D. W. 1994. Origin of somatic embryos from embryo rescue cultures of seedless grapes (*Vitis vinifera* L.). Inter. Symp. Table Grape Production. Special issue Amer. Soc. Enol. and Vit. pg 133-135.

Martinelli, L. and G. Mandolino. 1994. Genetic transformation and regeneration of transgenic plants in grapevine (*Vitis rupestris* S.). Theor. Appl. Genet. 88:621-628.

Mauro, M. C., S. Toutain, B. Walter, L. Pinck, L. Otten, P. Coutos-Thevenot, A. Deloire and P. Barbier. 1995. High efficiency regeneration of grapevine plants transformed with the GFLV coat protein gene. Plant Science 112:97-106.

May, P. 1995. A quarter century of mechanical grape harvesting in Australia. Austral. & New Zealand Wine Industry Journal 10(1):41-45.

Michelmore, R. W., I. Paran and R. V. Kesseli. 1991. Identification of markers linked to disease-resistance genes by bulked segregant analysis: A rapid method to detect markers in specific genomic regions by using segregating populations. Proc. Natl. Acad. Sci. USA. 88:9828-9832.

Moore, J. M. 1970. Cytokinin-induced sex conversion in male clones of *Vitis* species. J. Am. Soc. Hort. Sci. 95:387-393.

Mullins, M. G., P. C. A. Tang and D. Facciotti. 1990. *Agrobacterium*-mediated genetic transformation of grapevines: Transgenic plants of *Vitis rupestris* Scheele and buds of *Vitis vinifera* L. Bio/Techniolgy 8:1041-1045.

Nakano, M., Y. Hoshino and M. Mii. 1994. Regeneration of transgenic plants of grapevine (*Vitis vinifera* L.) via *Agrobacterium rhizogenes*-mediated transformation of embryogenic calli. J. Expt. Bot. 45:649-656.

Negi, S. S. and H.P. Olmo. 1971. Induction of sex conversion in male *Vitis*. Vitis 10:1-19.

Olmo, H. P. 1942. Storage of grape pollen. Proc. Am. Soc. Hort. Sci. 41:219-224.

Olmo, H. P. 1971. *Vinifera* x *rotundifolia* hybrids as wine grapes. Amer. J. Enol. Vitic. 22:23-29.

Pauquet, J., A. F. Adam-Blondon, S. Rousseau, J. P. Peros and A. Bouquet. 1998. Tagging a *Muscadinia rotundifolia*-originated powdery mildew resistance gene in *Vitis vinifera*. VIIth Inter. Symp. Grapevine Genetics and Breeding, Montpellier, France Abstract P2.10.

Pearson, H. M. 1932. Parthenocarpy and seed abortion in *Vitis vinifera*. Amer. Soc. Hort. Sci. 29:169-175.

Perl, A., O. Lotan, M. Abu-adied and D. Holland. 1996. Establishment of an *Agrobacterium*-mediated transformation system for grape (*Vitis vinifera L.*): The role of antioxidants during grape-*Agrobacterium* interactions. Nature Biotechnology 14:624-628.

Pommer, C. V., D. W. Ramming and R. L. Emershad. 1995. Influence of grape genotype, ripening season, seed trace size, and culture date on in-ovule embryo development and plant formation. Braganta, Campinas 54(2):237-249.

Pool, R. M., L. L. Creasy and A. S. Frackelton. 1981. Resveratrol and the viniferins, their application to screening for disease resistance in grape breeding programs. Vitis 20:136-145.

Ramming, D. W. 1990. The use of embryo culture in fruit breeding. HortScience 25:393-398.

Ramming, D. W. 1997. Families with 100% seedless progeny have been obtained from DOVine raisin grape. South African Inter. Table Grape Symp., Cape Town, Abstr poster 13.

Ramming D. W. and C. D. Fear. 1993. Dedication: John Weinberger - Fruit Breeder and Horticulturist . Plant Breeding Reviews 11:1-10.

Ramming, D. W., C. A. Ledbetter and R. Tarailo. 1990. Hybridization of seedless grapes. Vitis Special Issue, Proc. 5th Inter. Symp. Grape Breeding: 439-444.

Reisch, B. 1998. Molecular markers - the foundation for grapevine genetic maping, DNA fingerprinting and genomics. VIIth Inter. Symp. Grapevine Genetics and Breeding, Montpellier, France Abstract C2.1

Reisch, B. I. and C. Pratt. 1996. Grapes. p 297-369. In: Janick, J. and J. N. Moore, eds. Fruit Breeding Volume II Vine and Small Fruits. John Wiley & Sons, Inc., N.Y.

Rives, M. 1965. The germination of grape seeds, 1: preliminary experiments (in French, English summary).

Ann. Amelior Plantes 15:79-91.

Roy, R. R. and D. W. Ramming. 1990. Varietal resistance of grape to the powdery mildew fungus, *Uncinula necator*. Fruit Var. J. 44:149-155.

Sartorius, O. 1926. Zur Rebenselektion unter besonderer Berücksichtigung der Methodik und der Ziele aufgrund von 6-bis 14jährigen Beobachtungen an einem Klon. Z.f. Pflanzenz. 11:31-74.

Satorius, O. 1928. Über die wissenschafllichen Grundlagen der Rebenselektion in reinen Beständen. Z.f. Pflanzenz. 13:79-86.

Scholefield, P. B., P. May and T. F. Neales. 1977. Harvest-pruning and trellising of 'Sultana' vines. I. Effects on yield and vegetative growth. Sci. Hortic. 7:115-122.

Scorza, R., J. M. Cordts, D. W. Ramming and R. L. Emershad. 1995. Transformation of grape (*Vitis vinifera* L.) zygotic-derived somatic embryos and regeneration of transgenic plants. Plant Cell Reports 14:589-592.

Scorza, R., J. M. Cordts, D. J. Gray, D. Gonsalves, R. L. Emershad and D. W. Ramming. 1996. Producing transgenic 'Thompson Seedless' grape (*Vitis vinifera* L.) plants. J. Amer. Soc. Hort. Sci. 121:616-619.

Smith, C. A., C. H. Bailey and L. F. Hough. 1969. Methods for germinating seeds of some fruit species with special reference to growing seedlings from immature embryos. N J Agr. Expt. Sta. Bul. 823.

Snyder, E. 1937. Grape development and improvement. p. 631-644. In: USDA Yearbook of Agriculture.

Spiegel-Roy, P. N., N. Sahar, J. Boron and V. Lavi. 1985. In vitro culture and plant formation from grape cultivars with abortive ovules and seeds. J. Amer. Soc. Hort. Sci. 110:109-112.

Stamp, J. A. and C. P. Meredith. 1988. Proliferative somatic embryogenesis from zygotic embryos of grapevine. J. Amer. Soc. Hort. Sci. 113:941-945.

Staub, J. E., F. C. Serquen and M. Gupta. 1996. Genetic markers, map construction, and their application in plant breeding. HortScience 31:729-741.

Staudt, G. 1997. Evaluation of resistance to grapevine powdery mildew (*Uncinula necator* [Schw.} Burr., anamorph *Oidium tuckeri* Berk.) in accessions of *Vitis* species. Vitis 36:151-154.

Stewart, J. McD. and C. L. Hsu. 1977. In-ovulo embryo culture and seedling development of cotton (*Gossypium hirsutum* L.). Planta 137:113-117.

Stewart, J. McD. and C. L. Hsu. 1978. Hybridization of diploid and tetraploid cottons through in-ovulo embryo culture. J. Hered. 69:404-408.

Stout, A. B. 1936. Seedlessness in grapes. N.Y. Agri. Exp. Sta. Tech. Bul. 238.

Striem, M. J., G. Ben-Hayyim and P. Spiegel-Roy. 1994. Developing molecular genetic markers for grape breeding, using polymerase chain reaction procedures. Vitis 33: 53-54.

Striem, M. J., G. Ben-Hayyim and P. Spiegel-Roy. 1996. Identifying molecular genetic markers associated with seedlessness in grape. J. Amer. Soc. Hort. Sci. 121:758-763.

Thomas, M. R., P. Cain, S. and N. Scott. 1994. DNA typing of grapevine: a universal methodology and database for describing cultivars and evaluating genetic relatedness. Plant Mol. Biol. 25:939-949.

Thomas, M. R., T. K. Franks, P. Iocco, I. B. Dry, P. K. Boss, C. Davies, S. P. Robinson and N. S. Scott. 1998. Status and future direction of transgenic grapevines in Australia. In: Omura, M., T. Hayashi and N. S. Scott, ed. Breeding and biotechnology for fruit trees: Proc. 2nd Japan-Australia Inter. Workshop. National Institute of Fruit Tree Science pp 57-62.

United States Department of Agriculture, Agricultural Research Service. 1995. Notice to fruit growers and nurserymen relative to the naming and release of the DOVine raisin grape. 1 p.

United States Department of Agriculture, Agricultural Research Service. 1999 . Notice to fruit growers and nurserymen of the release of Summer Muscat raisin grape. 1 p.

United States Department of Agriculture, Agricultural Research Service and CSIRO. 1999. Notice to fruit growers and nurserymen of the release of Sunmuscat raisin grape. 2 pp.

United States Department of Agriculture, Agricultural Research Service. 2000. Notice to growers and nurserymen of the release of Diamond Muscat raisin grape. 2pp.

Wagner, R. 1967. Study of some segregation in progenies of Chasselas, Muscat Ottonel and small-berried Muscat (in French). Vitis 6:353-363.

Wang, Y., Y. Liu, P. He, J. Chen, O. Lamkanra and J. Lu. 1995. Evaluation of foliar resistance to *Uncinula necator* in Chinese wild *Vitis* species. Vitis 34:159-164.

Weinberger, J.H. and N. H. Loomis. 1974, "Fiesta" Grape. HortScience 9:603.

Woodham, R. C. and D. McE. Alexander. 1966. Reproducible differences in yield between Sultana vines. Vitis 5:257-264.

Xu, H., D. J. Wilson, S. Arulsekar and A. T. Bakalinsky. 1995. Sequence-specific polymerasee chain reactions markers derived from randomly amplified polymorphic DNA markers for fingerprinting grape (*Vitis*) rootstocks. J. Amer. Soc. Hort. Sci. 120:714-720.

Ye, G. N., G. Soylemezoglu, N. F. Weeden, W. F. Lamboy , R. M. Pool and B. Reisch, 1998. Analysis of the relationship between grapevine cultivars, sports and clones via DNA fingerprinting. Vitis 37:33-38.

Zhang, J. and D. W. Ramming. 1994. Seedtrace germination and plant formation from over ripening berries of seedless grapes. XXIV Inter. Horticultural Congress. Program and Abstracts Supplement. Abstr. 367.s

CHAPTER 5.

Natural Sun-Dried Raisin Production

Gregory T. Berg, M.S. and R. Keith Striegler, Ph.D.

Introduction

This chapter will cover cultural practices and their impact on raisin yield and raisin quality. Commercially available production systems will be emphasized. The practical aspects of each cultural practice included will be identified and described. Items not covered in this chapter, but discussed thoroughly in other chapters include raisin grape cultivars, new raisin vineyard development, major insect pests, and raisin grapevine diseases.

Dried-on-the-Ground Production

Fig. 5.1: The simple, single-wire raisin trellis system during the dormant period in the San Joaquin Valley (1998).

Trellis and Training Systems

Training systems for traditional dried-on-the-ground (DOG) raisin production consists of head training and cane pruning. The canes are tied to a single wire, a with two foliage support wires on a short crossarm [12 inches (0.3 m) or less], or two dual wires separated by a medium width crossarm [24 to 30 inches (0.6 to 0.8 m) wide]. The traditional practice of laying the grapes down on trays to dry between the vine rows has limited the development of taller or wider trellis systems due to problems with the trellis shading the trays during the drying period.

The earliest method of supporting Thompson Seedless vines in California was to tie the canes up vertically to a stake (Hussman, 1916). However, it soon became apparent that uneven bud burst, shading, and poor fruit bud differentiation were among the problems encountered when Thompson Seedless vines were trained in this manner.

By the early 1920s, growers began to devise new ways to spread fruiting canes out laterally by tying the canes to dual vertical wires at 34 and 48 inches (0.9 and 1.2 m) above the vineyard floor. At first, all canes were tied to the lower wire. Then, some growers saw improvement by using both wires. Ultimately, this early farm-level research led to using only the top wire for fruiting canes, because it gave the best yields (Lynn, 1965).

In the early 1960s, C. D. Lynn established several research trials to capitalize on successes in the table grape industry and extend them to the raisin industry. The traditional 48 inch (1.2 m) high, single-wire trellis was compared to a 48 inch (1.2 m) high, two-wire system, with the wires supported horizontally by a 24 inch (0.6 m) wide crossarm. Across all locations, Lynn's research showed that the crossarm style trellis systems increased vine size and leaf exposure, which led to significantly higher yields without sacrificing grape maturity.

As mentioned earlier, trellises with crossarms wider than 24 inches (0.6 m) or heights greater than 48 inches (1.2 m) can impose shading of the between-row raisin drying area. However, when Thompson Seedless vines were allowed to fill these larger systems, yields always increased (Weaver and Kasimatis, 1975; Kasimatis et al., 1975 and 1976).

In order to determine the optimum size and shape for a Thompson Seedless raisin grape trellis, nine different trellis systems were evaluated in a comprehensive study (Andris et al., 1991). All systems were established at the same time in the same soil and were also irrigated and balance-pruned similarly throughout the experiment. Increasing the height of a single wire trellis from 52 inches to 64 inches (1.3 m to 1.6 m) by using a 7 ft (2.1 m) stake in place of the traditional 6 ft (1.8 m) stake increased vine size by about 1 lb (0.45 kg) of pruning weight and increased raisin quality by about 3.5% without affecting yield, soluble solids, or percent substandard (**Figure 5.1**). Two-wire, narrow crossarm trellis systems, such as the 52 inch (1.3 m) high, 24 inch 0.6 m) wide crossarm compared to the 64 inch (1.6 m) high, 18 inch (0.5 m) wide crossarm trellis produced larger vines and higher yields, while the 52 inch (1.3 m) high, 24 inch (0.6 m) wide crossarm trellis displayed better soluble solids and raisin quality, when these two systems are compared to each other (**Figure 5.2**). Interestingly, the highest yields in the trial were achieved with a unique Australian trellis system, which consisted of tying the fruiting canes to wires on a 24 inch (0.6 m) wide crossarm mounted at 52 inches (1.3 m), with a single foliage wire mounted 12 inches (0.3 m) above the crossarm. The most similar system used in California is merely an upside-down version of the Australian trellis, with fruit canes tied to a single wire at 52 inches (1.3 m) and two foliage support wires mounted 12 inches (0.3 m) above on an 18 inch (0.5 m) wide crossarm (**Figure 5.3**). While the Australian system produced nearly 0.38 tons per acre (0.8 t/ha) of raisins more than the similar California system, its pruning weight, soluble solids, and raisin quality were significantly lower.

Probably the most practical and therefore popular system of the six trellis systems discussed here for raisin production, quality, and drying capability is the 24 inch (0.6 m) crossarm with dual fruit cane wires mounted on a 6 ft (1.8 m) stake at 52 inches (1.3 m) to 54 inches (1.4 m) above the vineyard floor (**Figure 5.2**). With this system, many growers have found success in controlling vine vigor while producing moderate to high yields with good raisin quality.

Pruning

Pruning of grapevines is performed annually during the dormant season in order to renew and limit fruiting nodes and to maintain vine shape and vineyard uniformity. The major raisin varieties grown in California are head trained and cane pruned. Some minor varieties, however, can be cordon trained and spur pruned. On one of the new overhead trellis systems described later in this chapter, many varieties can be cordon trained and cane pruned for fruit production. Varietal differences in relative fruitfulness of the basal nodes dictate which pruning-training method is most practical. Since most raisin varieties discussed in this treatise can be cane pruned, this practice will be discussed in further detail here.

For optimum productivity and fruit quality, cane selection while dormant-season-pruning is of great importance. As demonstrated by Christensen (1978), canes which grow on the outside of the vine canopy, or "sun canes," show an 11% higher bud burst, more clusters per node retained, and four times more nodes producing shoots with double clusters.

Perez and Kliewer (1990) further emphasized the importance of shading vs. sun-exposure on productivity of grapes. Grape leaves must be exposed to a certain amount and quality of sunlight for optimal photosynthesis and metabolism. In order to provide a surplus of carbohydrates, a mature grape leaf must be exposed to greater than 30 to 50 mol m-2 s-1 of photosynthetic photon flux density (PPFD). These researchers also found that as little as 15 days of shading resulted in a higher incidence of primary bud necrosis or death, causing decreased bud burst and/or increased burst of secondary and tertiary buds (which are often unfruitful).

How do the raisin grower and his field workers determine which canes grew in the sun vs. those that grew in the shade? If the outward position in the canopy is not apparent, then the presence of persistent woody laterals, as described by Christensen (1986), can be very good indicators of sun-cane characteristics. Persistent woody lateral canes themselves have been shown to produce more fruitful shoots across their entire lengths than the canes from which they originated or canes without woody laterals. The problem associated with retention of persistent woody lateral canes has been crowding of shoots and clusters due to the tightly spaced nodes on these canes.

However, entire woody laterals do not need to be retained to ensure good productivity. As Christensen and Smith (1989) have shown, the basal nodes of persistent woody lateral shoots are generally higher in percent primary bud burst, and are more fruitful than regular nodes. These nodes should be retained during pruning when persistent woody laterals are present.

Research has shown what to look for in selecting quality fruiting wood for raisin grapes in California. Vineyard experience and past performance, however, should be used to determine adequate cane, spur, and overall node numbers at pruning. Sun-canes should be kept whenever possible, along with occasional persistent woody lateral shoots or portions thereof in order to boost production without causing excessive canopy crowding. Sun-canes can be characterized by their round shape and relatively short internodes as compared to flat canes with long internodes, which often constitute shade canes (Kliewer, 1981).

Vegetation Management

Management of non-crop vegetation between vine rows (i.e. the row middle) is most often accomplished in DOG raisin grape vineyards by disking, herbicides, or other mechanical tillage means. However, some growers are returning to more sustainable methods by using temporary covercrops as a means of excluding weeds in their vineyard row middles. The use of cover crops in raisin vineyards offers many potential benefits such as reduced soil erosion, addition of nitrogen to the soil, addition of organic matter to the soil, improved soil structure, improved water infiltration, and enhanced pest management (Ingalls and Klonsky, 1998). Research by Striegler et al. (1997) has shown that vineyards with legume/insectary blend cover crops can perform as well as clean-cultivated vineyards while providing habitat for other beneficial organisms. Species selection is an important factor for DOG raisin growers who seek to cultivate the row middles in preparation for harvest terracing. The need for reseeding and costs of cover crop maintenance are equally important factors to consider. Very high biomass production or plant tissues that will not break down quickly can be detrimental to pre-harvest ground preparations.

In-row vegetation can be controlled through cultivation by French plowing or using rotary hoes or weeder blades (Sun-Maid, 1997). The majority of growers, however, use pre- and/or post-emergent herbicides to control in-row vegetation (Sun-Maid, 1997). New sprayer technologies are currently advancing and most serve to reduce herbicide usage while increasing effectiveness of weed control. These include hooded spray heads, controlled droplet applicators (CDA), and electronically-activated, chlorophyll-seeking spray nozzles. Roundup™ (a.i. glyphosate) remains the chemical tool of choice with these latter spray technologies, although new systemic herbicides are becoming available. For more information on herbicide usage and current recommendations, growers should consult their local farm advisor, weed specialist, or pest control advisor.

Other options for weed control without chemicals are also available to grape growers. Propane flamers are gaining popularity and are used to literally burn back young annual weed seedlings. Diesel or propane-fueled steam generators are other non-chemical weed control options. For either system, flame or steam nozzles are directed to the berms and operated at speeds determined adequate for weed control without damage to vine trunks or wooden stakes. Weeds must be burned or singed while they are very small (3 to 5 leaf stage). Repeat treatments are mandatory and only suppression of perennial weeds is possible. Many growers involved with non-herbicide weed control employ a systematic rotation of techniques in their cultural systems.

Vine Nutrition

Grapevines require certain nutrients for growth and development. These "essential" nutrients are grouped into two groups, according to the amount needed. The groups are known as the macro- and micronutrients. Macronutrients consist of carbon, hydrogen, oxygen, nitrogen, phosphorous, potassium, magnesium, calcium, and sulfur (California Fertilizer Association, 1990). Micronutrients required by grapevines are iron, zinc, manganese, boron, copper, molybdenum, and chlorine. Plants acquire carbon, hydrogen and oxygen primarily from the atmosphere and water, while other nutrients are supplied from the soil. All of these nutrients play key roles in certain vine physiological processes that impact yield, fruit composition and growth. However, fertilization of raisin vineyards in California typically only involves four of the essential nutrients (Christensen and Peacock, 2000). These nutrients are nitrogen, potassium, zinc and boron. An adequate supply of the other essential nutrients is generally available to vines from the weathering and breakdown of soil parent materials. This situation is due to the types of soils that raisins are grown on in the San Joaquin Valley. Raisin vineyards, for the most part, are planted on deep, well-drained soils that allow for good root growth and uptake of nutrients. Detailed descriptions of the nutrients required by grapevines and symptoms of deficiencies can be found in Christensen et al., 1978; Robinson, 1992; California Fertilizer Association, 1990; and Christensen and Peacock, 2000. The remainder of this section will cover vineyard fertilization practices.

Fertilization practices in raisin vineyards have changed significantly in recent years. Improvements have occurred in the areas of monitoring vine nutritional status, management of fertilizer application (timing and amount), and fertilizer application technology.

Proper monitoring of vine nutritional status is essential for efficient and profitable vineyard management. A good monitoring program allows growers to add nutrients in the correct amount to meet, but not exceed, vine requirements. Vine nutritional status can be estimated by soil or plant tissue analysis (Christensen and Peacock, 2000). Soil analysis has proven to be less reliable for determining vine nutritional status than plant tissue analysis. This is because soil analysis provides information on the nutrients found in soil but does not tell us whether nutrients are being removed from the soil and utilized by grapevines. Plant tissues used for analysis of nutrients in grapevines include the leaf petiole, leaf blade, fruit (analysis of juice) or dormant canes.

Bloom time petiole samples are the most widely used form of tissue analysis (Christensen et al., 1978; Christensen and Peacock, 2000). Samples should be collected from uniform areas of the vineyard and should not represent more than 10 acres. If the vineyard is not uniform (different soil types, uneven irrigation, presence of nematodes, etc.), more samples should be taken and sent to the laboratory. The size of the sample should be approximately 100 petioles. The method to collect petioles is to sample from a select group of vines or reference plot. Sampling from the same vines each year allows the grower to discern seasonal trends in vine nutritional status.

Samples should be collected as close to full bloom as possible. Full bloom is defined as the time when 75% of calyptras (caps) have fallen from the flowers. Petioles used for analysis should come from leaves opposite basal clusters (cluster nearest the cordon or trunk on a shoot). The leaf blade should be removed and discarded. Petioles are then placed in a clean, labeled paper bag. A record of all information regarding the sample should be retained by the grower to allow for sample identification and interpretation of results from the laboratory.

Petiole samples should be sent to the laboratory immediately. A delay in this process will reduce the accuracy of results. Samples should be kept in a dry and well-ventilated location until they are delivered to the laboratory. Also, growers should ask for their samples to be washed by the laboratory in order to avoid contamination from dust, pesticide residues, or recent foliar nutrient applications.

For accurate petiole analysis and interpretation of results, raisin growers should also be aware of the following guidelines: 1) Critical values for nutritional status of grapevines in California have primarily been developed from research on the Thompson Seedless variety. Other raisin varieties will have somewhat different nutritional requirements; 2) Some nutritional elements may be more accurately determined by analyses of other tissues or at other times of the growing season. For instance, post-veraison sampling of leaf blades from most recently matured leaves may give a more accurate picture of vine potassium status.

Most raisin growers in California are utilizing well-planned and consistent petiole sampling programs to acquire important information on vine nutritional status. Proper timing of application can lead to optimization of fertilizer use efficiency and vineyard profitability. This effort, in combination with a good fertilization program is essential for successful raisin production. Nutrients may be supplied to vines by application of inorganic fertilizer, application of organic fertilizer, in irrigation water originating from wells, or by use of leguminous cover crops. Growers long ago recognized growth and yield response to application of inorganic fertilizers, especially nitrogen. Consequently, historical use of fertilizers by raisin growers was directed toward insuring that

Fig. 5.2: The two-wire, narrow cross-arm raisin trellis system. Note canes divided evenly and tied to the two wires (1998).

Fig. 5.3: The three-wire raisin trellis system with narrow cross-arm for foliage support. Note canes tied to lower wire only (1998).

sufficient nitrogen was available to sustain grapevine growth and reproductive physiology. In general, the application of 80 to 120 lbs (36.36 to 54.54 kg) of nitrogen per acre before bud burst was the standard practice. Little attention was given to refinement of this practice since the cost of inorganic fertilizer was low in comparison to the response achieved or perceived to have been achieved. This situation has changed dramatically due to research primarily generated by L. P. Christensen and colleagues (Christensen and Peacock, 2000).

Nitrogen (N) is a major constituent of the organic compounds in plants. Amino acids, proteins, nucleic acids and many secondary metabolites are included in this group. As a result, N is the most commonly applied fertilizer element in vineyards.

Reduced growth is often the first N deficiency symptom observed in vineyards. Deficiency symptoms in leaves will not become evident until the deficiency is severe (Christensen, L.P. et al., 1978; or Gartel, 1993, for symptoms). Growth and yield of vines will usually be reduced significantly before deficiency symptoms in leaves are observed. However, it should be noted that vineyard pests (such as phylloxera and nematodes), soil physical problems, or poor irrigation management could also result in inadequate vine growth. A good program for monitoring N status is the only certain method to determine if N is a limiting factor in your vineyard.

Excessive use of N in vineyards seems to be a greater problem than deficiency of N in California. Application of N fertilizer should meet the N requirements of the vine. Use of N in excess of vine requirements is correlated with high nitrate in ground water, bud necrosis, reduced bud fruitfulness, excessive vine vigor, increased incidence of stem necrosis disorders, reduced fruit maturity and increased incidence of bunch rot. Also growers who apply more N than is needed are wasting time, money, and energy.

Improved efficiency in N fertilization can be achieved if N applications are done when active uptake of N is occurring. The seasonal pattern of N demand and allocation for grapevines in regions with long growing seasons is described in **Table 5.1**. Uptake of N occurs most rapidly from fruit set to veraison and from harvest to leaf senescence. Nitrogen used by vines during the period from budburst to fruit set comes primarily from reserves stored in permanent woody structures (roots, trunks, cordons, etc.) during the previous grow-

Table 5.1
Different Phases in the N-nutrition Cycle of the Grapevine*.

Phase in N-nutrition Cycle	Growth Stage	Specific Characteristics for N-uptake
I	Budburst to end of bloom	New growth dependent on reserve N accumulated during previous season(s). Little root growth and N uptake.
II a	End of bloom to end of rapid shoot growth	Active root uptake. Amount of "new" N sufficient to supply demand of new growth.
II b	End of rapid shoot growth to veraison	Leaves and clusters both important sinks for N.
III	Veraison to harvest	Root uptake may stop. Clusters main sink for N. Redistribution from roots, shoots and leaves to clusters.
IV	Harvest to start of leaf senescence	Active root uptake. Redistribution from shoots and leaves to permanent structure.

*Modified after Conradie, W.J.

ing season. During the veraison to harvest period, developing clusters become the dominant sink for N. Nitrogen is supplied to clusters at this time by uptake from the soil and redistribution from roots, shoots, and leaves. Applications of N in winter or early spring are not as efficient as applications during the growing season. Nitrogen applied during these times is subject to leaching and denitrification before active vine uptake begins.

The best timing for efficient use of N fertilizer appears to be from fruit set to veraison and then again post-harvest. In general, these periods coincide with flushes of root growth, when uptake of nutrients is optimal. Field experiments with Thompson and Flame Seedless vines in the San Joaquin Valley confirmed that N applied during fruit set, veraison, and post-harvest was most compatible with vine growth and fruit development. Post-harvest N application should be accomplished while vines still have a healthy, functional canopy.

Once it has been determined by petiole sampling and assessment of vine growth that N fertilization is needed, this important question must be addressed: How much N should you apply in your vineyard? N fertilization should meet the N requirements of your vines. N inputs are needed only to replace N that is lost from the vineyard soil-plant-atmosphere system. Yield, soil type, irrigation management, crop residue management, and use of legume cover crops influence the amount of N fertilizer required.

For a typical vineyard with crop residues (prunings, leaves, etc.) returned to the soil, the major source of N loss is harvested fruit. Research by Dr. Larry Williams, U.C. Davis and Kearney Agricultural Center, has shown that fruit removal from a "Thompson Seedless" vineyard results in a loss of 30 lbs (34 kg/ha) N per acre (**Table 5.2**). Conradie (1991) summarized the results of five N allocation studies done in different viticultural regions using various varieties. He found that the average amount of N lost at harvest was 3.2 lbs/ton of fruit (1.6 kg/t).

Table 5.2
Estimated annual N requirements and subsequent losses for 'Thompson Seedless' grapevines[z, y]

Nitrogen Status	Vine Part	Amount (lbs. per acre[x])
Requirements	Leaves	35
	Stems	35
	Clusters	35
	Total	**75**
Losses	Shoot trimming	
	-Leaves	5
	-Stems	3
	Fallen leaves	20
	Pruning	15
	Fruit Harvest	30
	Total	**73**

[z]Values were obtained by averaging the data collected over a 3-year period in the same vineyard.
[y]Modified after Williams, L.E., J. Amer. Soc. Hort. Sci. 112:330-333. (1987)
[x]Vineyard spacing was 8' by 12'; 454 vines per acre; avg. yield = 11.4 tons/ac.

This information can be used to estimate the N requirements of your vineyard. For example, a vineyard yielding 10 tons/acre (22.4 t/ha) would have 32 lbs N/acre (36 kg N/ha) removed at harvest. This vineyard would require an input of at least 32 lbs N/acre (36 kg N/ha) to maintain its productivity. Nitrogen inputs from all sources must be considered, however. Irrigation water, crop residues, atmospheric pollutants, and mineralization of soil organic matter may make significant contributions to the N supply in vineyards. Some vineyards in the San Joaquin Valley are supplied with irrigation water from wells that are high in nitrates. It is likely that many of these vineyards require little or no N fertilization.

This method of calculating vine N requirements is being used increasingly to optimize the amount applied. Rates are adjusted over time based on petiole test results and vine performance.

Irrigation

Three types of irrigation systems are prevalent in California's raisin producing areas. In districts where low-cost canal water is available, most growers will choose some form of flood irrigation. The most popular method has been furrow irrigation, where furrows of various number, shape, and size are formed between the vine rows. Furrows are used once or twice for channeling water from the ditch or underground pipe delivery system before weed growth requires cultivation for control. The furrows must then be formed again, prior to the next irrigation cycle.

One of the major drawbacks to furrow or flood irrigation systems is non-uniformity due to variability in soil characteristics (Prichard, 1996). With furrow irrigation, distribution uniformity (DU) usually ranges from 40 to 85%, with DU's of 70% or above being very rare (Prichard, 1996). Slopes should not exceed 2% and soil type homogeneity within the irrigation run is most desirable. Costs associated with furrow irrigation are low for initial setup, low for pumping requirements, but high for cultivation, and high for irrigation labor.

Sprinkler irrigation (solid set design) has also been used in grape and raisin vineyards where slope or soil variability factors limited the use of furrow irrigation. Solid set sprinkler irrigation requires adequate water supplies and larger pumping capacity to build and maintain system pressures at desired flow rates. Solid set sprinkler systems should be designed to deliver uniform flows at the lowest possible pump pressure in order to minimize operating costs. Advantages of solid set sprinklers include low labor requirements, frost protection and cooling capabilities, and their ability to overcome slope and soil variability problems. Furthermore, sprinkler irrigation systems often experience DU's of 75 to 85%. A disadvantage of overhead sprinkler irrigation of grapevines is the wetting of foliage and fruit, which can lead to increased disease pressure or burning under certain microclimatic conditions (Flaherty et al., 1992).

The latest innovations in sprinkler irrigation include microsprayers, fanjet sprinklers, and microsprinklers. These recent innovations have helped growers of grapes and other permanent crops to irrigate more acreage with less water and lower power requirements than with traditional solid set sprinklers. These systems use lower rates of flow and hence can be used where slope and/or soil characteristics limit infiltration or enhance runoff/erosion. Microsprinkler systems have a higher initial capital cost than solid set systems, but pumping costs are lower and DU's are better (80 to 95% range).

Trickle or drip irrigation systems can also be very efficient delivery systems with DU's commonly in the range of 80 to 95%. As with microsprinklers, these systems also provide grape growers with many advantages. Limited or high cost water supplies, steep slopes, poor infiltration rates and high labor costs can all be overcome through use of micro irrigation systems. However, proper system maintenance must be performed in order to ensure consistent distribution uniformity and to avoid clogged emitter orifices (Schwonkl et al., 1998). Further efficiencies in water savings and weed control costs can be achieved through use of subsurface drip irrigation due to reduced surface wetting (Phene, 1995).

Another way to possibly improve vine balance and water use efficiency in the vineyard may be a technique derived from split-root research introduced by Australian researchers in recent years (Loveys et al., 1997). Partial rootzone drying (PRD), as it has been named, is an irrigation technique, which serves to modify vine canopy growth and development by keeping alternating sides (50%) of the rootzone well-watered while the other side is allowed to dry down. This is believed to be caused in part by the production of abscissic acid (ABA) by the drying roots which in turn influences the leaf stomata to partially close in order to conserve water in the plant. The time period between alternating sides for irrigation/drying depends somewhat on the available water holding capacity of the vineyard soil and also on the rate of transpiration and evaporative demand (Et). This being said, the practical length of time has been found to be about ten to 14 days between switching sides for PRD, depending on vine growth stage and evaporative demand during the period.

It is likely that Thompson Seedless grapevines and other raisin varieties could also be grown using PRD irrigation techniques in the near future. In existing vineyards, modifications to the drip irrigation system will be required to achieve PRD. Some success has been achieved in Australia (and California) using conventional above ground drip irrigation. This consists of two complementary hoses per row, where hose A feeds the adjacent rootzones of vines A and B, and hose B feeds the adjacent rootzones of vines B and C. The same alternating pattern is repeated down the length of every row. PRD is achieved by opening valve A and closing valve B for 10 to 14 days, at which time the valves are switched, and the dry rootzone is irrigated and the wet rootzone is allowed to dry down. Many furrow-irrigated vineyards have probably been achieving the same effects for many years, although the corresponding poor distribution uniformity far outweighs the PRD benefit. Limitations to installation or retrofit of PRD in vineyards include the distance between vines or rows and the resulting ability to achieve true rootzone separation, as well as cost and availability of labor. For new plantings in California, some growers (primarily winegrape vineyards) have installed entire new systems with double lateral lines, which allow alternating sides of the rootzone to be irrigated by simply opening one valve and closing another for each irrigation block. Once vineyards are established using conventional above ground drip irrigation, these growers believe that optimum rootzone separation and water use efficiency will be achieved through the use of subsurface drip irrigation, with the placement of the hoses directly between the vine rows. Further research work and grower experience will be needed in order to optimize the use of PRD irrigation techniques for grape growing in California, whether for wine or raisin grape production.

Optimal irrigation scheduling for raisin grapes will vary according to soil type, water source, geographic location, age of the vines, and type of irrigation system. However, the amount of irrigation applied to raisin vineyards in California is approximately 24 to 36 inches (610 to 914 mm) of water (Peacock et al., 2000). Many growers have installed and are utilizing drip irrigation systems in their raisin vineyards. In addition, a substantial number of vineyards continue to use flood, furrow, or overhead sprinkler systems. However, grapevines will use the same amount of water whether irrigated by furrow, sprinkler, or drip irrigation. As mentioned earlier, the benefits of water use efficiency come with increased application uniformity and reduced wetted surface areas (reduced potential for surface evaporation). With furrow and sprinkler irrigation, there is a large wetted soil volume, so vines will take longer to deplete the soil water reservoir than they would if irrigated with drip or microsprinkler irrigation systems. With these latter systems, irrigations must be made more frequently, not to exceed 4 days between irrigations. Lysimeter research studies have shown optimal yield and fruit maturity levels are achieved on Thompson Seedless vines with sustained deficit irrigations of 60 to 80% of full grapevine evapotranspiration (Et) (Williams, 2000). Therefore, the proper amount of water to apply to a raisin grape vineyard can always be easily calculated by obtaining current daily potential grapevine Et data, multiplying by a sustained deficit factor of 60 to 80%, and finally multiplying by the number of days between irrigations. An excellent discussion of irrigation scheduling in California raisin vineyards is given by Peacock et al. (2000).

Missing the mark in irrigation scheduling can have negative impacts on vine physiology and productivity. Under-irrigation and over-irrigation can both have detrimental effects on vine performance. At no time during the growing season is it beneficial to allow more than 50% of the water holding capacity of the soil's wetted volume to deplete between irrigations (Peacock et al., 2000). Severe water stress results in a reduction in shoot growth, vine size (total vegetative growth as indicated by pruning weight), berry size, the number of clusters per vine, and yield. Application of excess irrigation water (generally greater than full vine Et) can result in excessive shoot growth and vine size, which leads to internal canopy shading. Shaded canopies have increased severity of fruit diseases such as bunch rot and powdery mildew, reduced sugar accumulation, fewer clusters per vine due to lower bud fruitfulness and bud necrosis, and lower yield. Good irrigation management is crucial to the successful and profitable production of raisins.

Harvest and Drying

Grapes to be dried into raisins are picked by hand labor and laid onto trays in the row middles, as described by Winkler, et al. (1974), Kasimatis and Lynn (1975) Enochian, et al. (1976), and Striegler, et al., (1996). Each vine row is associated with an adjacent row middle where its grapes are laid on trays to dry in the sun. Prior to picking, the row middles are disked to remove weeds, break up large dirt clods, and generally smooth them prior to terracing. The ultimate objective is to provide a smooth surface, which is moderately angled to increase the angle of incidence to sunlight and also to drain rainwater away from the raisins. Optimum sunlight interception should occur when most of the sun's rays reach the terrace at a 90 degree angle. Therefore, an east-west row orientation provides more direct sunlight exposure for drying fruit than does a north-south row orientation. Terraces are angled toward the south in east-west rows and toward the west in north-south rows. The minimum recommended slope of the terrace is 5% (Kasimatis and Lynn, 1975). Terracing the middles is done in a separate operation by a tractor-drawn implement. Terracing implements are available in various designs, but all generally perform the same function. These implements involve a blade, which pulls the soil into the desired surface angle and a float or roller, which firmly packs the terrace surface (**Figure 5.4**).

The selection of raisin tray material can be one of the most important management decisions raisin growers make, especially when there is precipitation during the drying period. There are essentially four basic types of raisin tray materials currently in use: regular strength paper, wet strength paper, poly coated paper, and specialty coated papers. The choice of material depends on many factors, including crop size, maturity date, expectation of rainfall, grower's level of risk adversity, and cost of the material versus cost of crop insurance (Striegler, 1992).

Regular strength paper is the most inexpensive form of material. However, regular strength paper readily absorbs water from rainfall and becomes very weak when wet. This material is recommended for use only by growers who are able to harvest early and avoid precipitation.

Wet strength paper costs are moderately higher than regular strength paper due to added ingredients. These added ingredients help wet strength paper remain stronger than regular strength paper during and after precipitation. Most raisin growers use wet strength paper tray material for normal harvest dates.

Poly coated paper and other specialty coated paper tray materials are the most expensive forms. However, these materials provide the most resistance to water penetration and absorption, as well as retention of strength when wet. The poly coating is usually applied to only one side of the paper, and this is the side to place down or toward the terrace. Growers who choose not to purchase crop insurance are more likely to use poly or specialty coated raisin trays in order to provide maximum protection in the event of precipitation during drying.

Again, selection of raisin tray material is as important as any other management decision in raisin production. The cost-effectiveness of a grower's raisin tray relies heavily on the material's ability to hold together during all operations that follow the actual harvest. The reliability of raisin tray material is severely tested during wet drying seasons. If all drying seasons were perfect, virtually any kind of paper could be used.

Natural Thompson Seedless raisin grapes should only be harvested after they have reached the proper maturity. High quality raisins will be made from grapes, which have reached 20 to 23 °Brix (percent soluble solids or sugar), however, minimum grade raisins can be made from grapes with about 19 °Brix (Jacob, 1942; Baranek et al., 1970; Kasimatis and Lynn, 1975; Kasimatis et al., 1977). Non-uniformity of ripeness in grape berries has been shown to vary widely between years as well as vineyards (Clary and Sawyer-Ostrom, 1991). The final raisin quality is affected by many factors leading to non-uniformity of sugar levels between individual grape berries and entire grape clusters, as well as between individual vines and areas across the entire vineyard. These factors include soil type, climatic conditions, irrigation and fertilization practices, pruning, etc. (Kasimatis et al., 1975, 1977). Furthermore, the drying ratio, defined as the amount of raisins that can be produced from a given amount of fresh grapes (fresh fruit weight: raisin weight), also becomes more favorable as grape maturity increases (Jacob, 1942; Winkler et al. 1974; Christensen, 1994b; Christensen et al., 1995a; Christensen et al., 1995b).

The majority of the raisins produced in California are harvested by hand. Hand harvesters distribute their own paper trays and pick and spread their own grapes onto the trays. Grapes are usually harvested from one row at a time and laid onto trays in the row middle to the north or east side of the vine row, depending on row direction. There is usually one person per row, however, some individuals choose to work in pairs. Grape harvest laborers are generally paid on a piece-rate (per tray) basis. Fresh grapes are cut from the vines and placed temporarily into picking pans, which hold approximately 20 to 25 lbs (9 to 11 kg) of grapes. When the pan is "full", the picker carefully spreads the grapes no more than one cluster deep onto a paper tray on the terrace surface (**Figures 5.5** and **5.6**). It is important that the integrity of grapes be monitored during hand harvest for raisin production. Excessive damage to grapes or "juicing" can reduce both the yield and quality of raisins produced.

Once the grapes are picked, they are allowed to dry naturally for approximately seven to 10 days, before they are handled again. Under good drying conditions, the grapes will have dried to a point where there are 20 to 25% of the crop dried into raisins on top and green grapes underneath. Turning the grapes hastens the drying process and helps the raisins to dry more uniformly (Fischer, 1959; Patterson and Clary,

Fig. 5.4: Typical terracing implement used to form flat angled surface for drying raisins on the ground (1995) (Insert shows close-up of terracing blade and roller.

Fig. 5.5: Harvest laborer spreading grapes onto an individual paper tray (1995).

1987). Without turning, some raisins will be too dry while others will be too moist, requiring a long curing period later. Turning fruit can be done in two ways.

The method, which has been in use for the longest time, involves turning the entire tray as described by Kasimatis and Lynn (1975). The first step is to partially slide an empty tray underneath the tray to be turned and then fold the empty tray up and over the full tray. The loose ends of both trays are grasped and the two trays are turned over, leaving the full tray empty and the empty tray full. After spreading the turned raisins out evenly, the worker(s) move(s) on to the next tray to be turned. The next tray will be turned onto the paper, which was just emptied, and so on. This method of turning is usually paid by the piece. The potential for spillage and loss of raisins is significant when this technique is used.

The second method, which has become more popular in recent years, involves turning only clusters (**Figure 5.7**). Only the clusters that need to be turned are turned (e.g., those that are green on the bottom). Clusters that are drying down satisfactorily are not turned. Turning fruit in this way is often paid by the hour. This method is believed to be less time consuming and with fewer losses than with the full tray turning method.

When the raisins are approaching average moisture content of 15 to 17%, the raisin grower can begin rolling trays. Additional drying and "moisture-curing" will occur inside rolled trays (under food drying conditions) unless the trays are of the poly coated type (Peacock and Christensen, 1997). The maximum moisture content allowed for incoming natural raisins at the processing plant is currently 16.0% (USDA 1990). While there are a few different types of raisin rolls (**Figure 5.8**), the most widely used type of roll has been the "biscuit" roll (**Figure 5.9**). This roll is made by folding the tray lengthwise into thirds, forming the "cigarette roll", and then rolling the cigarette roll along its length into a biscuit roll. The biscuit roll is then placed toward the high side of the terrace, in the sun, to finish drying and curing, or evening-out (Kasimatis and Lynn, 1975). Another type of roll has become known as the "flop" roll, which is similar to the cigarette roll, with the exception that the tray is folded crosswise into thirds. Since the biscuit roll is not easily made following a flop roll, these rolls are usually retrieved as is. Open-ended "flop" or cigarette rolls can be used to roll the fruit a little bit early, as drying will continue at a steady pace (Peacock and Christensen, 1997). This technique may help the grower in times when inclement weather is possible, but must be retrieved in a timely manner to avoid over-drying and caramelization.

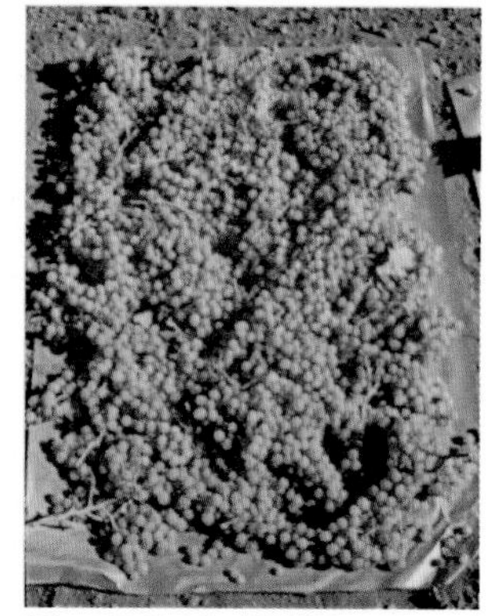

Fig. 5.6: A "full" tray of Thompson Seedless fruit (1995).

If the raisins are nearly dry (all berries at least brown in color) and a rainfall event is predicted, the trays can be rolled to protect them from the rain, as mentioned above. Research by Patterson and Clary (1987) has shown that cigarette rolling the trays causes the least amount of damage, i.e., fewer embedded capstems and cluster stems than if the trays were biscuit rolled under high fruit moisture conditions. The flop roll, while not tested, should prove to be less damaging than even the cigarette roll due to lower handling requirements.

If the raisins are dry, biscuit rolls provide the best protection from adverse weather (**Figure 5.9**). However, unless poly or other specially coated trays are used, even biscuit rolls are susceptible to mold and insects if high moisture weather conditions persist. The rolls should be picked up as soon as possible.

Within approximately five to seven days after rolling, assuming good weather, the rolled raisin trays should be ready to be retrieved. Most growers use tractor-drawn bin trailers and a crew of four or five people for each setup. Each tray is picked up by hand and dumped into bins on the trailer. Usually, two people are picking up trays, one person drives, and one or two people help fill the bins and dispose of used papers. Papers are often collected on spikes and then burned at the end of the row. Some growers simply disk the used trays into the soil, while others are looking at recycling used trays. Recycling may not be feasible, however, depending on the composition of the tray.

Precipitation at any time during the raisin drying period

Fig. 5.7: Two trays of Thompson Seedless raisins approximately nine days after harvest. "Green" clusters have been turned on the tray to the right (1995).

Fig. 5.8: Three common types of raisin tray rolls: the cigarette (right), the flop (center), and the biscuit (left) (1995).

will damage the crop and cause substantial revenue losses (RAC, 1993). As little as 0.25 inches (6.3 mm) of precipitation can cause substantial damage to the crop. The damage is most severe when rain falls on grapes in the green or brown stage of drying (Striegler et al., 1992). Rain during these stages of drying leads to rapid mold formation and decay. Sand particles may be splashed by raindrops onto the drying berries, leading to the problem known as embedded sand. Sand becomes embedded as the wrinkles are forming in the shriveled berry skins. Removing embedded sand is feasible, but costly in terms of economic and yield losses.

Methods of accelerating natural sun drying of raisin offers the opportunity to reduce the time the raisin trays remain in the field and reduces chance of exposure to rain. One method that proved successful was to apply a spray containing a mixture of methyl oleate and potassium carbonate directly on the fruit laying on trays in the vineyard row. The application of these materials changes the water retaining nature of the grape berry by dissolving the wax cuticle and allows moisture to evaporate from the fruit at an accelerated rate. This effect is discussed further in **Chapter 7** (**Figure 7.1**) and in **Chapter 8** (**Figure 8.10**). Spray-on-the-tray (SOT) raisin production was developed at California State University, Fresno and is described by Petrucci et al. (1978). The similarity of this process to the conventional method of sun drying raisins facilitates grower acceptance with the advantage of reducing the time in the field and exposure to rain.

Raisins produced using the SOT method are yellow to light amber in color (**Chapter 3 Plate 3.3.19**) compared to the dark purple brown color of the natural sun-dried raisin.

Fig. 5.9: Close-up of the raisin tray biscuit roll (1995).

Therefore, it is important for the grower to consult with their packing house to be sure the raisins are acceptable for delivery. In order to spray the grapes on the trays, the terrace should be prepared with a furrow on each side so that a tractor and sprayer can pass over the terrace (**Chapter 3 Figure 3.3.16**). The furrows reduce dust exposure to the grapes on the trays and serves as a drain for water in the event of rain. It is important that the fruit is distributed evenly on the trays with a depth of not more that 3 to 4 inches (7.62 to 10.16 cm) so the spray can penetrate and contact as much of the fruit as possible. The spray is applied immediately after fresh fruit harvest and after the trays are turned, which is after about four days.

A 2% spray formulation consists of 4 gal (15.2 L) of methyl oleate and 32 lbs (14.4 kg) of potassium carbonate per 100 gal (380 L) of water and is applied at 200 gal (760 L) per acre at harvest. A 1% solution is applied after turning the trays at a rate of 100 gal/acre (935.4 L/ha). Latest research shows that it is possible to use effectively concentrations of spray mix as low at 0.6%. A 40 to 42 inch (101.6 to 106.68 cm) wide spray boom is recommended using four flat fan nozzles operated at 30 to 40 lbs/in^2 (2.1 to 2.8 kg/cm^3).

Depending on timing, there are a few strategies raisin growers may take to salvage the crop. Trays are often "slipped," a practice used to break the water seal between the raisin tray and the soil, or turned (full tray method), which achieves the same effect plus exposes the underside of clusters. In any case, unless sunshine and dry, moving air follow the rain, the decay will continue.

A small percentage of the industry is using mechanized raisin handling equipment to turn and retrieve raisins (Berg et al., 1992 and 1996). By using continuous tray materials instead of individual trays, tractor-drawn machinery can be used to perform the turning and retrieval processes. Some growers have seen up to 50% savings in labor hours and 25% savings in costs for the entire raisin harvest process (Berg et al., 1996).

Most raisin growers choose or are required to perform some sort of rudimentary cleaning of the raisins at the farm before delivery. In the best of drying seasons, sand, leaves and other materials become part of the raisins as they are retrieved from the field. To remove this debris, the raisins are dumped onto a screen shaker. Usually, from two to four people man the screen shaker, looking for foreign material and breaking up large clumps of raisins or leaves. The shaker sifts out sand, some substandard raisins, and other debris. The "cleaned" raisins continue down the shaker and are recombined into bulk bins.

The bulk bins are often stored on the farm in covered stacks to "sweat out." Sweating out assures that the moisture levels of all raisins in the bins have been allowed to equalize, which makes a better product for delivery to the processing plant. This process occurs naturally over a period of two to three weeks while covered in warm weather. If insect problems arise, the covered stacks can also be fumigated prior to delivery.

Following a sufficient curing period at the farm (two weeks to one month), raisins can be delivered to an appropriate processing facility. Raisins from the farm are usually transported in bulk bins via truck.

Before the processor takes responsibility for any raisins, USDA officials inspect incoming raisin deliveries for the following defects: excessive moisture, fermentation, uncured berries, mold, damaged raisins, foreign material, and immaturity. Foreign material in the raisins may arise from many sources. For example, seeds from noxious weeds such as longspine sandbur (*Cenchrus longispinus*) and common cocklebur (*Xanthium strumarium* L.) can cause serious problems in processing if not discovered and removed beforehand. If a raisin delivery does not meet with USDA inspection, it can be returned to the producer or held for reconditioning (USDA, 1990). Upon meeting USDA incoming inspection, raisins are stored according to quality and moisture content until the processor is ready to begin processing. Through use of fumigation or low-oxygen atmosphere modification units, unprocessed raisins can be successfully stored for up to one year.

The traditional process for harvest and handling of natural sun-dried raisins is expensive and laborious. The California raisin industry requires the use of some 45 to 50,000 seasonal workers to perform the raisin grape harvest during the short three to four week period in late August and early September (Alvarado et al., 1992). For the future, the availability of labor for the harvest of raisin grapes is in question. Existing and proposed government regulations on labor use, immigration, and health care will continue to put pressure on raisin grape producers to reduce the use of labor. If labor shortages occur, increases in the cost of harvest labor could be substantial. Methods of mechanically handling raisins after hand harvest have been developed, but are not widely adopted (Berg et al., 1992).

Although mechanical harvest of fresh raisin grapes has not yet proven successful, Studer and Olmo (1971) developed a system, which uses harvest cane severing in conjunction with a mechanical harvester. This system allows the fruit to partially dry on the vine before the fruit is mechanically harvested and laid onto a continuous paper tray in the vine row. After about seven to 10 days of drying, the raisins are mechanically retrieved using a machine developed by the same researchers (Petrucci et al., 1983). Currently, some California raisin growers are taking a new look at complete dried-on-the-vine (DOV) raisin production systems (see the following section in this chapter for Mechanized Raisin Production) in order to remain competitive.

Mechanized Raisin Production

Each year, about 345,000 tons (313,000 t), or 95% of all raisins produced in the U.S. are grown within a 60-mile (96 km) radius of the city of Fresno, CA (RAC, 2001; CA Agr. Stat. Serv., var.). Each ton of raisins requires about 4 tons (3.6t) of green grapes. Consequently, some 1.4 million tons (1.3 million tonnes) of fresh fruit (95% "Thompson Seedless") is harvested annually and made into raisins. Of the California raisin crop, 90% is produced naturally, i.e., the grapes are dried in the sun without chemical drying aids on paper trays between the vine rows. The other 10% is artificially dehydrated as either dipped seedless or golden seedless raisins (RAC, 2001).

With the current California raisin production, the industry requires about 14.5 to 18 million labor-hours per season for harvest and handling operations alone. Timeliness of harvest operations can be critical to successful production of natural sun-dried raisins. As discussed earlier, Thompson Seedless grapes reach optimum maturity (21 to 23 °Brix) in late August or early September. Once placed on trays, the grapes require 14 to 21 days of low relative humidity and afternoon temperatures of 90 to 100 °F (32 to 38 °C) to reach 14 to 16% moisture content. Expected drying conditions are usually optimal until mid-September, when the day lengths become shorter, morning dews become heavier, and the chance of precipitation increases substantially.

As stated earlier, 40,000 to 50,000 seasonal workers are employed in the California raisin harvest each year (Alvarado et al., 1992). The passage of the Immigration Reform and Control Act of 1986 (IRCA) raised concerns that the supply of seasonal workers available to raisin growers would be reduced. Increased enforcement of IRCA regulations could also reduce the available labor supply significantly (Alvarado et al., 1992; Cuhna, 1996). In the past, the labor supply for the harvest of raisin grapes has generally been sufficient, although recent harvests have been somewhat challenging for some growers. During most adverse (cool, wet) drying seasons, as well as in normal years, some raisin growers have perceived that a shortage of labor existed.

To avoid the effects of future labor shortages, several progressive raisin growers have or are developing alternatives to the traditionally labor-intensive raisin harvest and handling methods. Three of these include the RAMEC, Unruh, and Rocca systems for partial or complete mechanical raisin harvesting (current sources for these systems are listed in the references section at the end of this chapter).

Mechanical Handling

The RAMEC and Unruh systems are examples of mechanized raisin handling systems, which utilize hand harvest onto continuous tray material, mechanical fruit turning, and mechanical raisin retrieval. As with traditional raisin grape

Fig. 5.10: Thompson Seedless raisin grapes harvested onto a continuous paper tray (1992).

Fig. 5.11: Mechanically turning raisins on a continuous paper tray (1992).

harvesting, terraces are created in the vine row middles and hand laborers harvest the fresh fruit clusters from the vines. One difference of note is that a raised-bed terrace is required for machinery clearances after fresh fruit harvest. Grape harvest laborers are usually paid a piece-rate wage and, therefore, collect fruit and unroll the continuous tray at their own pace. Fresh fruit is collected into grape picking pans and then spread out onto a continuous tray in 20 to 22 lbs (9 to 10 kg) quantities. When possible, the groupings of fresh fruit are separated on the continuous tray by a short empty space, which allows the pickers to be compensated by the standard piece-rate method (**Figure 5.10**).

The process of fruit turning can then be performed mechanically by specialized attachments within these systems. For either system, similar tractor-drawn implements turn the fruit by lifting the continuous tray and raisins over a hydraulically driven belt, through a series of rollers, and then back onto the ground. The turning machines are mounted to the three-point hitch of a standard 30 to 45 hp (22 to 34 kW) farm tractor (**Figure 5.11**). Raisin turning may be performed once or several times during the drying period, depending on drying conditions. Precipitation events may require use of the turning machines in order to break the seal between the soil and the tray (a.k.a. slipping the tray). Special considerations should be made with regard to soil type and probable soil conditions following rain events where raisin handling machinery are to be used.

The RAMEC and Unruh systems utilize specially designed machinery to retrieve the raisins from the continuous trays. Unlike traditional individual tray raisin drying, the continuous trays are not rolled and then left to cure in the vineyard. Instead, machines are brought in to remove the fully dried raisins from the continuous tray, sift out loose sand, leaf particles, and other debris, and then convey them over the row to a bin trailer towed by another tractor (**Figure 5.12**). Both the UNRUH and RAMEC retrieval machines require the use of a 60 hp (45 kw) tractor. Raisins are conveyed over the row to a separate bin trailer being towed by a 30 hp (22 kw) tractor. Depending on the conditions of the vineyard, the continuous tray, and the raisins, a support crew of one to three additional persons may accompany the retrieval machine and/or bin trailer.

Fig. 5.12: Mechanically retrieving raisins from a continuous paper tray (1992).

These systems were evaluated and compared to traditional hand harvesting under field conditions during the 1992 raisin grape harvest at California State University, Fresno (Berg et al., 1996). Data was collected in order to describe the procedures involved with each mechanized system, the estimated costs of operation, and the labor requirements for each system from pre-harvest practices to the "boxed," farm-level product.

Performance data was used to compare the three systems on the basis of labor-hours per acre. Reductions in labor-hours per acre and per ton were realized for the mechanized systems, even though the trial vineyard was very low yielding. Labor hours were reduced by the mechanized systems during all harvest operations. The processes of rolling trays and shaking/boxing the raisins were eliminated by the mechanized systems. Using traditional methods, harvest and handling of raisins required 40 to 50 person-hours per ton (about 36 to 45 person-hours per tonne) of raisins produced. The UNRUH system saved 26% of the labor-hours required by the traditional methods. The RAMEC system cut additional labor hours during pre-harvest and picking, leading to a 47% overall labor-hour savings from traditional raisin harvest and handling methods. Other factors, such as raisin drying rate, raisin quality, and yield losses were not found to be affected by the systems.

Fig. 5.19: Dormant season view of the Sun-Maid South Side DOV trellis system in a Caruthers, CA Thompson Seedless vineyard. Note fruiting canes tied to two wires held by the arm extending toward the south (left) side. Double cross-arms above will guide the growth of renewal shoots away from the fruiting zone (1999).

DOV raisin production research began in Australia in the late 1950s and early 1960s (May and Kerridge, 1967). The most significant development of this early research was the practice of harvest pruning, or cutting the fruiting canes upon fruit maturity (**Figure 5.13**). This practice was used in combination with either "(1) [grapes] picked after wilting, dipped, and dried on drying racks; (2) [grapes] shaken off the vines after wilting and dried on ground sheets; (3) [grapes] shaken as dried fruit off the vines; or, (4) [grapes] sprayed with dipping emulsion on the vine and shaken off as dried fruit" (May and Kerridge, 1967). Due to Australia's climatic conditions and market preferences, the methods of raisin production, whether conventional or DOV, use oleate sprays or dipping emulsions to accelerate drying and reduce browning. Please refer to **Chapter 7** in this treatise for more information on current practices related to DOV Sultana production in Australia.

In recent years, new trellis systems have been developed for DOV production in Australia. The Irymple system, which is described in detail by Gould and Whiting (1987), the Shaw system as described by Shaw (1986), and the Swing-Arm system described by Clingeleffer and May (1981) are examples. These trellis systems follow the general principle of separating the vine canopy into fruiting and non-fruiting zones. These systems were developed to facilitate mechanization of harvest pruning.

Parallel research efforts were undertaken in California during the same time period, although the resulting developments were somewhat different. DOV raisin production was evaluated using the Black Corinth and Thompson Seedless cultivars. DOV Zante Currant production was successfully accomplished (Christensen et al., 1970) but, Thompson Seedless had some limitations due to later maturity, higher vigor, and larger berry size (Studer and Olmo, 1973). Without oil

Fig. 5.20: Sun-Maid South Side DOV trellis with Thompson Seedless after mechanical cane severance and leaf blowing (1996).

emulsion sprays, harvest-pruned Thompson Seedless grapes dried to 25 to 35% moisture after six to nine weeks of vine drying on a standard trellis system (Studer and Olmo, 1973). Attempts to expedite the Thompson Seedless DOV process included oleate sprays (Petrucci et al., 1974; Petrucci, 1996), which reduced the drying time to 15 to 18 days after harvest pruning, but, as in Australia, adequate emulsion spray coverage was difficult to achieve and the final product was different from the natural Thompson Seedless raisin. Trellis systems such as the Alternating Duplex (Studer, 1984) have also been studied for mechanized raisin production. This system also serves to separate fruiting and renewal zones, but was not proven for use in Thompson Seedless DOV production.

Innovations and improvements to DOV production systems continue to evolve both in Australia and the United States (California). Ivan Shaw and Peter Clingeleffer are currently working on the design and improvement of their own DOV production systems in Australia. Their latest version does away with the stationary trellis design and now uses a swinging crossarm system, which separates fruiting and renewal zones and allows for the mechanization of cane severing and harvest.

Sun-Maid Growers, Inc. of California have developed and patented a DOV raisin production system (Sun-Maid DOV Manual, 1996). Their system involves a specially designed trellis, which is very adaptable to the conversion of existing vineyards. The Sun-Maid South Side DOV trellis system separates the vine canopy into distinct fruiting and renewal zones, which facilitates harvest cane severing and promotes a good environment for the growth of renewal canes. The fruiting zone, which later becomes the drying zone, is oriented toward the south side in east-west directional rows (**Figures 5.19** and **5.20**). This orientation is designed to enhance the drying rate, thereby making it possible to dry Thompson Seedless grapes on the vine without oils or other aids in most seasons. Several companies are involved with the design of necessary machinery, such as cane severing units (**Figure 5.21**), leaf blowers/removers, and harvesters (**Figures 5.22a** and **5.22b**). All fruit produced on a Sun-Maid South Side DOV system must be marketed through the Sun-Maid Growers' Cooperative.

Another unique DOV system, originally designed for Thompson Seedless grapes, has been developed as a modifi-

Fig. 5.21: Prototype mechanical cane cutter developed by L & H Manufacturing (Selma, CA) for the Sun-Maid South Side DOV trellis system (1996).

Fig. 5.22a: Ag-Right DOV raisin harvester developed by Ag-Right, Inc. (Madera, CA) for use in various DOV raisin trellis systems (1996).

Fig. 5.22b: Close-up view of the Ag-Right DOV harvester's picking head (1996).

cation of the K. Hiyama cross trellis. The Hiyama trellis has been used in California for DOV production of Zante Currant raisins for many years. Joe Epperson, a Madera, California trellis system supplier/installer is credited with the development of the Epperson DOV system. His system wraps the fruiting canes to two wires spaced 8 inches (20 cm) apart on opposite ends of a 4 ft (1.2 m) crossarm (for a total of four fruiting wires). Fruiting canes can originate from either head trained cane pruned vines or from cordon trained vines

Fig. 5.23: Cordon trained, cane pruned Fiesta vines on the Epperson DOV trellis system at the California State University, Fresno farm laboratory (1996).

(**Figure 5.23**). Wooden posts of 12 ft (3.7 m) length are driven one-third of their distance into the ground with the remainder used to support the fruiting and foliage wires. The crossarm is mounted at 50 inches (1.3 m) above the vineyard floor and two sets of foliage wires are secured on the posts at 22 inches (0.5 m) and 46 inches (1.1 m) above the crossarm. A third set of movable foliage wires lie on top of the crossarm, near the inside fruiting wires.

As the renewal shoots push upwards from the head of the vine in early spring, the movable foliage wires are manually pushed toward and secured at the upright posts. This contains and directs the following year's renewal shoots to the vertical portion of the trellis and separates them from the horizontal fruit zones. At the same time, clusters and weak shoots originating in the head of the vine are removed. Additional passes are sometimes needed to maintain adequate canopy separation during the growing season.

Upon proper sugar accumulation, fruiting canes are severed by hand to initiate the drying process. Approximately three to five days after severing, dry leaves are blown out of the drying zone with leaf blowers to enhance drying rates. Under good drying conditions, the raisins can be harvested in six to eight weeks after cane severing.

This system is unique in that it is easily adaptable to

Fig. 5.24: Dual-sided fruiting zones of the Epperson DOV trellis system after cane severance (1996).

Fig. 5.25: Head-trained, cane pruned Fiesta vines on the Simpson DOV trellis system (1997).

Fig. 5.26: Fruiting canes of two adjacent rows trained to a single row middle (center), while shoots from renewal spurs will be oriented toward the open row middles (either side of center) (1997).

Fig. 5.27: Fiesta raisins drying on the Simpson DOV raisin trellis (1995).

Fig. 5.28a: A Simpson DOV harvester entering a fruiting row (1995).

Fig. 5.28b: Simpson DOV harvesters upon exit. Note the simple design of the rotating bar picking mechanism (1996).

existing vineyards as well as to being used for new plantings. Fruiting zones located on both sides of the vine row (**Figure 5.24**) provide a balanced system, which facilitates canopy spraying and harvest practices. Post and crossarm assemblies are placed at 14 ft (4.3 m) intervals to avoid excessive wire sagging. This system also appears promising for full mechanization in the future.

DOV Overhead Trellis Systems

Overhead trellis systems, such as arbor and pergola, have been in existence for centuries throughout the grape producing regions of the world, i.e. Spain, Italy, Argentina, Chile, South Africa, Taiwan, etc. Several overhead trellis systems are now emerging in California vineyards for the production of table, raisin and wine grapes. Two systems particularly adapted to raisin production described below are the Simpson system and the Pitts System, which both utilize mechanization to a great extent and completely eliminate ground preparation and use of paper trays as described for natural sun drying of raisins on the ground. These systems promote the ideal system of drying raisins on the vine (DOV), thus coping with the constant threat of rain during the drying season under the climatic conditions of California's San Joaquin Valley. DOV raisin production can be most successful with the earlier maturing seedless grape varieties such as Fiesta, DOVine, Summer Muscat, etc. The Thompson Seedless, a later ripening variety, in certain years will not fully dry on the vine and will need to be artificially finish-dried in a dehydrator.

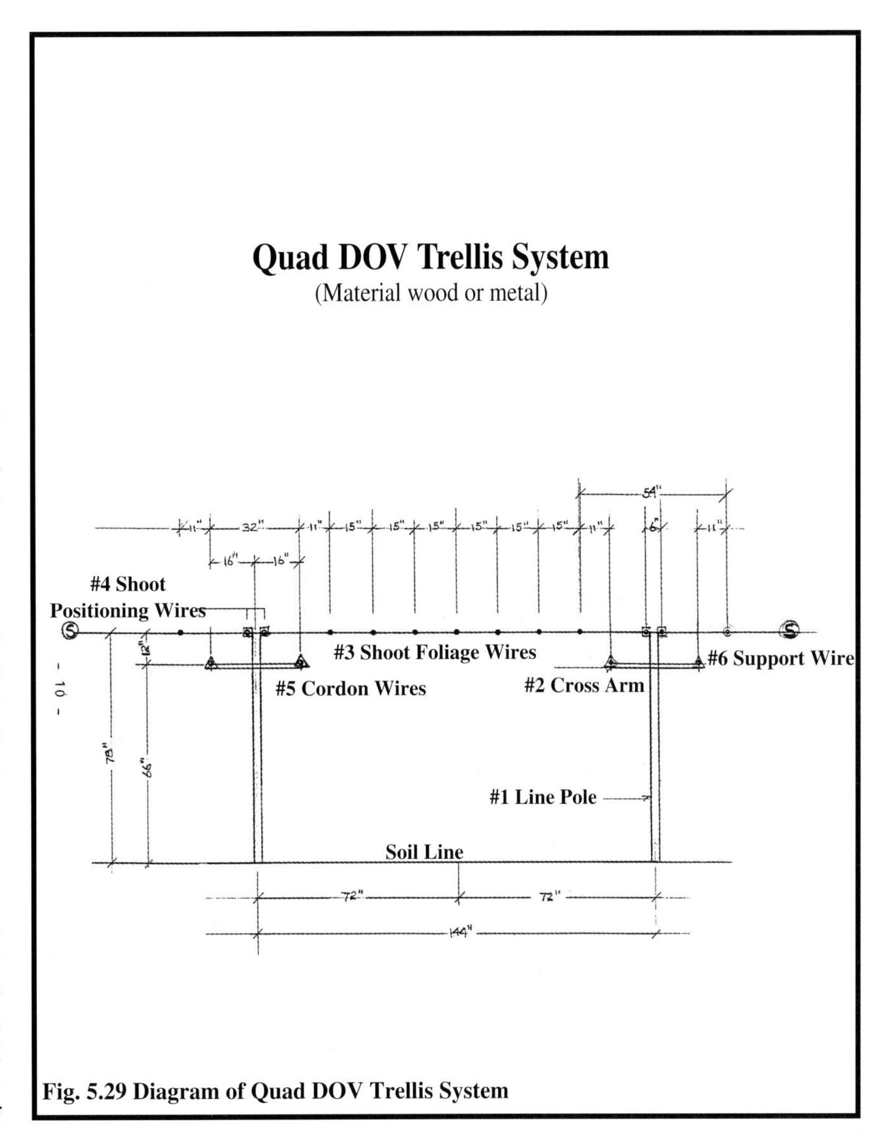

Fig. 5.29 Diagram of Quad DOV Trellis System

Simpson Overhead DOV

Lee Simpson, Simpson Vineyards, Madera, California, has completely redesigned his vineyard from the ground up for DOV overhead raisin production. He has developed a total systems approach to raisin production, which involves the selection of cultivar, training system, planting density, irrigation method, and mechanization. One-hundred-sixty acres (64 ha) of Fiesta vines were planted as cuttings in 1992 on a 5 ft by 8 ft (1.5 m x 2.4 m) vine by row spacing, which allows 40 ft^2 (3.6 m^2) per vine and 1,089 vines per acre (2,691 vines per ha). Subsurface drip irrigation is used to supply water and necessary nutrients to the closely spaced vines, and also substantially reduces the possibility of surface moisture and high humidity under the canopy. An overhead trellis system, pergola, is used and the vines are head trained and cane pruned (**Figure 5.25**). Fruiting and renewal zones are placed in alternate row middles from year to year (**Figure 5.26**). Canes are severed by hand and the raisins are allowed to dry naturally (**Figure 5.27**). Custom built tractor-mounted harvesters are used to gently knock the raisins off into bins at harvest time (**Figures 5.28a** and **5.28b**). Simpson's operation keeps a small crew of laborers (five or six) working all year long performing vine training, irrigation system maintenance, trellis and equipment repairs, and pest management. A moderate crew of contract labor is hired at the time of cane severing to assist the permanent crew. Simpson is consistently producing yields of Fiesta raisins at two times the regional average (see **Chapter 6**).

Pitts Quad DOV System

Gary Pitts of Fowler, California, a table and raisin grape producer, has recently patented an overhead DOV growing method. The Quad DOV Trellis System, (**Figure 5.29**) for the Pitts Growing Method (U.S. Patent 5,711,109) is described as follows, bearing in mind that the distance between line posts, endposts, and anchors is dependent on planting density, soil type and fertility of the vineyard.

Please refer to **Figure 5.29** and the following key which describes the various components of the Quad, DOV Trellis system:

1. Line poles or stakes.
2. Cross arm.
3. Foliage wires placed above cordon branch to aid in canopy and fruit support
4. Movable shoot positioning wire (13 gauge) placed above cordon branch to position shoots and canes.
5. Cordon wires (12 gauge), which support the permanent cordon quadrilateral trunks and arms (these wires are parallel with the vine row).
6. Support wires run perpendicular or across the vine row.

Fig. 5.30a: Quad DOV Trellis System in a newly planted vineyard (2001).

Fig. 5.30b: Quad DOV Trellis System in an established vineyard retrofit situation (2001).

Fig. 5.31: Fully developed quadrilateral cordon system on the Pitts Quad DOV raisin trellis (2001). Note 45 degree angle of fruit bearing canes to the cordon.

7. Cable (5 lbs galvanized) on perimeter of the trellis on both ends of the vineyard block is for maintaining tension on all wires running parallel with the vine row, (Fig 5.30b arrow).
8 Screw anchors are used to prevent movement of endposts, (Fig 5.32 with arrow).

When developing a new vineyard, the vine is trained as a single trunk which is topped to create two trunks that are trained outward toward the cross arm wires to form a wishbone (**Figures 5.30a** and **5.30b**). These secondary trunks are attached to the crossarm and are again topped to form a bilateral cordon on each of the two crossarm wires spaced 32 inches (0.8 m) forming a quadrilateral cordon head. This quadrilateral cordon, when fully developed, (**Figure 5.31**) establishes the base for implementing the alternate row DOV bearing principle (See description of alternate row raisin production above in the Simpson DOV section). Another feature of the Pitts growing method is that it lends itself to efficient mechanical or hand severance of the fruit-bearing canes. The Pitts Quad DOV trellis system employs a crossarm that is attached 14 inches (0.35 m) below the top trellis support wires. This enables the fruit-bearing canes to rise from the cordons at about a 45 degree angle (**Figure 5.31** and **5.32**) and in a fairly uniform position, which facilitates mechanically severing the canes. The Pitts method is said to substantially reduce the number of clusters in the renewal zone of the vine.

Overhead trellis systems employed for producing raisins (regardless of variety) are optimized when cane-pruned to retain uniformity of the fruit-bearing area and to renew the fruit-bearing area annually. Because the quadrilateral cordon training system uses long fruit-bearing units (canes), the non-bearing side of the vine can be pruned to basal buds arising from the cordon on that side of the vine. These buds in turn produce shoots, which are supported by the trellis wires. When mature, these shoots will provide the following year's fruit-bearing canes. This alternate bearing system bears full crop on one side of the quadrilateral cordon covering a row middle while the other side covering the other row middle produces fruit canes for the following year (**Figure 5.32**, Note the insert showing the crop).

Fig. 5.32: Alternate row production of the Quad DOV raisins trellis system (2000). Insert at right shows grape clusters in fruiting side of the canopy.

Fig. 5.33a: Amaro DOV 2000 raisin harvest system (left-front view) [2000].

Fig. 5.33b: Amaro DOV 2000 raisin harvest system (right-front view) [2000].

Fig. 5.33c: Amaro DOV 2000 raisin harvest system (rear view) [2000].

The Pitts growing method and Quad DOV Trellis System are also well adapted to machine harvesting. The Amaro grape and raisin harvester, model DOV 2000, is the latest mechanical grape harvester that can harvest fresh grapes as well as raisins from overhead trellis systems (**Figures 5.33a, 5.33b** and **5.33c**). This machine harvests mature grapes and/or raisins from vineyards trained to an overhead trellis. According to the machine's designers and manufacturers, "All product (grapes/raisins) is simultaneously collected and delivered to a shuttle unit traversing

Fig. 5.34a: Experimental propane flaming of leaves after cane severance on a Pitts Quad DOV trellis (2000).

Fig. 5.34b: Close-up view of fruit zone after experimental propane flaming on a Pitts Quad DOV trellis (2000).

through the center of the DOV 2000 machine, eliminating the need to coordinate bulk storage with harvesting." Such overhead trellis systems are currently being commercialized in California vineyards (as of the 2000-2001 season) and they appear to have become the trend in raisin, table grape and perhaps wine grape production.

Growing grapes on overhead trellis systems provides maximum leaf exposure to the sun, thus maximizing light interception by the canopy as well as within the canopy. Therefore, photosynthesis is enhanced when the largest percentage of the canopy is exposed to direct sunshine (Perez and Kliewer, 1990). Overhead trellis systems afford the greatest sunshine exposure to the newly forming buds on the green shoots that, when mature, will be used as the following year's fruiting canes. Because of the exposure of these buds to the sun, fruit bud differentiation is greatly enhanced. This in turn increases vine yields because of the larger number of clusters produced. As mentioned earlier, "sun canes" are generally more productive than "shade canes" (Christensen, 1978, 1986 and 2000; and Kliewer, 1981 and et al., 2000).

Later maturing varieties may require leaf blowing or removal by other means (e.g., propane flaming) to provide maximum drying rates. On a purely experimental basis, Fowler raisin grower John Paboojian has used propane fueled flames to disable leaves just prior to cane severing, followed by propane flaming of dried leaves several days after cane severing (**Figures 3.34a** and **3.34b**). More research is needed to determine the effect of propane flaming on actual drying rate, disease incidence during inclement weather, economics, and raisin quality.

For any dried-on-the-vine growing system, the following

Table 5.3
Effect of maturity level[z] on fruit composition[y], raisin quality, and yield of DOV raisins from Thompson Seedless grapevines. 1995. Caruthers, CA.

	Treatment (Targeted °Brix Level)			
	16°	18°	20°	22°
Berry Weight (g)	1.6c[y]	2.0a	1.8ab	1.7bc
Soluble Solids (%)	17.2c	19.0bc	20.2ab	22.2a
Raisin Yield (tons/acre)	2.6	3.3	3.0	3.4
Percent Grade B or Better	35.6%b	56.6%a	69.1%a	65.3%a
Percent Substandard	20.3%a	9.1%b	7.2%b	6.8%b

[z] DOV canes were severed on 17 Aug, 31 Aug, 14 Sep, and 21 Sep 1995, lrespectively for the 16, 18, 20, and 22° Brix targeted treatments.
[y] Means in the same row followed by the same letter do not differ significantly at the 0.05 level. Means separated by Duncan's Multiple Range Test. n.s. = not significant.

guidelines may help interested growers fine tune their selections. Canopy separation between fruiting and renewal zones is critical in order to provide easy access to fruiting canes during the cane severing process and to keep live canes from shading the drying zone. Where mechanical cane severing is desired, trellis system design and vine training must allow minimal interference with cutter mechanisms and maximum access to the fruiting canes.

As with dried-on-the-ground raisin production, fruit maturity is the key to raisin quality, although it is possible that the aerodynamic, bullet shape of DOV raisins helps to improve airstream sorter raisin grades by a few points. **Table 5.3** presents the results of data collected in 1995 on a Caruthers area Thompson Seedless vineyard where DOV practices were used. As can be seen, percent grade B or better raisins and substandard raisins were largely unaffected by maturity levels of 18 °Brix or higher, when grown and dried on the Sun-Maid DOV production system. Overall, total raisin yields were not statistically different for any maturity level. It should be noted here, however, that the 1995 season was one of abnormally high temperatures, which had a negative impact on the vines. Primarily, we believe that the abnormal weather conditions experienced in 1995 contributed to delayed fruit maturity and to the relatively high substandard raisins across all treatments.

Irrigation method and management are also very important considerations when choosing a DOV production system. For example, complete overhead systems on moderately well drained soils could encourage problems caused by high surface soil moisture unless subsurface drip irrigation is used. Climatic conditions during the drying period will also be a factor with regard to the potential need for finish drying in some years or losses due to high winds (which may affect some systems more than others). When managed with attention to detail, DOV raisin production can be a very successful way to produce high quality natural sun-dried raisins with fewer defects and by using far less labor at harvest time.

It is likely that conversion of acreage from standard raisin production to DOV production will continue in California as DOV production systems undergo further development and improvement. The time frame within which this occurs will be influenced by factors such as labor availability, labor costs, government regulations related to labor, and raisin prices.

Photo Credits

Figures 1-12, 15a-17, 19-28b: G. T. Berg
Figure 13: V. E. Petrucci
Figure 14, 18, 30a-31, 33a, b and c: Randy Vaughn-Dotta
Figure 29: Pitts Farms
Figure 32: Paboojian Vineyards
Figure 34a and b: Gary Pitts

References

Alvarado, A., H. O. Mason, and G. Riley. 1992. The labor market in the central California raisin industry: five years after IRCA. Center for Agricultural Business, CSU Fresno. Employment Development Department, California Agricultural Studies Bulletin # 92-4.

Andris, H., L. P. Christensen, and F. L. Jensen. 1989. Evaluation of raisin grape trellis systems. California raisin Advisory Board, 1989 Raisin Research Reports. pp. 33-40.

Baranek, P. P., M. W. Miller, A. N. Kasimatis, and C. D. Lynn. 1970. Influence of soluble solids in Thompson Seedless grapes on airstream grading for raisin quality. Amer. J. Enol. Vitic. 21:19-25.

Berg, G. T., R. K. Striegler, and M. Salwasser. 1992. Survey and economic analysis of mechanical raisin harvesting systems. California Raisin Advisory Board 1992 Raisin Research Reports. pp. 157-167.

Berg, G. T., R. K. Striegler, and D. K. Smith. 1996. Reducing production costs and labor requirements by use of mechanized raisin handling systems. In: Proceedings of First Vincent E. Petrucci Viticulture Symposium. California Agricultural Technology Institute Publication # 980702. pp. 141-153.

California Agricultural Statistics Service. California fruit and nut statistics. P.O. Box 1258, Sacramento, CA, 95812. Annual Crop Reports (1988-2000).

California Fertilizer Association, Soil Improvement Committee. 1990. Western fertilizer handbook. Horticulture edition. Danville, IL, USA; Interstate Publishers. 338 p.

Christensen, P. C. Lynn, H. P. Olmo, and H. E. Studer. 1970. Mechanical harvesting of Black Corinth raisins. Calif. Agric. 24(10):4-6

Christensen, L. P. 1978. Pruning for "sun-canes" in Thompson Seedless. Blue Anchor 55:9-11.

Christensen, L. P. 1986. Fruitfulness and yield characteristics of primary and lateral canes of Thompson Seedless grapevines. Amer. J. Enol. Vitic. 37(1):39-43.

Christensen, L. P. and R. J. Smith. 1989. Effects of persistent woody laterals on bud performance of Thompson Seedless fruiting canes. Amer. J. Enol. Vitic. 40(1):27-30.

Christensen. L. P., M. L. Bianchi, C. D. Lynn, A. N. Kasimatis, and M. W. Miller. 1995a. The effects of harvest date on Thompson Seedless grapes and raisins. I. Fruit composition, characteristics, and yield. Amer. J. Enol. Vitic. 46(1):10-16

Christensen. L. P., M. L. Bianchi, M. W. Miller, A. N. Kasimatis, and C. D. Lynn. 1995b. The effects of harvest date on Thompson Seedless grapes and raisins. II. Relationships of fruit quality factors. Amer. J. Enol. Vitic. 46(4):493-498.

Christensen, L. P., A. N. Kasimatis, and F. L. Jensen. 1978. Grapevine nutrition and fertilization in the San Joaquin Valley. Berkeley: University of California Division of Agricultural Science publication 4087.

Chistensen, L. P., and W. L. Peacock. 2000. Mineral nutrition and fertilization. In: Raisin Production Manual. L. P. Christensen (ed.) Berkeley: University of California Division of Agricultural Science publication 3393. Pp. 102-114.

Christensen, L. P. 2000. Vine pruning. In: Raisin Production Manual, L. P. Christensen (ed.). University of California Agriculture and Natural Resources Publication #3393 oo. 97-101.

Clary, C. D. and G. A. Sawyer-Ostrom. 1991. Effect of fresh fruit maturity on raisin quality. California Raisin Advisory Board 1991 Raisin Research Reports. pp. 167-177.

Clingeleffer, P. R., and P. May. 1981. The Swing-arm trellis for Sultana grapevine management. S. Afric. J. Enol. Vitic., 2(2):37-44.

Conradie, W. J. 1991. Translocation and storage of N by grapevines as affected by time of application. Proc. International Symposium on Nitrogen in Grapes and Wine. Pp. 32-42.

Cuhna, M. 1996. The grape industry's most vital resources are labor and mechanization. In: Proceedings of First Vincent E. Petrucci Viticulture Symposium. California Agricultural Technology Institure Pulication #980702. pp.123-127.

Enochian, R. V., M. D. Zehner, S. S. Johnson, and V. E. Petrucci. 1976. Production costs and consumer acceptance of dried-on-the-vine raisins. Agr. Econ. Rprt. No. 337, ERS-USDA.

Flaherty, D. L., L. P. Christensen, W. T. Lanini, J. J. Marois, P.A. Philips, and L. T. Wilson. Eds. 1992. Grape pest management. Second edition. University of California Division of Agriculture and Natural Resources Pub. # 3343. 400pp.

Fischer, C. D. 1959. Experiments on possible quality influencing factors of natural sun-dried raisins. Twenty Years of Raisin Research. California Raisin Advisory Board pp. 39-40.

Gartel, W. 1993. Grapes. In: Nutrient deficiencies and toxicities in crop plants. W. F. Bennett (ed.) APS Press. St. Paul, MN, USA. Pp. 177-183.

Gould, I. V., and J. R. Whiting. 1987. Mechanization of raisin production with the Irymple trellis system. Trans. of the ASAE-1987, pp 56-60.

Jacob, H. E. 1942. The relation of maturity of the grapes to yield, composition and quality of raisins. Hilgardia. 14:321-345.

Husmann, G. C. 1916. The raisin industry. USDA Bulletin # 349, Washington, D.C. March 17, 1916.

Ingalls, C. A., and K. M. Klonsky. 1998. Historical and current uses. In: Cover Cropping in Vineyards - A Growers Handbook; Eds. C. A. Ingalls, R. L. Bugg, G. T. McGourty, and L.P. Christensen. Univ. of Calif. Div. of Agric. and Nat. Resour. Publication # 3338. pp. 3-7.

Kasimatis, A. N. and C. D. Lynn. 1975. How to produce quality raisins. University of California, Berkeley, CA.

Kasimatis, A. N., E. P. Vilas, F.H. Swanson, and P. P. Baranek. 1975. A study of the variability of Thompson Seedless berries for soluble solids and weight. Amer. J. Enol. Vitic. 26:37-42.

Kasimatis, A. N., L. A. Lider, and W. M. Kliewer. 1975. Influence of trellising on growth and yield of Thompson Seedless. Amer. J. Enol. Vitic. 26:125-129.

Kasimatis, A. N., L. A. Lider, and W. M. Kliewer. 1976. Increasing growth and yield of Thompson Seedless vines by trellising. Calif. Agric. 30(5):14-15.

Kasimatis, A. N., E. P. Vilas, F.H. Swanson, and P. P. Baranek. 1977. Relationship of soluble solids and berry weight to airstream grades of natural Thompson Seedless raisins. Amer. J. Enol. Vitic. 28(1):8-15.

Kliewer, W. M. 1981. Grapevine physiology: How does a grapevine make sugar? Div. Of Agri. Sci. Univ. of Calif. Leaflet No. 21231.

Lynn, C. D. 1965. Two-wire horizontal trellis for Thompson Seedless raisin production. Amer. J. Enol. Vitic. 16:237-240.

Loveys, B., J. Grant, P. Dry, and M. McCarthy. 1997. Progress in the development of partial rootzone drying. The Australian Grapegrower and Winemaker. July 1997.

May, P., and G. H. Kerridge. 1967. Harvest pruning of Sultana vines. Vitis 6:390-93.

Patterson, W. K. and C. D. Clary. 1987. Methods tested to determine quality control in raisins. California Agricultural Technology Institute Publication # 871202. 6 pp.

Peacock, B. and L. P. Christensen. 1997. Influence of tray type, roll type, and tray filling on the rate of raisin drying. In: Proceedings of San Joaquin Valley Grape Symposium. December 16, 1997. University of California Cooperative Extension.

Peacock, W. L., L. E. Williams, and L. P. Christensen. 2000. Water management and irigation scheduling. In: Raisin Production Manual; Ed. L.P. Christensen. University of California Agriculture and Natural Resources Publication No. 3393, pp.127-133.

Perez, J. and W. M. Kliewer. 1990. Effect of shading on bud necrosis and bud fruitfulness of Thompson Seedless grapevines. Amer. J. Enol. Vitic. 41:168-175.

Petrucci, V. E., C. D. Clary, and M. O'Brien. 1983. Grape harvesting systems. In: *Principles and Practices for Harvesting and Handling Fruits and Nuts*, M. O'Brien, B. F. Cargill, and R. B. Fridley (eds.), AVI Publishing Co., Westport, CT, pp. 525-574.

Petrucci, V. E., N. Canata, H. R. Bolin, G. Fuller, and A. E. Stafford. 1974. Use of oleic acid derivatives to accelerate drying of Thompson Seedless grapes. J. Amer. Oil Chem. Soc. 51(3):77-80.

Petrucci, V. E., S. D. Foster, C. D. Clary, T. W. Thorsen and D. J. Bavaro. 1978. An alternative method of field drying raisins. ASAE paper 78-1544.

Petrucci, V. E. 1996. DOV raisin research at California State University, Fresno: a historical perspective. In: Proceedings of First Vincent E. Petrucci Viticulture Symposium. California Agricultural Technology Institute Publication # 980702. pp. 19-30.

Phene, C. J., 1995. The sustainability and potential of subsurface drip irrigation. In: Proceedings of Fifth International Micro-irrigation Congress, Orlando, FL, April 2-6. pp. 359-367.

Prichard, L. 1996. Unpublished information provided in course materials for Raisin Production Short Course. University of California Kearney Agricultural Center, Parlier, CA. February 6-8, 1996.

Raisin Administrative Committee (RAC). 12001. RAC marketing policy: 2001-2002 marketing season. 3445 N. First Street, Suite 101, Fresno, CA, 93726 USA.

Robinson, J. B. 1992. Grapevine Nutrition. IN: Viticulture, Volume 2, Practices. B. R. Coombe and P. R. Dry (eds.). Wine Titles, Adehide. pp.178-208.

Schwankl, L., B. Hanson, and T. Prichard. 1998. Micro-irrigation of trees and vines: a handbook for water managers. University of California Irrigation Program. Water Management Series Publication # 3378. 142 pp.

Shaw. I. 1986. Development of specialized trellis drying systems. Hanging cane. In: Trellis drying and mechanical harvesting of grapes. Eds. Ballantyne, I., and I. Macrae. Department of Agriculture and Rural Affairs. Conference Proceedings Series # 7, pp 93-4.

Striegler, R. K. 1992. Selection of raisin trays. American Vineyard. 1(7):3,6.

Striegler, R. K., C. D. Clary, D. R. Wineman, G. A. Wagy, and S. Glossner. 1992. Effect of simulated rainfall on raisin quality. HortScience 27(6):189.(Abstr.)

Striegler, R. K., G. T. Berg, and J. R. Morris. 1996. Raisin production and processing. In: Processing Fruits: Science and Technology, Volume II; Major Processed Products; Eds. L.P. Somogyi, D.M. Barrett, and Y.H. Hui. pp. 235-263.

Striegler, R. K., M. A. Mayse, W. O'Keefe, and D. R. Wineman. 1997. Response of Thompson Seedless grapevines to sustainable viticultural practices. . California Agricultural Technology Institute Publication # 970102. 6pp.

Studer, H. E. 1984. The Alternating Duplex: A vine training system for mechanized raisin production. Trans. of the ASAE-1984, pp. 986-89.

Studer, H. E. and H. P. Olmo. 1971. The severed cane technique and its application to mechanical harvesting of raisin grapes. ASAE vol 14(1):38-43.

Studer, H. E., and H. P. Olmo. 1973. Vine-drying of Thompson Seedless grapes. California Raisin Advisory Board Raisin Research Reports, 1973, pp. 960-70.

Sun-Maid Growers of California. 1997. California raisins best management practices handbook. First edition. 59pp.

Sun-Maid Growers of California. 2001. Succesful Raisin

Production for the 21st Century - using existing trellis, trellis retrofits, and new plantings. 18 pp.

USDA Dried Fruit Inspection Division. 1990. Handbook for inspecting and receiving of natural condition raisins.

Weaver, R. J. and A. N. Kasimatis. 1975. Effect of trellis height with and without crossarms on yield of Thompson Seedless grapes. J. Amer. Soc. Hort. Sci. 100:252-253.

Williams, L. E. 1987. Growth of "Thompson Seedless' Grapevine: II. Nitrogen Distribution. J. Amer. Soc. Hort. Sci. 112:330-333.

Williams, L. E. 2000. Grapevine water relations. In: Raisin Production Manual; Ed. L.P. Christensen. University of California Agriculture and Natural Resources Publication #. 3393, pp.121-126.

Winkler, A. J., J. A. Cook, W. M. Kliewer, and L. A. Lider. 1974. Raisins. In General Viticulture. University of California Press, Berkeley, CA. pp. 622-656.

Additional Sources of Information

AG RIGHT Dried-on-the-Vine Mechanical Raisin Harvester — Ag Right Enterprises, 12657 Road 28, Madera, CA 93637 USA

A&P Ag Structures DOV Raisin Trellis Systems — A&P Ag Structures, Inc., 11266 Avenue 264, Visalia, CA 93277 USA

Cal-Pac (Pitts Quad DOV Raisin Trellis Systems) – P.O. Box 577406, Modesto, CA 95357 USA

Epperson Cross DOV Production System — Epperson's Market, Inc. Farm Service and Supply, 2590 N. Madera Avenue, Madera, CA 93630 USA

Korvan DOV Raisin Harvesters — KORVAN Industries, Inc. 270 Birch Bay Lynden Road, Lynden, WA 98264 USA

L & H Manufacturing Mechanical DOV Cane Severing System — L & H Manufacturing, Inc. 9739 E. Manning Avenue, Selma, CA 93662 USA

Ramec Mechanical Raisin Handling Systems — RAMEC Harvest Systems, Inc., 4275 N. Chateau Fresno Avenue, Fresno, CA, 93722 USA

Rocca Mechanical Raisin Harvesting System — Rocca Ranches, Inc. 5342 W. Dakota Avenue, Fresno, CA 93722 USA

Simpson Vineyards Overhead Dried-on-the-Vine Raisin Trellis and Mechanical Harvesting System — Simpson Vineyards, Inc., 6708 Road 26, Madera, CA 93637 USA

Sun-Maid South-Side DOV Raisin Production System — Sun-Maid Growers of California, 13525 S. Bethel Avenue, Kingsburg, CA 93631 USA

Unruh Mechanical Raisin Handling System — Johnny's Welding, Inc., 5295 S. Clovis Avenue, Fresno, CA 93725 USA

Table 6.1
Capital Budgeting Example Costs and Returns.

Year	Cost/Return for Investment 1	Costs/Return for Investment 2
Year 0*	($100)	($100)
Year 1	$25	$50
Year 2	$25	$40
Year 3	$25	$20
Year 4	$25	$10
Year 5	$25	$5

** This Year*

Capital Budgeting Example

The analysis portion of the capital budgeting process includes the calculation of the measures discussed previously, namely the payback period, NPV, IRR, and MIRR. These measures can most readily be understood through a simple example. Let's imagine two investments that each cost a one time payment of $100. The $100 must be paid this year and the investments will return the amounts shown in **Table 6.1**. Notice that each investment costs the same, and at first view, each investment returns a total of $125. The question is whether there is any difference between the two investments or are both the same from a financial feasibility perspective?

Payback Period

The payback period tells us how long it will take us to recover our initial investment, in this case how long it will take the investor to recover the initial $100. To do this we simply add up the returns for each year until we get to $100. For investment 1 this number would be equal to the $25 from year 1 plus the $25 from year 2 plus the $25 from year three and finally, plus the $25 from the fourth year for a total of $100. Thus, the payback period for investment 1 is equal to four years. Investment 2 is slightly different.

The first year returns to the investor $50 plus the return from the second year of $40 gets us to $90, leaving us needing only $10 more to payback the initial investment. However, the third year returns us $20, thus the investor would only need half of the third year to get to a total of $100. Therefore, the payback period for the second investment is 2.5 years or two years and six months. Typically, we believe the faster an investment returns our initial cost to us the better. Using that rationale, investment 2 would be preferred according to the payback criteria. However, the payback period measure suffers from two shortcomings.

First, it does not consider any returns after the payback period. Thus, concerning investment 1 the $25 in year five is not considered and with investment 2 the returns from the second half of year three and all of the returns from years 4 and 5 are not considered. Secondly, payback period does not consider the time value of money. The time value of money relates to the cost associated with the investor having to wait five years for each of the investments to completely pay out. These two shortcomings are addressed by using net present value to evaluate the two investments.

Net Present Value

Net present value (NPV) takes into account both the timing and magnitude of all cash outlays and returns. The calculations involve discounting. Discounting is the process economists use to estimate the current value of money that will not be received until some future date. This is accomplished by dividing the return due in the future by a discount factor. The discount factor is equal to $(1 + r)^n$, where r is the discount rate and n is the number of years. In most financial analyses the discount rate, r, is equal to the cost of capital or the prevailing interest rate. To illustrate let's assume your savings account pays an interest rate of 5%. If your bank were to offer you a payment of $1.05 payable one year from now the present value of this amount would be equal to $\$1.05/(1 + 0.05)^1$ or $1.05/1.05 which equals $1.00. Thus, you wouldn't care whether your bank paid you $1.00 today or $1.05 a year from now. The basic assumption in NPV calculations is that you can re-invest returns at the cost of capital or the discount rate. In our simplistic example above, if your bank gave you the $1.00 today you could put it in your savings account and earn $0.05 in interest giving you the $1.05 one year from now. The generic equation for NPV is:

$$(1)\text{NPV} = -\text{INV} + \sum_{n=0}^{N} \frac{R_n}{(1+r)^n}$$

where INV is the cost of the investment, R_n is the net return in year n, r is the discount rate, and N is the number of years in the life of the investment. Using this information we can re-write equation (1) for the two hypothetical investments in **Table 6.1** as:

$$(2)\text{NPV} = -\text{INV} + \frac{R_1}{(1+r)^1} + \frac{R_2}{(1+r)^2} + \frac{R_3}{(1+r)^3} + \frac{R_4}{(1+r)^4} + \frac{R_5}{(1+r)^5}$$

where: NPV is net present value, INV is the initial cost of the investment, R_n is the return in year n, and $(1+r)^n$ is the discount factor in year n, with r being the discount rate. Using hypothetical investment 1 from above and a discount rate of 5% we re-write equation as follows:

$$(3)\text{NPV} = -\$100 + \frac{\$25}{(1.05)^1} + \frac{\$25}{(1.05)^2} + \frac{\$25}{(1.05)^3} + \frac{\$25}{(1.05)^4} + \frac{\$25}{(1.05)^5}$$

In turn equation (3) becomes:

$$(4)\text{NPV} = -\$100 + \frac{\$25}{1.05} + \frac{\$25}{1.10} + \frac{\$25}{1.16} + \frac{\$25}{1.22} + \frac{\$25}{1.28}$$

Recall, the initial $100 investment cost is not discounted since it is paid in current dollars. The subsequent $25 dollar payments are not received until the end of each of the next five years. Thus, they must be reduced or discounted to what their value would have been if the payments were each received in the present. Continuing the process, and rounding to the nearest dollar, equation (4) reduces to:

(5) NPV = -$100 + $23.81 + $22.68 + $21.60 + $20.57 + $19.59

Table 6.2
Summary of Example Capital Budgeting Calculations.

Measure	Example Investment 1	Example Investment 2
Payback (years)	4.0	2.5
NPV (dollars)	$8.24	$13.32
IRR (%)	7.93%	11.97%
MIRR (%)	6.68%	7.66%

Thus, NPV for investment 1 is equal to $8.24. Following the same procedure for the second hypothetical investment yields a NPV of $13.32.

Net present value decisions are based first on whether NPV is positive or not, then on magnitude. As long as investments have positive NPVs, the investments should be undertaken. Given that each investment have positive NPVs, as is the case in this example, then the larger of the two is ranked highest. Therefore, as with the payback period, investment 2 is more valuable than investment 1.

Internal Rate of Return & Modified Internal Rate of Return

The internal rate of return (IRR) is the discount rate that sets NPV equal to zero. To find this return, a trial and error procedure is used on the NPV equation to find the IRR. Again, when comparing alternative investments, larger IRRs are preferred to smaller rates of return. IRRs also place an upper limit on the cost of capital for investments that must be undertaken with borrowed funds. In the case of financing investments, the investment should only be made if the IRR is greater than the interest rate that is paid on the borrowed funds.

For our example the IRRs for investments 1 and 2 are, 7.93% and 11.97%, respectively. Using the "bigger is better" criteria investment 2 is, once again, the preferred investment. As upper limits on the interest rate that can be paid to borrow funds to finance the investments, the IRRs indicate that if financing were to cost more than 7.93% then investment 1 would be ruled out and if interest rates exceeded 11.97% then both investments would not be profitable. A key underlying assumption with IRR is that the returns may be re-invested at the internal rate of return for the entire life of the investment. Often times IRRs are large enough to render this assumption unrealistic. This limitation is overcome by using the modified internal rate of return.

Despite a conceptual preference for NPV in financial analyses, most business managers also like rate of return measures. To this end analysts have modified the IRR measure to compensate for it's shortcomings. The result of these alterations is the modified internal rate of return (MIRR), equation (6).

$$(6)\text{MIRR} = \left[\frac{FV_{CI}}{PV_{CO}}\right]^{1/N} - 1$$

where: FV_{CI} is the future value of cash inflows (returns), PV_{CO} is the present value of cash outlays (costs), and N is the number of time periods (usually years) in the planning horizon.

The future value of the cash inflows is equal to:

$$(7)FV_{CI} = \sum_{n=0}^{N} R_n(1+r)^{N-n}$$

where R_n is the return in year n, and r is the cost of capital. Using example investment 1 to illustrate, equation (7) becomes:

$$(8)\ FV_{CI} = \$25(1.05)^{5-1} + \$25(1.05)^{5-2} + \$25(1.05)^{5-3} + \$25(1.05)^{5-4} + \$25(1.05)^{5-5}$$

which reduces to:

$$(9)\ FV_{CI} = \$25(1.05)^{4} + \$25(1.05)^{3} + \$25(1.05)^{2} + 25(1.05)^{1} + \$25(1.05)^{0}$$

$$(10)\ FV_{CI} = \$25(1.22) + \$25(1.16) + \$25(1.10) + 25(1.05) + \$25(1)$$

$$(11)\ FV_{CI} = \$30.39 + \$28.94 + \$27.56 + \$26.25 + \$25.00 = \$138.14$$

The present value of cash outlays (PV_{CO}) is generated by the following equation:

$$(12)PV_{CO} = -INV + \sum_{n=0}^{N} \frac{-R_n}{(1+r)^n}$$

where -INV is the cost of the investment, and $-R_n$ is any cash outlays in year n. In our example the summation portion of this equation is met as the cash outlays are all captured in the -INV part of the equation since the only money being expended is in the present time period.

However, one can see how this would not be the case in the establishment of a raisin vineyard. Establishment costs are spread out over, usually, a three-year period so some discounting would be necessary. For our example, investment 1 has a present value of cash outlays simply equal to $100. Incorporating the generated information into equation (6) yields:

$$(13)\text{MIRR} = \left[\frac{138.14}{100}\right]^{1/5} - 1$$

Thus, the MIRR for investment 1 is equal to 6.68%. Following the same process for example investment 2 yields a MIRR of 7.66%.

The results of our example financial analysis are summarized in **Table 6.2**. The standard rules of thumb are for the payback period, smaller is better, for each of the other measures bigger is better. Keep in mind that, concerning rates

Table 6.3

Projected Per Acre Revenue and Costs per Production System.

	Production System			
	DOG	Rocca	Simpson	Pitts
Revenue				
Yield (tons)	2.0	2.6	5.0	6.0
Price	$1,025	$1,025	$1,025	$1,025
Total Revenue	$2,050	$2,665	$5,125	$6,150
Operating Costs				
Insecticide, Herbicide, Fungicide	$164	$251	$184	$250
Vine Pruning & Brush Control	$209	$608	$856	$600
Water and Fertilizer	$153	$213	$105	$250
Harvesting and Machinery	$590	$487	$140	$480
Miscellaneous	$169	$43	$117	$100
Total Operating Costs	**$1,285**	**$1,602**	**$1,402**	**$1,680**
Income Above Operating Costs	**$765**	**$1,012**	**$3,723**	**$4,470**

Table 6.4

Net Present Value, Payback Period, Internal Rate of Return (IRR), and Modified Internal Rate of Return (MIRR) of Production Systems.

	Net Present Value			
Discount Rate	DOG	Rocca	Simpson	Pitts
3%	$5,441	$7,387	$50,736	$64,784
6%	$2,127	$3,213	$33,932	$47,383
9%	$73	$667	$23,310	$38,324
12%	($1,245)	($939)	$16,314	$29,672
15%	($2,117)	($1,981)	$11,528	$23,717
18%	($2,709)	($2,673)	$8,139	$19,427
21%	($3,119)	($3,142)	$5,667	$16,217
24%	($3,408)	($3,464)	$3,815	$13,737
27%	($3,614)	($3,688)	$2,395	$11,771
30%	($3,761)	($3,843)	$1,284	$10,176
33%	($3,866)	($3,951)	$401	$8,858
36%	($3,940)	($4,024)	($312)	$7,749
39%	($3,991)	($4,313)	($895)	$6,803
42%	($4,025)	($4,307)	($1,377)	$5,985
45%	($4,046)	($4,294)	($1,778)	$5,267
Payback	10 yrs 7 mos	14 yrs 7 mos	4 yrs 1 mos	3 yrs 1 mos
IRR	9.14%	10.09%	34.61%	46.36%
MIRR	6.48%	6.90%	13.09%	16.03%

of return, the existence of other goals may play a role. For instance, in this example if a company or individual was to require a minimum return of 12%, then neither of the investments would be deemed worthy.

Economic Analysis

The economic analysis of raisin production systems includes establishment costs, estimated costs and returns and subsequent net returns, the aforementioned capital budgeting results, breakeven analysis, and a discussion of risk. Establishment costs, costs and returns and net returns per acre are necessary inputs for the capital budgeting process. The breakeven analysis and discussion of risk highlight sensitivity of the production systems to price and yield risk as well as weather and labor availability concerns.

Initially, the traditional, dried-on-the-ground (DOG) system is discussed followed by a financial feasibility study of three alternative systems developed by San Joaquin Valley California growers. The Rocca system (Rocca) involves retrofitting a DOG system and combines dried-on-the vine (DOV) and DOG production practices with mechanical harvesting. The Simpson (Simpson) and Pitts (Pitts) systems are mechanically harvested alternatives that utilize higher planting densities of Fiesta and Thompson Seedless vines and self-designed trellising systems, respectively. Each of these systems are described in greater detail in **Chapter 5**. It is not intended for this to be an exhaustive list of alternative production systems and practices. Much of the research and development of alternative systems is being undertaken by either individuals or small groups of growers, rendering the data from those experimentations proprietary and, therefore, unavailable to the public. However, these alternative system analyses will give the reader an appreciation for the overall direction of the American raisin industry.

Establishment Costs & Projected Costs & Returns

Using a cost basis, the most significant difference between the systems is in establishment costs and in a trade-off between purchasing machinery and hiring labor. The traditional DOG system costs about $2,600 per acre to establish. Retrofitting to the Rocca system requires an additional $1,200. Establishment costs of the Simpson system is approximately $7,000 per acre over three years. The Pitts System establishment costs are $4500 per acre. The economic viability of these production alternatives then rests with the ability to offset these higher initial costs with greater revenues generated by increased yields and lower exposure to risk. Rocca has achieved about a 25% increase in yields, while Simpson and Pitts Systems have been able to nearly triple the DOG yields via increased canopy area.

Four alternative production systems are used to illustrate current economic conditions in the raisin industry. A comparison of the projected costs and returns for the four alternative production systems is presented in **Table 6.3**. The characteristics of each system have been discussed previously, therefore this section concentrates on per acre costs and returns. The comparative budgets use a projected raisin price of $1,025 per ton. Total projected revenues are $2,050, $2,665, $5,125, and $6,150 for the DOG, Rocca, Simpson, and Pitts systems, respectively.

Operating costs have been aggregated for ease of presentation, but do represent actual budgeted amounts. As expected, significant differences exist among the operating costs for the four alternatives with the Pitts system operating costs being the highest. Income above operating costs is a simple calculation subtracting total operating costs from total projected revenue. Income above operating costs are $765, $1,012, $3723, and $4,470 for the DOG, Rocca, Simpson,

Table 6.5
Net Returns Per Acre by Price Per Ton and Production System.

	Production System			
Price	*DOG*	*Rocca*	*Simpson*	*Pitts*
$/ton				
$400	-$485	-$582	$598	$720
$600	-$85	-$72	$1598	$1920
$800	$315	$438	$2598	$3120
$1000	$715	$948	$3598	$4320
$1200	$1115	$1458	$4598	$5520
$1400	$1515	$1968	$5598	$6720

and Pitts systems, respectively. It is important to note that while year-to-year yields are quite consistent, prices per ton may vary significantly from season to season.

Capital Budgeting Results

Net Present Value

Positive NPVs indicate that the investment returns more than it costs in current dollars. Thus, both the DOG and Rocca systems make economic sense at discount rates of 9% or less. Meanwhile, Simpson is still economically viable at discount rates of up to 33%. The Pitts system maintains a positive net present value at discount rates over 45%. The considerable difference is attributable to yield variations. The standard DOG system is assumed to yield two tons of raisins, while the Rocca system has averaged slightly more than 2.5 tons per acre for the time frame of the available data. By contrast, Simpson and Pitts have verified yields in the 5.5 to 6 tons per acre range.

Payback Periods

A significant concern to growers is the time it will take to recover establishment costs. This time frame is referred to as the payback period. Rounding to the nearest year indicates that the DOG, Rocca, Simpson, and Pitts systems would take approximately 10.5, 14.5, 4 and 3 years to payback establishment costs, respectively (**Table 6.4**). The figure for Rocca includes the cost of retrofitting a vineyard after five years under a DOG regime. The shorter time frames help to reduce risk and lower interest charges in the event of financing the origination of the vineyard.

Internal Rate of Return & Modified Internal Rate of Return

Internal rates of return represent the discount rate at which NPV is equal to zero. Modified internal rate of return calculations require making assumptions as to rate of return available. In this case a re-investment rate of 5% was assumed. The DOG and Rocca systems have comparable internal rates of return (**Table 6.4**). However, the Simpson and Pitts system's IRRs are from three to five times that of the DOG and Rocca system's IRRs. Given the comparable nature of cash costs between the systems these differences in rates of return are attributable to increasing yields.

When reinvestment is considered, MIRR, the results are quite similar. The DOG and Rocca systems are quite comparable with Simpson and Pitts, earning about twice the rate of return of the other two systems. The authors believe the modified internal rate of return is the more reasonable of the two measures. The internal rate of return assumes the profits can be reinvested at that rate. However, consistent returns of the magnitude calculated here are unlikely, hence our preference for the more conservative MIRR measure.

Table 6.6
Breakeven Yield by Price Per Ton and Production System.

	Production System			
	DOG	Rocca	Simpson	Pitts
Price				
$400	3.2	4.0	3.5	4.2
$500	2.6	3.2	2.8	3.4
$600	2.1	2.7	2.3	2.8
$700	1.8	2.3	2.0	2.4
$800	1.6	2.0	1.8	2.1
$900	1.4	1.8	1.6	1.9
$1000	1.3	1.6	1.4	1.7
$1100	1.2	1.5	1.3	1.5
$1200	1.1	1.3	1.2	1.4
$1300	1.0	1.2	1.1	1.3
$1400	0.9	1.1	1.0	1.2

Table 6.7
Breakeven Raisin Prices by Yield and Production System.

	Production System			
Yield	DOG	Rocca	Simpson	Pitts
1.0	$1285	$1602	$1402	$1680
2.0	$643	$801	$701	$840
3.0	$428	$534	$467	$560
4.0	$321	$400	$350	$420
5.0	$257	$320	$280	$336
6.0	$214	$267	$234	$280
7.0	$184	$229	$200	$240

This rate of return analysis allowed for an estimation of a lower bound on fixed costs (machinery and management would be the primary fixed costs in raisin production). The bound on fixed costs is estimated at levels allowing the internal rate of return to remain about 2.5 to 3%. We estimate that the DOG and Rocca systems could sustain fixed costs of $300 per acre. The Simpson and Pitts systems could support per acre fixed costs of $2800 and $3500 per acre per year for the life of the investment and still have comparable rates of return with the other production alternatives. The reader should be aware that these figures are fixed costs per acre per year with a twenty five year planning horizon.

Breakeven Analysis and Risk

Net Returns

Breakeven analyses provide a useful management tool when the producer is evaluating alternative prices and potential yields. When examining the establishment decision, the projected net returns at different prices can provide useful planning information. Potential net returns for the four alternative systems, given a range of raisin prices and based on the yields indicated in the original budgets, are presented in **Table 6.5**. For example, at a price of $800 per ton (given the respective yields in the budget) the DOG, Rocca, Simpson, and Pitts systems project net returns (profits) of $315, $438, $2,598, and $3,120 per acre, respectively. Even more revealing is the trend in net returns as prices increase. Net returns by price are displayed in **Figure 6.1**. Notice how the rate of increase changes for the different systems. The slope of the trend lines for the Simpson and Pitts Systems indicates that for those systems the net returns increase at a faster rate as the raisin price increases. **Figure 6.1** also provides a view of when profits might become negative. For the DOG and Rocca Systems, profits would be below breakeven (zero profit) at prices near $600 per ton while the same breakeven for the Simpson and Pitts Systems would be less than the $400 shown.

Tables 6.6 and **6.7** demonstrate breakeven yields and prices for the production alternatives, respectively. The breakeven yields bear out the result that Rocca, Simpson, and Pitts systems all have higher production costs than the traditional system, thereby creating the need for greater yields in order to breakeven. However, with prices averaging nearly $1000 per ton in the 1990s one can see that the breakeven yields at this price are sufficiently low enough for each of the alternatives to breakeven even in low yielding years.

When breakeven prices at varying yields are considered (**Table 6.7**), as would be expected, the Rocca and Simpson systems require higher breakeven prices than the traditional DOG system. This is attributable to higher annual production costs. However, at historical production levels for these alternative systems, it is clear that price levels to date have been high enough to cover the additional production costs.

Yields

A similar analysis can be constructed looking at breakeven yields necessary to cover operating costs given different

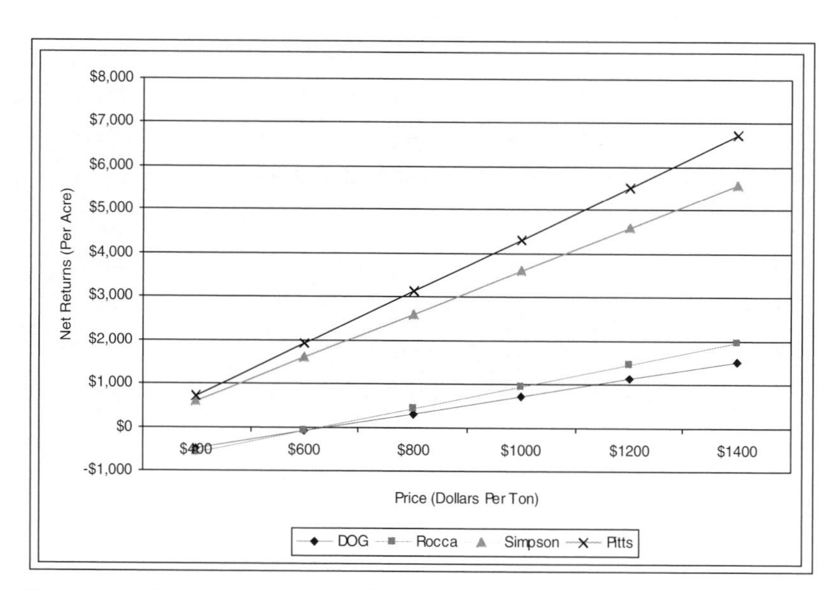

Fig. 6.1: Net Returns at Historical Yields by Price and Production System.

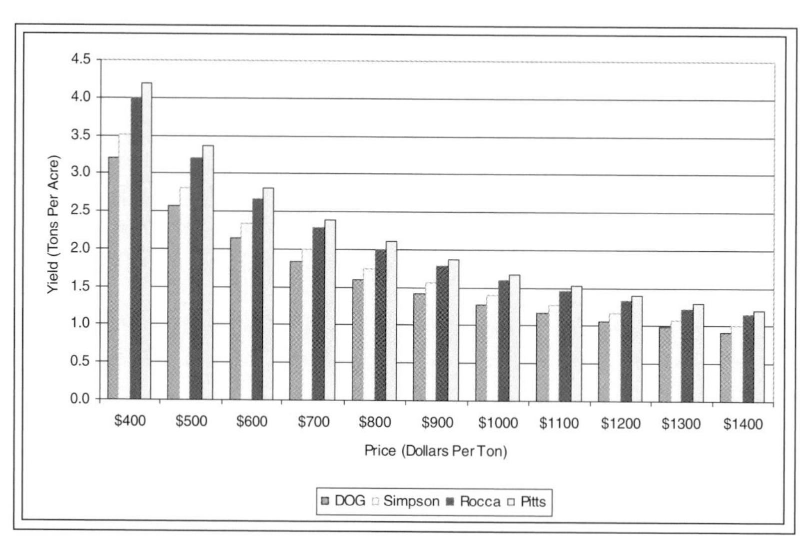

Fig. 6.2: Breakeven Yields by Price and Production System.

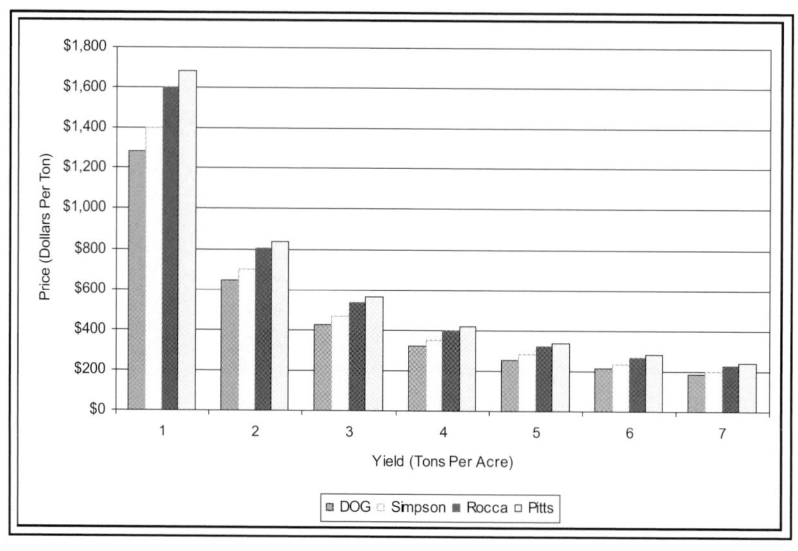

Fig. 6.3: Breakeven Prices by Yield and Production System.

raisin prices. Table **6.6** shows the breakeven yields necessary to cover operating costs at several different raisin prices. For example, if the raisin price is $900 per ton, the DOG System would require a yield of 1.4 tons per acre to cover operating costs. Similarly, the Simpson, Rocca and Pitts Systems would require 1.6, 1.8, and 1.9 tons per acre, respectively. **Figure 6.2** is a graphical view of the same results.

Prices

Based on the projected costs and returns shown in **Table 6.3**, the Simpson and Pitts systems have returns above operating costs of $3,723 and $4,470. These operating costs are significantly higher than the $1,012 and $765 for the Rocca and DOG Systems, respectively. These returns are based on a raisin price of $1,025 per ton. The returns would vary as the price of raisins vary. Given the yield estimates for each of the systems, a breakeven price can be computed where the returns above operating costs would be zero, or stated alternatively, the minimum price the producer could receive and to just cover the operating costs. For the DOG system, the required minimum price to cover operating costs would be $642.50 per ton. For the Rocca, Simpson and Pitts Systems, the breakeven prices are $628.20, $280.40, and $280 per ton, respectively. Under the Simpson and Pitts systems, the proportional increase in yield is much greater than the proportional increase in operating costs, thus allowing the breakeven price to be significantly less.

Table 6.7 shows the breakeven prices necessary to cover operating costs at several different yields and **Figure 6.3** provides a graphical view of the same results. Interestingly, as the yields increase the breakeven prices tend to converge, but only the Simpson and Pitts systems seems to have the capability of reaching the higher yields. Producers could use this analysis to estimate the prices needed based on their projected yields. At the yields projected in the budgets (**Table 6.3**) the DOG and Rocca systems required raisin prices in the range of $640 to almost $800 per ton. However, the Simpson and Pitts Systems would only require prices in the range of $280 per ton in order to cover per acre operating costs.

Risk

What is not captured explicitly in this analysis is the reduced level of risk faced by those who adopt alternative, mechanized systems. Directly, this risk reduction manifests itself in the, almost complete, elimination of weather risk. The Rocca system expedites the on-ground drying process by starting the drying process on the vine and then more evenly distributing the crop on trays to finish drying. This appears to be especially true with large crops. The Simpson and Pitts Systems, being a fully dried-on-the-vine systems, all but eliminate adverse weather effects. Indirectly, this reduces the risk of price discounts based on quality and embedded sand problems.

Finally, these two alternative systems reduce labor requirements, especially at harvest, while increasing yields. With perceived uncertainty about adequate supplies of labor and with most all growers demanding labor at the same time, reducing individual producer's labor requirements can speed up the harvest process. This, in turn, further abates weather-related problems. With increased yields the pressure for high prices is also reduced. This will allow growers who adopt these alternatives to more evenly distribute year to year cash flows.

References

Alvarado, A., H. O. Mason, and G. Riley. 1992. *The labor market in the central California raisin industry: five years after URCA*. Center for Agricultural Business, California State University, Fresno. Employment Development Department, California Agricultural Studies Bulletin 92-4.

Barry, Peter J., Paul N. Ellinger, John A. Hopkin, and C.B. Baker. 1995. *Financial Management in Agriculture.* Fifth Edition Interstate Publishers, Inc. Danville, IL

Mamer, J. W. and A. Wilkie. 1990. *Seasonal Labor in California agriculture, labor inputs for California crops.* Employment Development Department, California Agricultural Studies Bulletin 90-6.

Mason, Bert. 1998. *The Raisin Grape Industry.* Paper presented at the Immigration and Changing Face of Rural California Conference. University of California Kearney Research Station.

Pitts, Gary. Owner/Operator Pitts Carbonics, 9827 S, Clovis Ave. Fowler, CA 93625.

Raisin Administrative Committee. http://www.raisins.org/newsletter/rac_news.html.

Rocca, Earl. Owner/Operator Rocca Ranches, 5342 W. Dakota Ave, Fresno, CA 93722.

Russell, Kiley. Worries about Health of California Raisin Industry Persist. Available at http://www.uniontrib.com/news/state/20010731-1457-wst-agr-rais.html.

Simpson, Lee. Owner/Operator Simpson Vineyards, 6708 Road 26, Madera, CA 93637.

Simpson, L., D. Delaney, G. T. Berg, and R. K. Steigler. The Simpson DOV Raisin Production System: Development and Commercial Application. Available at http://www.dovraisins.com.

University of California Cooperative Extension. http://www.agecon.ucdavis.edu/outreach/crop/cost-studies/97Raisins.pdf.

CHAPTER 7.

Sultana Raisin Production

Peter R. Clingeleffer, B.S.

History

Light colored sultana raisins are produced after treatment with alkaline drying emulsions in a number of countries including Australia, Turkey, Greece, Iran, South Africa and India. The sultana raisins are distinctly different from the purplish-brown 'naturals,' frequently referred to as Thompson Seedless raisins, from the sulphur-bleached sultanas produced in Turkey, Iran, U.S.A. and South Africa and from the light-colored, greenish-yellow 'naturals' or 'Soyagi' which are dried in enclosed structures built from mud in Afghanistan (Grncarevic, 1969).

The treatment of grapes with emulsions to enhance water loss and, hence, minimize climatic risks associated with raisin drying, have their genesis in Mediterranean countries. It is reported that the first drying oils, which consisted of olive oil and wood ash, were used in early Roman times. Other techniques involved dipping the fresh grapes in hot lye solutions to crack the skin and promote drying (Grncarevic and Lewis, 1976). The traditional approach adopted in Turkey involves dipping of fresh Sultana grapes in an emulsion of 0.5% olive oil and 5% potassium carbonate (K_2CO_3) at ambient temperature (Kerridge, 1970). The fruit is ground dried on a puddled clay straw mixture, although more recent developments have included the use of paper trays, concrete slabs and plastic tents to protect the drying fruit from rain (Kerridge, 1970). When ground dried in this manner, emulsion treated fruit dried twice as fast as untreated fruit (i.e. seven days compared to 14 days) and produced light, red colored raisins.

When grape drying commenced in Australia in the early 1900s, brown sultana raisins were produced using the hot dip method. The lye dip of wood ash, caustic soda and quicklime was heated to 185 °F (85 °C) and the fruit dipped for five seconds to split the berry and partially remove the bloom and the fruit dried on wooden trays (Grncarevic and Lewis, 1976). The trays were gradually abandoned in favor of drying on locally developed tiered racks of 'wire netting' to reduce labor inputs and to provide protection of the fruit during rain. Racks were in general use by 1911. The perforated 'dip tin' was also a very early development used for dipping and as a harvesting container (Grncarevic and Lewis, 1976).

In the early 1920s the Turkish method of drying was introduced in Australia and adapted to rack drying. Although drying was slower than if the grapes were hot dipped, the technique produced a light, golden-amber colored sultana raisin rather than the amber to brown raisin obtained with hot dipping. By experimentation, the correct proportions of the dip components necessary for successful drying were ascertained, and by 1925 light type sultanas from Australia were well received on the London market. Further development of the cold dip took place in the 1940s when, stimulated by the shortage of olive oil, dipping oils consisting mainly of ethyl esters of fatty acids and emulsifiers were developed and soon became commercially available. Further research led to the identification of esters of fatty acids as accelerants of the drying rate (Grncarevic and Lewis, 1976). Animal tallow was then used as a cheap source of esterfied fatty acid. However, modern drying oils are now based entirely on vegetable oils.

This chapter will focus on modern techniques for the production of light colored, amber sultana raisins, which are produced in Australia and to a lesser degree in South Africa and India. Recent technological advances have shown that economically sustainable raisin production can be achieved through integrated approaches to vine management, which combine high productivity and quality and minimize inputs and environmental impacts (Clingeleffer, 1998). Mechanization of the drying process based on the concept of trellis drying (May and Kerridge, 1966) is a key component of this integrated approach (Clingeleffer, 1998).

Modern Sultana Management Features

In Australia there has been a consistent trend to improved productivity of dried sultana raisins from around 0.3925 tons/acre (2 t /ha) in the 1950s to current levels of 2.23 to 2.676 tons/acre (5 to 6 t/ha) (Clingeleffer, 1994 and 1998). Producers on all but the most marginal soils have the potential to achieve 3.345 tons/acre (7.5 t/ha) in most years. Growers who have adopted improved technology have achieved 4.46 to 5.352 tons/acre (10 to 12 t/ha) of sultana raisins. Contributing factors to improved productivity have been the adoption of high yielding clones, nematode tolerant rootstocks, larger trellises which support lighter pruning techniques and improved irrigation and soil management practices (Clingeleffer, 1994 and 1998). There has also been a trend to the adoption of systems for integrated pest and disease management in Sultana vineyards.

High Yielding Clones

The adoption of high yielding Sultana clones such as H5, H4 and M12 (Antcliff and Hawson, 1974) has significantly increased yields. Compared to unselected material, these clones increased productivity by 15 to 25%. A further increase in productivity of about 10% may be expected from eliminating virus diseases from improved clones (Clingeleffer and Krake, in press). Small plantings of the high yielding variety, Merbein Seedless (Antcliff, 1981), are also used to produce sultana raisins.

Rootstocks

The increasing adoption of nematode tolerant, vigorous rootstocks, in particular the *Vitis champini* selection, Ramsey has contributed to increased yields (Sauer, 1977 and Clingeleffer, 1998). Replanted vines grafted to these rootstocks are inherently higher in vigor and productivity over a range of soil types. Typically, yields of Ramsey vines were 100 to 250% higher in replant situations in nematode infested, lighter sandy loam soils. When planted in virgin soils, 40% higher yields were also reported for Ramsey vines (May et al., 1973).

High vigor rootstocks, such as Ramsey and 1103 Paulsen, will increasingly play an important role in sustainable raisin production. Considerable benefits in Sultana productivity have been demonstrated with Ramsey rootstock under saline conditions compared to own-roots (Walker et al., 1997). The higher salt tolerance of Ramsey rootstock compared to own-roots was attributed to its higher inherent vigor, development of bigger canopies, lower leaf chloride concentrations levels and maintenance of normal photosynthesis rates under salt stress. Furthermore, recent observations indicate that the inherent vigor of Ramsey may in part be attributed to the development of more extensive root systems deeper in the soil profile (Myburgh et al., 1998). Thus, rootstocks may access more of the available water and nutrients and hence may be used for more efficient water and nutrient uptake compared to ungrafted vines.

Improved Trellises

Increased yields have resulted from the adoption of improved (tall and/or wide) trellis systems, which reduce shoot crowding and maximize light interception, potential carbon dioxide (CO_2) assimilation, and vine fruitfulness. These larger trellis systems are suited to light pruning techniques to utilize the full potential of vigorous vines resulting from changes in management techniques and the adoption of improved clones and rootstocks (Clingeleffer, 1981 and 1985). Compared to own-rooted vines trained on a small 0.99 ft (0.3 m) T-trellis and severely pruned to nine canes May et al. (1973) obtained a 100% yield increase from the combined responses to Ramsey rootstock, 3.96 ft (1.2 m) wide T-trellis and light pruning treatments. Tall trellis systems used for mechanization are cordon-based systems and utilize hanging canes (Clingeleffer, 1998). They facilitate light pruning, upward vertical growth of replacement shoots and reduce shoot crowding leading to improved fruitfulness and productivity (Clingeleffer, 1998). These trellis systems also provide reasonable shade for the developing bunches to minimize sunburn and color deterioration (Walker and Clingeleffer, 1993; Uhlig, 1998).

Improved Cultural Practices

The adoption of improved cultural practices for management of soil, water and vine nutrition has contributed to the upward trend in productivity (Clingeleffer, 1994). These improvements include changes in irrigation techniques to improve water distribution, the introduction of demand-oriented water scheduling and identification and correction of nutrient-limited vine performance. They also have a major impact on land, water and nutrient use efficiency and reduced environmental impact, including problems associated with salinity, water logging and nutrient leaching.

Irrigation and Water Management

Traditionally, Australian raisin vineyards were flood irrigated from an open channel supply system. Improvements in the supply system, including the extensive use of pipes, has reduced evaporation losses and facilitated the availability of water on demand. Consequently there has been a trend to move from flood irrigation to overhead or undervine sprinkler systems and drip irrigation (Clingeleffer, 1996). The characterization of water use by vines will further provide growers with a scientific basis to manage irrigation and improve water use efficiency. For example, Yunusa et al. (1997) found under drip irrigation that Sultana grafted to Ramsey had a higher (double) crop water use efficiency (i.e. yield per unit of water applied) than own-rooted Sultana. Compared to own-roots, the higher yielding rootstock vines developed larger canopies and had higher levels of transpiration, but had lower levels of soil evaporation due to the higher levels of light interception by the larger canopy.

Soil Management

Recent studies indicate soil management is important to the achievement of sustainable raisin production. The studies of Myburgh et al. (1998) used soil pits and soil penetrometer techniques to study root distribution and soil structure and identify problems with irrigation and soil management in Australian raisin vineyards. They showed that major problems in vine health and growth could be attributed to soil compaction limiting both root growth and water movement in the profile. While natural compaction was observed at various depths within the soil profile, particularly in lighter sands, other contributing factors included compaction caused by wheel tracks and post ramming. Cultivation with rotary hoes was identified as a serious problem leading to poor infiltration of irrigation water due to soil "glazing" just below the soil surface. The use of deep "ripping" and other soil management practices including the use of cover crops to rehabilitate soils in problem sites was recommended.

Nutrition

The use of fertilizers by growers in Australian raisin vineyards is very variable, ranging from none to very high applications, particularly of nitrogen. Optimal use of fertilizers must balance supply with vine requirements to optimize productivity without contributing to excessive environmental contamination. Currently, petiole nutrient levels determined at flowering, are used by many growers to assess vine nutrient status and as a guide to the application of fertilizers (Robinson et al., 1997). However, care must be taken when assessing fertilizer needs as optimal petiole nutrient levels, particularly of nitrogen may vary for different rootstocks, e.g. between own-rooted Sultana and Sultana grafted on Ramsey (Robinson et al., 1997). Furthermore, excessive nitrogen levels have been implicated in reduced sultana quality during processing and storage (Clingeleffer and Tarr, unpublished).

Integrated Pest and Disease Management

Climatic conditions for sultana raisin production lead to low pest and disease pressures which, in most seasons in Australia, are easily controlled. Insecticides are rarely used. Growers prefer to rely on natural predators for control of light brown apple moth and mealy bug (Kelly and MacGregor, 1997, see **Chapter 10**). The main fungal pathogens, powdery mildew (oidium, *Uncinula necator*) and downy mildew (*Plasmopara viticola*) (see **Chapter 11**) are controlled with protectant sprays of wettable sulphur and copper based fungicides. Many dried fruit growers (Kelly and MacGregor, 1997; Glenn et al., 1997) practice integrated pest and disease management (IPDM). Emphasis is placed on correct diagnosis, monitoring, understanding the control options, effective timing of control and targeting the use of chemicals. Biocontrol technologies have been introduced. They include *Bacillus thuringiensis* (Bt) and the parasitic wasp, *Trichogramma* for the control of light brown apple moth and *Cryptolaemus* ladybirds for mealy bug.

The widespread adoption of IPDM is likely to be linked to changes in irrigation practices. The widespread adoption of flood irrigation is an impediment as it can restrict access to vineyards for targeted spraying at a critical period for control (Clingeleffer, 1998). The adoption of the taller and more open trellises carrying a high number of smaller bunches should also facilitate the adoption of IPDM due to reduced incidence of pests and diseases and improve spray coverage as has been demonstrated for winegrapes (Emmett et al., 1994).

Principles of Drying: Effects & Mode of Action of the Drying Emulsion

Thermodynamic Principles

The rack, trellis and ground drying techniques used to produce light colored sultanas rely on solar radiation energy either directly or indirectly through its influence on air temperature and humidity. Inputs of solar radiation decrease rapidly over the drying period due to the combined effects of decreasing light intensity and shorter days, causing slower drying (Clingeleffer, 1984). During drying, water movement from the berry is dependent on its availability at the berry surface, the rate of transfer being governed by differences in vapor pressure between the fruit , and the surrounding air, i.e. vapor pressure differential (VPD) (Szulmayer, 1973; Clingeleffer, 1986). For an emulsion-treated berry there are three stages of drying. In stage one, water evaporates from the berry surface at a constant rate as if from a body of water. The only decrease in drying rate is due to berry shrinkage. In the second stage, water movement within the berry becomes the limiting factor and the drying rate decreases as the berry approaches 'dryness.' Under normal ambient conditions on either the rack or trellis the VPD is small at the latter stage of drying and 'finishing off' is necessary to reduce the moisture level to the 13% required for packing. In the third stage of drying 'bound water' is lost by volatilization of sugars, normally at a slow rate.

Differences between rack, trellis and ground drying and dehydration can be related to thermodynamic parameters (Clingeleffer, 1986). On the rack, drying occurs under ambient conditions. The berry temperatures are not elevated above air temperature except briefly in the early morning and evening when fruit is exposed to direct solar radiation. Good drying conditions are achieved with high air and fruit temperature and low humidity, which create a large VPD. During periods of high humidity or rain, uptake of moisture can occur, in particular when air temperatures are quite high. In the trellis drying system, more of the fruit may be exposed to direct solar radiation for a longer period than on the rack. The direct input of solar radiation raises berry temperatures above ambient, increases the VPD and, consequently, the drying rate. Berries within the bunch may also receive some solar radiation as it is transmitted through a number of berry layers (Clingeleffer, 1984). Thus more rapid drying rates may be achieved with trellis drying, provided emulsion application has been satisfactory.

When berries are dried on the ground, interception of solar radiation is high and the resulting high berry temperatures and low air humidity result in large VPD values. At night little movement of moisture occurs, even under dewy conditions because air and fruit temperatures are both low and consequently VPD values are small (Clingeleffer, 1984).

The Drying Emulsion and Its Effect

The premium golden amber color of Australian sultanas is achieved by dipping or spraying the grapes with a drying emulsion to hasten drying. A standard emulsion used for dipping is an alkaline (pH 11) oil in water mixture of food grade potassium carbonate (2% K_2CO_3) and 1.6% drying oil. There are several commercially available drying oils developed from a vegetable oil base of food grade quality. They consist of about 70% of ethyl esters of C_{16} and C_{18} fatty acids with free oleic acid (5%) and emulsifiers (Grncarevic and Lewis, 1976). The active components of the emulsion are the esters and the potassium carbonate, which neutralizes the free oleic acid in the oil to form soap (Grncarevic and Lewis, 1976). The alkalinity of the emulsion also protects it from fermentation when damaged fruit is being dipped.

Untreated grapes lose water very slowly as they are covered by a hydrophobic layer of epicuticular wax (bloom) which consists of densely packed, irregular, intersecting and overlapping wax platelets (Chambers and Possingham, 1963' Possingham, 1972; Uhlig et al., 1996). The epicuticular wax layer, which is 3 to 4 *u*m thick (Uhlig et al., 1996), provides a barrier to water movement (Grncarevic and Radler, 1971). The mode of action of the alkaline, oil-in-water emulsion appears to be both a physical and chemical modification of the structure of the outer layer of wax platelets so that its permeability to water is increased (Possingham, 1972 and Uhlig et al., 1996) (**Figure 7.1**). The effect is also reversible as the emulsion is easily washed off by rain (Grncarevic and

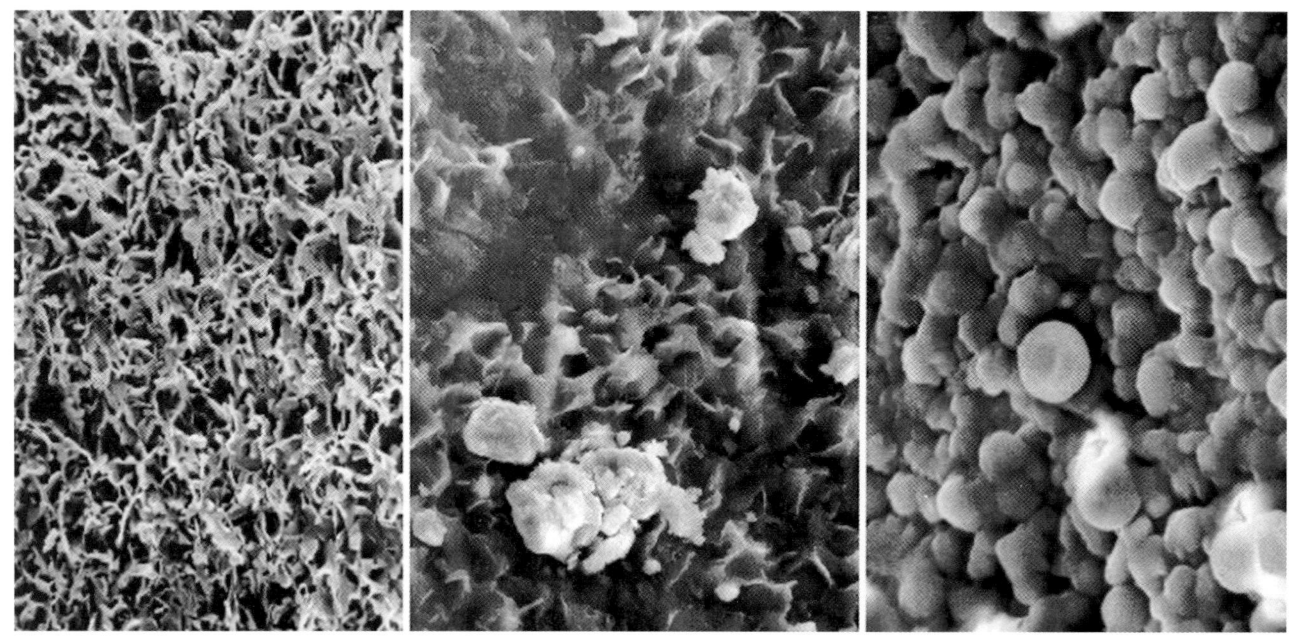

Fig. 7.1: Scanning electron micrographs of surface wax structures of dried sultanas (x 2500). *Left*: Untreated wax platelets. *Middle*: Modified surface wax structure after drying emulsion spray application. *Right*: Modified surface wax structure after dipping in drying emulsion.

Lewis, 1976). Consequently, re-spraying after heavy rain may be necessary, particularly during the early stages of drying.

Studies of the transcuticular movement of the potassium cation following dipping, using X-ray microanalysis and a rubidium tracer, demonstrated cation penetration of both the epicuticular wax layer and the cuticle but not the cell wall (Uhlig et al., 1996). Furthermore, substitution of the potassium cation in the emulsion indicated that drying rate was influenced by the size of the hydrated ion radius, which followed the Lyotropic series. In other words, rubidium carbonate produced a faster drying rate than potassium carbonate (Uhlig et al., 1996). Trials with alternative anions proved that the carbonate ion was the most effective (Uhlig et al., 1996). Further studies in which the pH of the drying emulsion was adjusted indicate that the cuticle acts as a weak acid, highly cross-linked cation exchanger (Uhlig and Walker, 1996). Over a pH range from four to 14, a linear increase in drying rate was determined as the pH of the emulsion increased (Uhlig and Walker, 1996). The most effective pH values were found between 9.5 and 11.5 because at higher levels negative effects on quality were shown (i.e. taste and processing damage).

The drying emulsion accelerates the rate of water loss from drying sultanas resulting in lighter colored dried grapes. Browning of grapes during drying is principally due to the action of polyphenol-oxidase, which is mainly located in the skin (Radler, 1964; Rathjen and Robinson, 1992). Faster drying minimizes reaction of the polyphenol-oxidase enzyme with phenolic substrates, keeping browning to a minimum. When dried, the high sugar concentration inhibits the enzymatic browning process (Grncarevic and Hawker, 1971). As a result, light golden amber sultana raisins are produced. Wet or humid weather reduces the rate of water loss from drying grapes allowing greater opportunity for polyphenol-oxidase to mix with the phenolic substrates and browning to occur.

Key Factors in the Use of Drying Emulsions

Research and commercial experiences have identified a number of key factors in the use of drying emulsions required to produce light colored sultanas.

Harvest at Optimum Fruit Maturity

An optimum period to harvest Sultanas for drying has been established with respect to fruit maturity. Immature fruit tends to give lightweight berries with a green tinge, due to the presence of chlorophyll (Clingeleffer, 1986; Uhlig and Clingeleffer, 1998). High maturities of around 22 °Brix are required to consistently produce well-filled (plump) and light colored berries with a good drying ratio (Grncarevic, 1973; May et al., 1983; Uhlig and Clingeleffer, 1998). Over-mature, sun-exposed fruit tends to dry to a darker color (Uhlig, 1998). Good irrigation management to avoid water stress and give uniform application through the vineyard is important to ensure uniform maturation of the crop and evenness in color (Walker and Clingeleffer, 1993; Uhlig and Clingeleffer, 1998). Furthermore, May et al. (1983) observed that emulsion spread from treated areas to untreated parts of mature berries. The spread of emulsion was linearly related to sugar content, with best results around 22 °Brix. This effect helps to explain the excellent results that can be achieved with both rack spraying and trellis drying, in particular with the high volume recycling spray units which apply sufficient emulsion to move through the bunches.

Complete Emulsion Cover

Grncarevic and Hawker (1971) and May et al. (1983) showed that complete cover of the berry surface with drying

Fig. 7.2: Bulk dipping of sultanas that were hand harvested into perforated 'dip tins'.

Fig. 7.3: Placement of sultana fruit onto the traditional drying rack.

Fig. 7.4: View of sultana bunches after careful spreading on the drying rack.

emulsion is essential to produce a high quality product and a fast drying rate. Poorly covered berries dry more slowly and become 'blobs' of higher moisture content (Clingeleffer et al., 1980). Excessive 'blob' numbers increase the number of dark berries and cause darkening of the surrounding light color fruit due to moisture equilibration in storage. Blobs may also contribute to sugaring and excessive damage during processing.

Emulsion Formulation and Application Level

The original drying emulsion formulations used for dipping consisted of 2.5% K_2CO_3 and 2% drying oil, although commercial recommendations have now been reduced to 2% K_2CO_3 and 1.6% oil. May et al. (1983) demonstrated that the amount of emulsion to produce light colored fruit is much less than that adhering to a fully immersed berry ('dipped'). Field studies with both rack and trellis drying have shown that high emulsion levels resulting from multiple or high strength formulations have little further effect on final drying rate. However, they contribute to rapid moisture uptake and darkening during periods of rain or high relative humidity and during storage and processing and to increased levels of processing damage, stickiness, sugaring and compaction in the final product (Clingeleffer, 1993). For both drying methods there was no advantage in exceeding 1.25% potash and 1% oil. For trellis drying, 0.6% potash and 0.5% oil produced satisfactory, cost effective results. The use of multiple spray applications should be avoided except after rain during the early stages of drying.

Rack Drying Methods

Overview

In the traditional method of production of light colored sultanas described by Grncarevic and Lewis (1976), the fruit is hand-picked using a short-bladed knife and placed in perforated containers, known as 'dip tins' (**Figure 7.2**), that hold 15.4 to 22 lbs (7 to 10 kg) of fresh grapes. The fruit is transported in the perforated containers to a central drying yard and spread on drying racks after treatment by dipping (**Figures 7.3** and **7.4**). The emulsion hastens drying and under favorable climatic conditions, produces light, golden-amber dried fruit. Drying is complete when moisture levels reach about 16% on a wet weight basis, usually after 10 to 14 days. The fruit is then shaken mechanically from the racks and spread on black, polypropylene sheets on the ground where direct solar radiation increases berry temperatures above ambient, finishing the drying process (Clingeleffer, 1984). When delivered to packers the final moisture content of the fruit should be below 13%. This moisture retards darkening in storage, facilitates capstem removal and minimizes processing damage (Simmons et al., 1979). A financial penalty is applied to fruit delivered above 13% moisture.

The Drying Rack

Drying racks are usually 165 to 330 ft(50 m or 100 m) long, 7.92 to 9.9 ft (2.4 to 3.0 m) high and have eight to 12 horizontal tiers which are spaced vertically, about 9.2 inches (23 cm) apart. Racks may or may not be roofed. Each tier is formed from 3.96 ft (1.2 m) wide, 2 inches (5 cm) mesh galvanized wire netting which is reinforced along both edges by a strand of heavier wire and fitted on the underside with wooden 'spreaders' spaced 4.95 ft (1.5 m) intervals to prevent sagging when loaded. A pair of heavy posts at each end of the rack supports end cross pieces from which the tiers are strung and strained, while pairs of lighter intermediate posts at about

Fig. 7.5: Trailer load of sultanas that were hand harvested into solid plastic containers prior to placement on the drying rack.

9.9 ft (3 m) intervals along the rack carry cross pieces supporting the netting (Grncarevic and Lewis, 1976). Before fruit is loaded onto a rack, a 'berry hessian' of polypropylene mesh is placed and pinned on the bottom tier to catch berries.

Rack Drying Processes

Drying emulsion may either be applied by bulk dipping or by rack spraying (Grncarevic and Lewis). Bulk dipping involves placement of 75 to 100 dip tins onto a frame fitted to the transport trailer. The loaded frame is mechanically lifted from the transport trailer and the fruit completely immersed in a large, 780 gal (3,000 L) tank of emulsion (i.e. 2% potash and 1.6% oil) for one to three minutes (**Figure 7.2**). After immersion the fruit is lifted from the emulsion, allowed to drain briefly and spread by hand as a single bunch layer on the drying rack (**Figures 7.3** and **7.4**). A second application of 0.6% potash and 0.5% oil may be applied to minimize blobs or if rain occurs within three days of the initial application.

Solid plastic containers (**Figure 7.5**) are now commonly used for hand-picking to avoid contamination of the fruit with soil particles which produce 'grit' in the dried fruit and for which financial penalties are imposed if detected (Grncarevic and Lewis, 1976). In this case the fruit is placed directly onto the drying racks where the drying emulsion is applied by spraying with a hand-held, multiple-nozzle forked wand (**Figure 7.6a**) or directly from a tractor-mounted fixed boom fitted with high volume, 'flooding' nozzles (**Figure 7.6b**) (Clingeleffer, 1994 and 1998). Bunches are spread onto the rack as single layers with the larger bunches placed to the outside of the rack wires and, where necessary, cut in half to facilitate application of emulsion and give complete berry coverage. Levels of 1% oil and 1.25% potash are now used to formulate the emulsion for rack spraying which is applied at a rate of about 15.84 gal/ton (60 L/t) of fresh fruit. A second application of 0.6% potash and 0.5% oil may be applied to minimize blobs or if rain occurs within three days of the initial application.

Drying on the rack is usually completed to a moisture level of 16 to 18% within 10 to 14 days, depending on the ambient conditions. Removal of excess moisture to meet the delivery

Fig. 7.6b: Rack Drying - Rack spraying of sultana bunches with a high volume fixed boom.

Fig. 7.6a: Rack Drying - Rack spraying of sultana bunches with a hand-held 'wand'.

specification of 13% may be achieved by rack dehydration in-situ on the rack or after removal from the rack. Clear reinforced polyethylene rack curtains may also be placed on racks at the latter stages of drying (i.e. below 25% moisture content) to prevent re-absorption of moisture at night or during unfavorable, humid weather (Fuller and Redding, 1988; Schache, 1990). These solar curtains elevate daytime berry temperatures by about 50 °F (10 °C) and increase the drying rate. Solar curtains also protect the fruit from darkening as a result of rain and sunburn on the edge of the rack. The use of curtains in combination with simple solar collectors, with or without rock-pile storage of heat for night drying has also been developed but not widely used.

The dried sultanas are shaken from the rack using a tractor mounted mechanical shaker (**Figure 7.7**) fitted with long, horizontal fingers, which pass between each tier (Grncarevic and Lewis, 1976). During shaking the fingers oscillate in a vertical plane and lift the netting from below. The mechanical shakers are fitted with catch trays to receive the fruit as it falls. Before this process takes place a 5.94 ft (1.8 m) wide length of hessian is run out along each side of the rack to catch spillage of sultana raisins.

Removal of Excess Moisture

Ground Drying

Once the fruit is removed from the rack, it is spread out in the sun for one or two days on black 165 ft (50 m) poly-

Fig. 7.7: Removal of dried sultanas from the drying rack with a tractor mounted 'rack shaker'.

Fig. 7.8: Ground drying sultanas to remove excess moisture after removal of the fruit from the rack.

propylene sheets ('finishing off') (**Figure 7.8**). The sheets may be folded over the fruit if necessary to protect the fruit from rain or heavy night dew (Clingeleffer, 1984). Ground drying may reduce green tinge related to the amount of chlorophyll present in berries (Bottrill and Hawker, 1970). The placement of fruit on the ground sheets in direct sunlight raises the fruit temperature above ambient. On clear sunny days in early March, April and May the fruit reach temperatures levels of 149, 131, 113 ºF (65, 55, and 45 ºC), respectively (Clingeleffer ,1984). Drying is normally completed in one to two days. Following hot days fruit should not be boxed until the late evening or preferably the next morning (Clingeleffer, 1984). It is recommended that sultanas should not be boxed directly from the rack, as the ground drying process appears to harden the skin and facilitates cap stem removal and reduces damage during processing.

Mechanical aids have been developed to unroll and tip the fruit onto the ground sheets and when drying is complete, to elevate the fruit into bulk bins and roll up the sheets. After placement on ground sheets the fruit is racked out by hand to give uniform fruit distribution to promote accelerated drying, even out the color and remove blobs (Clingeleffer, 1984). Stalks and other contaminants may be removed at this stage. The moisture content of fruit drying on groundsheets should be monitored to ensure fruit is not over-dried.

Fig. 7.9: Sultanas in bin dehydrator used to remove excess moisture after removal from the rack or trellis.

Dehydration

Cost effective rack and bin dehydration techniques are also extensively used to remove excess moisture before or after removal from racks, respectively (Grncarevic and Lewis, 1976; Hayes, 1989). Liquid propane gas (LPG) is preferred as a clean-burning energy source for dehydration (Clingeleffer, 1994).

For rack dehydration, the rack is covered and made reasonably airtight along the sides and ends with heavy reinforced plastic curtains (Grncarevic and Lewis, 1976). A portable LPG burner and fan unit capable of discharging about 79.576 gal/min (280 m^3 per minute) is used to heat the air to 140 ºF (60 ºC). Perforated ducting is used to convey and distribute the heated air along the length of a rack. Optimum drying conditions and control of the rate of drying are achieved by manipulating the fuel supply, fan speed, recycling of air and venting of humid air by raising the curtain opposite the ducting. Compared to ground drying, rack drying enables finish drying of high moisture fruit prior to rack shaking, saves time and reduces the risk of fruit contamination. When dry the fruit is allowed to cool to ambient temperature, it is shaken from the rack and immediately boxed.

Many variants of the bin dehydrator are used (Grncarevic and Lewis, 1976). They are all based on the principle of attaching a LPG burner and fan dehydration unit to the end of a series of metal-sided bins. The fruit is placed in the bin to a depth of 4 to 6 inches (100 to 150 mm) on woven polypropylene mesh placed over a wire mesh floor located about 6 inches (150 mm) above the ground (**Figure 7.9**). The bins are normally about 19.8 x 7.92 ft (6 x 2.4 m) and 1.65 ft (0.5 m) high. An airtight cover is clamped to the bins and held on by a number of frames placed along the bins (**Figure 7.10**). The dehydration unit blows hot air [122 to 140 ºF (50 to 60 ºC) into the space between the cover and the fruit. This air is forced down through the fruit to discharge the moisture-laden air from vents in the sides of the bins near ground level. Increased efficiency has been achieved by recirculating a proportion of the heated air instead of venting it. Mechanical tipping of the fruit into the bins directly from rack shaker trays or from bulk storage bins is used to facilitate mechanization of bin dehydration (**Figure 7.11**). Similarly, individual bins are designed

Fig. 7.10: View of covered bin dehydrator and heating unit during operation.

to accommodate standard, tractor mounted 'forklift' equipment which is used to tip the fruit directly into bulk bins once the fruit is cooled.

Boxing and Storing Fruit for Delivery

Sultanas are placed into 1,100 lbs (500 kg) bulk bins for delivery to processors. To reduce insect and rodent infestation and to avoid moisture uptake during periods of high humidity, long-term on-farm storage by growers is rarely practiced. Fruit delivered above 13% moisture attracts a penalty applied as a weight deduction depending on the moisture level, e.g. a 2.5% weight deduction is applied for fruit with moisture contents between 13.5 and 15%. A maximum delivery temperature of 86 °F (30 °C) has been set by industry to minimize rapid darkening and excessive fruit compaction in storage (Clingeleffer, 1994). Consequently, ground dried fruit should not be boxed until the late evening or preferably the next morning when fruit temperatures are at ambient levels (Clingeleffer, 1984). Fruit temperatures should be carefully monitored after bin or rack dehydration. If necessary, the dehydrator fan may be run for a few hours after dehydration is complete to reduce fruit temperature. Prior to delivery, growers should store fruit in bulk bins in the shade as fruit temperature may rapidly increase on the surface or close to the sides of the bulk bins.

Trellis Drying

The Concept

Partial mechanization of the sultana harvest, based on the concept of trellis drying, has been adopted by many of Australian growers. Trellis drying was introduced by CSIRO in the late 1960s (May and Kerridge, 1967). It involves drying of the fruit *in-situ* on the trellis after severance of fruit bearing canes and application of the drying emulsion and mechanical harvesting of the dried product. Mechanical removal of the dried sultanas has a low energy requirement compared to fresh grapes. Consequently, growers may use small, tractor-mounted, purpose-built mechanical harvesters or larger, self-propelled or tractor drawn wine grape harvesters. Final drying of the trellis-dried raisins is required to reduce the moisture level to below 13%. This is fully mechanized using modified systems of bin dehydration with liquid propane gas (LPG) heating.

Fig. 7.11: Mechanical placement of dried sultanas into storage bins from dehydrator bins prior to delivery to packers.

Advantages of trellis drying over rack drying include lower costs, improved risk management, reduced handling damage, elimination of soil contamination (grit), reduced opportunities for weed seed, stone and mould contamination, and improved quality (Clingeleffer, 1998). Considerable cost savings may be achieved with trellis drying compared to hand harvesting and rack drying, particularly with higher yielding vines, i.e. greater than 2.23 dried tons/acre (5 dried t/ha). Cost reductions are achieved in the areas of labor recruitment and management, wages, workers compensation insurance and record keeping, although these savings are offset in part by the cost and operation of equipment. More advanced mechanized production systems also involve investment in improved plant material, irrigation and trellis systems. These systems increase yields and minimize the labor required for both winter and harvest operations. With respect to risk management, trellis drying enables growers to commit large areas and reduce rain damage caused by bunch rots (see **Chapter 11**) and berry drop during the harvest period. It also enables growers to maximize yields by delaying the commencement of harvest to improve drying ratios. Trellis dried fruit is more

Fig. 7.12: Trellis drying of sultanas on a standard T-trellis after application of drying emulsion and hand cane cutting.

Fig. 7.13: Application of drying emulsion to trellis dried sultanas by high-volume recycling spray equipment.

robust and less likely to be damaged in processing than rack dried fruit (Clingeleffer, 1994). Consequently, bonus payments are now provided to growers of trellis dried fruit.

Since its introduction, the trellis drying concept has continued to evolve with developments to facilitate it use and optimize quality when used for conventional, narrow T-trellis systems and recently in combination with tall, high yielding management systems to achieve almost full mechanization of sultana production.

Conventional Trellis Drying Systems

Research and commercial experiences have shown that trellis drying is ideally suited to vigorous, high yielding vines, not only to maximize savings by mechanical harvesting but also to ensure retention of photosynthetic capacity (i.e. 50% leaf area) when the fruit bearing canes are cut (Scholefield et al., 1977a). Trellis drying can reduce yields in low vigor situations (May and Scholefield, 1972 and Scholefield et al., 1977a). Excessive defoliation when cane cutting produces bunches with fewer flowers in spring (Scholefield et al., 1977b) and compound problems of low vigor. Low vigor vines producing less than 2.453 dry tons/acre (5.5 dry tonnes per ha) should not be trellis dried. Installation of foliage wires, either vertically above the T-trellis or in a V or U formation may be used to encourage vertical shoot growth, particularly of replacement canes and retention of a larger canopy after cane cutting (**Figure 7.12**). Delaying cane cutting, application of post-harvest nitrogen and careful scheduling of pre- and post- cutting irrigation is used to minimize the effects of defoliation and maintain vigor. Vineyard design may also limit its suitability for trellis drying. Headlands should be sufficiently wide, 15.84 to 19.8 ft (4.8 to 6.0 m) to provide access for spraying and harvesting equipment. A minimum trellis height of about 3.96 ft (1.2 m) is necessary to provide access for spray and harvesting equipment below the bunch zone.

Key operations involved in the trellis drying process are severance of fruiting canes, 'crown bunch' removal, application of drying emulsion, harvesting and removal of excessive moisture from the fruit.

Severance of Fruiting Canes

Severance of fruiting canes, 'cane cutting' initiates the drying process. It is usually undertaken just prior to or within a day of the application of drying emulsion. The process involves hand-cutting the fruiting canes with hand held secateurs at the wire just beyond any replacement canes required for next season. Installation of foliage wires which encourage vertical growth of replacement shoots, manual attachment of replacement canes to foliage wires in late spring and careful placement of canes during winter pruning facilitate both the location and manual cutting of fruiting canes and retention of more than 50% of the leaf canopy.

To achieve the high fruit maturity required for quality and optimum drying conditions canes are usually cut when sugar levels exceed 20 °Brix, around February 20 up to March 7.

'Crown Bunch' Removal

Bunches on shoots arising from the crown of vine are not severed during the cane cutting operation and must be removed manually. These berries dry very slowly if left attached to the vine, leading to dark berries of high moisture content. During mechanical harvesting, berries from crown bunches may contaminate the fruit sample and contribute to stickiness and adhering leaf problems, lead to down-grading of the fruit sample and contribute to excessive damage during processing (Clingeleffer et al., 1980). Crown bunches may be removed by hand in the spring or at harvest. In the latter case, crown bunch removal should be undertaken within three to five days of cutting. They may be placed on the fruiting wire in lighter crops or removed and rack dried.

Application of Drying Emulsion

To be completely effective and ensure a high quality product, free of dark berries, 'blobs' and 'bloomy' fruit, the emulsion must cover all berry surfaces (Clingeleffer et al., 1980; Clingeleffer, 1985). Grapes are sprayed with emulsion either just before or soon after the canes are cut, when the leaves on the severed canes have wilted. High-volume, recycling spray units are used for emulsion application, or 'wetting' (**Figure 7.13**). Nozzles are located to maximize bunch coverage and minimize application onto the foliage. Applica-

Fig. 7.14: Mechanical removal of trellis dried sultanas with a small tractor-mounted mechanical harvester.

tion rates vary between 2,354 and 3,531 gal/acre (22,000 and 33,000 L /ha) at ground speeds of about 0.93 miles/hour (1.5 km / hour). About two-thirds of this emulsion is caught as runoff and recirculated leaving 749 to 1,187.7 gal/acre(7,000 to 11,000 L/ha) on the vines, depending on crop size. The process of wetting is facilitated by foliage wires, placement of replacement canes on foliage wires, and attachment of canes in winter to spread the fruit and minimize clumping. Spreading of the emulsion around the berries is improved if sugar levels are above 20 °Brix (May et al., 1983). The most cost effective drying emulsion formulation has been shown to be 0.5% drying oil and 0.6% potash although many growers use slightly higher levels, up to 0.8% oil and 1% potash (Clingeleffer, 1993). A second application is only applied if it rains heavily within three days of the initial application, or a touch up with a hand directed nozzle is used to more adequately cover clumped, tight bunches likely to result in blobs.

Harvesting

The dried fruit is usually mechanically harvested after two to three weeks of drying (**Figure 7.14**) with moisture levels between 16 and 18%. Under exceptional drying conditions lower moisture levels may be achieved. The fruit is most suited to removal when the fruit bunch stems are most brittle in the afternoon and evening.

Removal of Excess Moisture

Final drying of trellis dried sultanas is required to reduce the moisture level to below 13%. This may be achieved by ground drying although more commonly, bin dehydration as described above is used. Final drying costs are in the order of $8-10/t. The final drying process reduces green tinge effects, promotes more uniform berry color and usually leads to enhanced fruit quality (Bottrill and Hawker, 1970; Clingeleffer, 1986).

Trellis Design and Vine Management

Trellis design and equipment development have now evolved almost to the point where it is possible to completely mechanize all the steps of raisin production, including winter

Fig. 7.15: Tall, Shaw swing-arm trellis used for management of high vigor grafted vines and mechanization of trellis drying processes.

pruning, and while maintaining high productivity (Clingeleffer, 1998). These developments undertaken by innovative growers are largely adaptations of hanging cane techniques (Clingeleffer, 1981) and the swing-arm system (Clingeleffer and May, 1981). Trellis design and vine management has been modified such that the non-fruiting replacement shoots and the fruit bearing canes are positioned in separate zones. When trellis drying, this facilitates mechanical cane severance, drying emulsion, the retention of leaf canopy and the mechanical harvesting operation (Clingeleffer, 1998).

Vine management involves the use of tall trellis systems, 4.95 to 6.6 ft (1.5 to 2 m), and permanent cordons with hanging canes to separate the fruiting and renewal zones and the retention of high cane numbers to maximize productivity (Clingeleffer, 1994). Each season, production is rotated between the two sides of the trellis. The hanging canes are supported below the cordons by additional wires placed in the form of an inverted V-Trellis. The evolution of systems for mechanization described by Clingeleffer (1998) has led to a simplification of the original Shaw trellis system based on a split, 0.99 ft (0.3 m) quadrilateral cordon. They include a single bilateral cordon (Shaw Mark 2 trellis) and the replacement of the inverted V-trellis with a sloping, swing-arm trellis (Shaw swing-arm trellis) (**Figure 7.15**). The swing-arm concept (Clingeleffer and May, 1981) is used to alternate the cropping side between seasons by rotation of the trellis. The modified sloping system facilitates the growth of vertical

Fig. 8.8: Workers scraping trays to recover raisins.

the trays (**Figure 8.8**). The trays are inverted and the raisins are collected in bins and are ready for processing. The empty trays are washed and sent to the dipping line to be loaded with fresh grapes.

The process flow described for raisins is referred to as counter flow dehydration. The reason is the movement of cars in the tunnel is in the opposite direction of the flow of air. Grapes are dried using counter-flow because of the difficulty of obtaining acceptably low final moisture content. The driest fruit must be exposed to the highest temperature. Prunes and other larger fruits are usually dried using concurrent flow. Since the harvest season for plums for production of prunes does not interfere with the grape harvest, many dehydrators will dry prunes in the same tunnels used for grapes. To accomplish concurrent flow, the cars of fresh prunes are loaded into the hot end of the dehydrator tunnel.

Continuous conveyer dryers are used extensively in drying food and many other products. This type of dryer consists of an endless belt that carries the product through a tunnel and has the advantage of essentially automatic operation that minimizes labor requirements. Initial equipment cost is higher than tunnels and cannot be justified in seasonal fruit drying operations.

Tunnel Dehydrator Optimization

Air Flow

Dehydrators move a high volume of air across the drying fruit to remove water vapor. Grapes and prunes contain 70 to 80% water and therefore release a substantial amount of water vapor during the initial stages of dehydration. A dehydrator tunnel holding 9.9 tons (9 t) of prunes with a dry ratio of 2.5 to 1 will require removal of over 12,000 lbs (5,400 kg) of water in the 24-hour period, or over 8 gal/min (3.7 kg /min) (Cruess and Christie, 1921). This is equal to about 1,400 gal (630 hL) of water.

A heated airflow of about 500 to 600 m^3/minute (17,637 to 21,164 ft^3/min) results in a drying time of 18 to 24 hours for grapes (Cruess and Christie, 1921). Nichols and Christie (1930) define air velocities of 22,532 and 28,392 ft^3/min (638 and 804 m^3 per minute), respectively for prunes and grapes, and research conducted by Thompson et al. (1981) indicates

Fig. 8.9: Photo micro-graphs of grape epidermis before and after treatment with methyl- esters (Eissen, W. and W. Muhlbauer, 1983). Note how the wax platelets are disassociated after treatment (*lower photo*).

that prune dehydrators operate at 179.6 °F (82 °C) using an air velocity of 2,575 ft^3/min (73 m^3/min).

Dehydrator Air Temperature and Humidity

Air temperature should not exceed 165 °F (74 °C) for drying prunes and grapes. Research suggests that the level of relative humidity in the exhaust air of a counter-flow tunnel should be between 35 to 40% (Nichols and Christie, 1930; Eissen and Muhlbauer, 1983). Some research has shown that relative humidity of exhaust air could approach 60% although the wet bulb temperature should not exceed 125 °F (50 °C) (Thompson et al., 1981).

Dehydrator Energy Use and Losses

Evaluation of an existing prune dehydrator showed the main areas of heat loss are in the exhaust air, burner inefficiency and air leaks (Thompson et al., 1981; Brown et al., 1983). As a result, about 59% of the energy is available to vaporize water from the fruit. Survey information (Clary and Moso, 1983) indicates an energy usage range of 241 to 552 therms and 137 to 393 kW·h per dry ton to dehydrate grapes into raisins (24.9 to 64.2 GJ/t and 151 to 433 kW·h/t). Based on an electric rate of $0.135 per kW·h, and a gas rate of $0.565 per therm, this amounts to about $0.036/lb water removed ($0.079 per kg of water) from grapes, or about $0.365 per kg of dried product.

Alternative Methods of Preparing Grapes

Heat is an important factor affecting the rate of reaction of in enzymes (Singleton et al., 1985). Indiscriminant exposure of fruit to moderate levels of heat can increase the rate of reaction of various enzymes, some of which can cause discoloration. It is suspected that the heat used to crack the grape skin in the dipping process contributes to discoloration of the grapes. The dipping treatment exposes the grapes to heat and

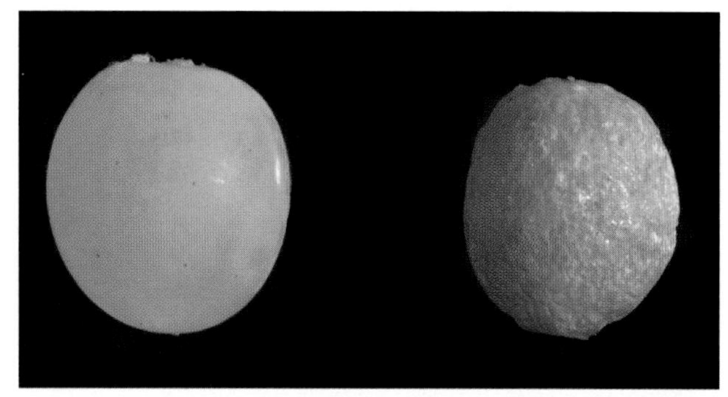

Fig. 8.10: Thompson Seedless comparison of the fresh grape berry (left) and the Grape Puff™ (right). Note similarity.

induces discoloration from activation of enzymes, oxidation and caramelization of sugars. The sulfur treatment commonly used must reverse this discoloration, as well as produce a residue sufficient to preserve color for an extended time.

Ethyl- and methyl-esters of fatty acids have been used as a treatment for grapes in Australia (Grncarevic, 1963; Ponting and McBean, 1970). Use of these compounds originated in traditional Greek and Turkish sultana production, and consisted of dipping grapes in an emulsion of potash and olive oil in water, followed by drying the grapes in covered racks. Refined methyl oleate and potassium carbonate mixed in water disassociates the hydrophobic platelets on the epidermis of the fruit (**Figure 8.9**). This permits more rapid moisture transfer from the fruit tissue to the surrounding atmosphere during rack drying. It has proven effective in reducing drying time of sun-dried raisins from 14 to seven days (Petrucci et al., 1983). Use of esters of fatty acids reduced drying time of grapes in a solar dryer [140 °F (60 °C) air at 99 ft/min (30 m/min)] from 50 to 25 hours (Eissen and Muhlbauer, 1983), and has been used with other waxy fruits including cherries, blueberries, and prunes (Ponting and McBean, 1970). Formulations of this fatty acid are available for use on grapes and is also sold for application to alfalfa to hasten drying.

This chemical treatment has been adapted to for treatment of grapes for tunnel dehydration. Petrucci et al. (1981) describe methods of using esters of fatty acids to treat grapes for production of golden raisins. Clary and Schmidtgall (1997) found the use of a solution of 1% EE-Muls-Oyle[1] and 1.5% potassium carbonate formulation with water proved effective in treating grapes for production of golden seedless raisins. Although reduced levels of sulfur were used in this study, all raisins produced exhibited good color. Grapes treated with EE-Muls-Oyle and 6.2 or 9 lbs (2.8 to 4 kg) SO_2 / fresh ton produced golden raisins with a color comparable to those made by conventional means, and their sulfur content was considerably lower. Golden raisin samples held in storage for a year exhibited good color retention and appearance. The benefit of this treatment is production of a brightly colored golden raisins with a reduction of sulfur residue.

[1]EE-Muls-Oyle is a formulation of ethyl oleate and potassium carbonate manufactured by Victorian Chemical Company, Australia.

Fig. 8.11: Red Grape Puffs™.

Low Pressure Dehydration

Low pressure dehydration is used to dry food products that are sensitive to heat. Although raisins are not produced commercially using low pressure, the following principles may some day have applications to raisin production. Low pressure systems often referred to as vacuum dryers use an operating pressure lower than atmospheric pressure to dry foods. Applications include freeze drying, vacuum band drying, vacuum pans and microwave vacuum dehydration. In all these applications, low pressure contributes to preservation of fresh product character.

Absolute pressure has a significant effect on the properties of water during dehydration. Pressure less than atmospheric conditions induces water to vaporize more rapidly and the boiling point is depressed below 212 °F (100 °C). Fruit juice is concentrated using vacuum pans. Vacuum band dryers consist of evacuated chambers that contain heat exchangers to heat and dehydrate food products. Water vapor is removed from the drying environment by the vacuum pumping system.

Use of vacuum for production of golden raisins is described by Clary (1996) and Clary and Petrucci (1991). This method uses a liquid media to conduct heat to grapes for dehydration. Since the process is completed under vacuum, sulfur dioxide is not required to produce golden raisins. The liquid media is usually a stabilized vegetable oil which is removed from the raisins during processing. Absence of sulfur residue does impact an extended shelf life.

Freeze drying uses very low pressure to remove water from a food product by direct sublimation (King, 1973). Because water is either ice or vapor at the operating pressure of a freeze dryer, the dried product maintains shape and much of its original color and flavor. The absence of heat contributes to reaching very low levels of final moisture content without damaging the product.

Microwave applied to fruits warms the product uniformly dehydration heats the product to cause water to vaporize, without causing undesirable change. Microwave energy penetrates deeply into food products and can reduce process time by 90% (Decareau and Peterson, 1986). It offers opportunity to process foods in ways not possible by conventional means. Microwave heating alone offers distinct benefits in dehydra-

Fig. 8.12: Green Grape Puffs™.

Fig. 8.13: Both green and red Grape Puffs™.

tion because of the penetration of energy and uniform heating resulting in water vaporizing from throughout the product. Some applications have combined microwave heating with conventional air heating for dehydration (Shivhare et al., 1991).

Application of microwave energy results in an increase in product temperature. However, if microwave energy is applied in a vacuum, the temperature rise is limited to the boiling point of the water at the lowered pressure. Since water boils at 72 °F (22 °C) at a pressure of 20 mm Hg (3 kPa), the product temperature would be limited to that temperature during dehydration. The potential benefits of microwave vacuum dehydration include rapid dehydration using reduced temperature to preserve color, flavor, and nutritional value. The uniform heating of microwaves induces an inner pressure that maintains shape of the dried product.

Microwave vacuum dehydration was first used for concentration of citrus juice in France (Decareau, 1985). Microwave vacuum drying of agricultural commodities has included grain (McKinney et al., 1977) and rice (Wear, 1982). This technology was adapted to grapes by McKinney et al. (1983) for production of Grape Puffs™ using zoned microwave vacuum dehydration patented by McKinney and Wear (1987) and described by Petrucci and Clary (1989) (**Figures 8.10 through 13**). This method is superior to fruits dried using dried fruit freeze drying. Microwave vacuum dehydration avoids freezing the fruit and disruption of the cell structure. Therefore, the dried product maintains better color and flavor, nutritional value and good texture when reconstituted.

Reconditioning Raisins

Contributed by: Wayne Albrecht, Albrecht Farms, Del Rey, CA

History of Reconditioning

The San Joaquin Valley of California is generally hot and dry in the fall and lends itself to good drying conditions for the production of sun-dried raisins. Until recently, little was known of El Nino or the wet or drought cycles. Better understanding of changes in weather patterns correlates closely with the construction of drying facilities to recondition raisins.

Since most of the California raisin crop is sun dried in the field (**Chapter 5**), it is susceptible to damage due to rain during the harvest and drying season. Rainfall in excess of 0.5 inches in a 24-hour period presents serious potential of mold infestation. In addition to mold, sand is a potential defect. Some research has been conducted identifying the molds and how to remove mold from rain damaged raisins. Tilden et al. (1958) conducted one of the initial evaluations of molds present in rains damaged raisins. Further research was conducted by Erke and Dokoozlian in 1983. Fisher (1958) described methods for removing rain damaged raisins from sound fruit using mechanical and chemical methods.

The reason for the research was that in rain years, the partially dried crop was collected from the field to prevent exposure to more rain. It was transported to a dehydrator facility and reconditioned. The process of reconditioning varies depending on history and the condition of the raisins. However, the intent is recover sound raisins and to dry them to a stable moisture content.

The recovery of rain damaged sun-dried raisins dates back to the turn of the century. Initially, drying the wet or high moisture raisins was the extent of what is now called reconditioning. The concept of reconditioning started when farmers tried to save their rain-damaged crop by drying their wet raisins. Some of the early methods were very simple in design. One method was to blend dry raisins with wet raisins in sweatboxes. This method allowed the wet and dry raisins to equilibrate or sweat, thus the name sweatboxes. Another method was to place half-full sweatboxes along a barn wall and allow the sun's warm reflection to dry the wet fruit. The sweatboxes were covered at night so morning dew would not re-moisten the fruit. Other methods included filling sweatboxes half full and placing sticks between the stacked boxes so air could flow between them. The stacked sweatboxes were then set in a row and covered with a tarp. A fan was used to draw daytime warm ambient air or heated air through the stack of sweatboxes. Sometimes the farmer moved the fan to the other end of the stack in order to dry the raisins more uniformly. Many of the old drying structures can still be seen around the valley today.

The raisin crop has been exposed to rainfall at 20-year intervals including the early 1930s, the late 1950s, the late 1970s and early 1980s, and 1992 and 1996. Based on these patterns, we are now realizing that El Nino type patterns have effected the California Raisin Industry for years. As a result, dehydrators such as Melikian, Chooljian, and Sanger were built in the 1930s, Enoch, Metzler, Lamanuzzi and Pantaleo and Lone Star were built in the late 1950s, then in the late 1970s Albrecht, 4 Bar C, Rosenthal and Victor were constructed because of rain years. The purpose of these dehydrators was to dehydrate grapes during the harvest season, and finish dry rain-damaged raisins collected from the field during rain years.

Commercial Reconditioning

Commercial reconditioning began in 1976 when Valley Welding and Machine Works began experimenting with separating rain-damaged raisins prior to drying them in the tunnels. This consisted of conveying rain-damaged raisins through a hot water bath using augers. These augers were mounted in a trough and set at an angle in a tank of hot water. This allowed the raisins to be dumped into the tank and be funneled into the bottom of the trough where the augers transported the fruit at an upward angle through the hot water. Experiments indicated that a stiff wire screen on the bottom of the trough was needed to separate debris, such as sand and infestation. In addition, the screen is abrasive enough to cause the raisins to tumble and rub against each other. The raisins infected with putrid mold were softened to the point they are pushed through the screen. The nodular mold was loosened by the hot water and the rubbing action between the fruit and the screen. This caused the mold to rub free of the fruit. Within a short time, commercial reconditioning began on a large scale. The auger trough was 15 ft (20.32 cm), 8 inches (4.545 cm) long, which allowed 15 ft (20.32 cm) of screen on the bottom side. The majority of dehydrator operations elected to use two pairs of these augers in two tanks so the water temperature could be controlled to allow for the best results between tanks.

In 1978, the raisin industry was devastated by one of the industries worst rains hit on Labor Day. This early rain, and several rains that followed, marked the need to go one step further than the reconditioning augers. The augers handled the light mold and wet raisins, but loads of heavily molded raisins were doomed. Many began working on ideas to separate the moldy berries from the good berries. When moldy raisins were dipped in hot water, they became sticky. It was found that moldy berries stuck to a standard conveyor belt. Tilting the belt at an angle caused the good fruit to roll off the side of the belt and the moldy fruit stuck. The addition of a vibrator to conveyor belt further improved the performance.

In a matter of weeks, many variations of this concept were developed. The Chooljian method used a large frame to hold a series of short vibrating belts running at an upward angle. The belts ran in series and the fruit dropped from one belt to another. As the belts ran up hill, the moldy raisins were carried over the top and the good raisins would drop to the next belt. The moldy raisins were scraped off of the underside of the belts. This method had two problems: one was that the unit weighed so much that anchoring it to the ground nearly impossible because of the weight of the shaker. Second, by the time the fruit reached the third or fourth belt, the fruit had cooled and no longer stuck to the belt.

The second method was the Melkonian belt. This method used the same theory as the Chooljian method, but the belts were tilted to the side. This allowed the sound fruit to slide off the side of the belt. The benefit to this method was that it provided more belt surface area for the moldy fruit to stick. A formed sheet-metal pan placed against the bottom side of the top belt was used to shake the belt and loosen the good fruit. As with the Chooljian belts, this unit was difficult to anchor.

Valley Welding and Machine Works and Bogoshian Raisin developed the third method. This method used one inclined belt, but used a series of beaters under the belt to vibrate the belt more aggressively. This method required three to four units to provide enough surface area. Although these units were easy to anchor, the angle of the incline belts and speed of vibration were critical.

Since these methods of reconditioning used a hot water dip, the raisins absorbed moisture. Therefore, the separated fruit was loaded onto wood trays and dried in gas fired tunnels.

State of the Art Reconditioning

By the time the rains of 1982 came, the reconditioning units were fully developed. An array of equipment is available to recondition raisins based on the condition of the rain-damaged raisins. These methods include:

1. Dry reconditioning consists of running raisins over a shaker with or without the use of a scalper, vacuum, or dry capper. No water is used in this method. This method is effective in removing foreign material, bunch rot and sub-standard raisins.
2. Raisins collected from the field that have high moisture as the only defect are reconditioned by loading the wet raisins on wood trays and drying them down to acceptable moisture content in tunnel dehydrators.
3. The wash and dry method consists of running rain damaged raisins through augers or a series of augers in hot or cold water to remove defects including moisture, embedded sand, low molds, micro, fermentation, sub-standard and maturity.
4. Mold belts augment the wash and dry method. Following treatment in the hot water and augers, the fruit is transported onto mold belts. This method is the most effective in removing defects from rain-damaged raisins including high levels of mold and damage.
5. All methods that include the use of water must be processed at dehydrators because the fruit must be dried back down to meeting moisture levels.

Since 1982, refinements have included the use of cold water to retain bloom. However, developments in optical electronic sorting, raisin processing now incorporates state of the art technology with new and improved electronic sorters. These sorters were originally installed to detect stems. As sorter technology improved, they began to have other capabilities such as separating by color. The sorters are capable of removing some of the white putrid and green or black nodular molds. Although raisin reconditioning was highly effective in salvaging raisin crops in the 1970s and 1980s, it has become apparent that the future of reconditioning lies in the computer age of sorters and scanners. Although these devices are effective in removing defect raisins, the industry will continue to depend on reconditioning and dehydrator operations to remove the moisture Mother Nature does not when the rains come.

Photo Credits

Figures 3, 4, 8 and 9: Courtesy of Albrecht Farms, Del Rey, CA.

Figures 5 and 6: Courtesy of Del Rey Packing, Del Rey, CA.

References

Baranek, P., M. W. Miller, A. N. Kasimatis, and C. D. Lynn. 1970. Influence of Soluble Solids in Thompson Seedless Grapes on Airstream Grading for Raisin Quality. Am. J. Enol. Viticult. Vol. 21:1, pp 19-25.

Barbosa-Canovas, G. V., H. Vega-Mercado. 1996. *Dehydration of Foods*. Chapman and Hall, New York.

Brown, D. E., R. P Singh, and J. F. Thompson. January 12, 1983. Energy Conservation in Raisin Drying. Proceedings: 55th Rural Energy Conference, University of California, Davis.

Christensen, P. February 6, 1985. Factors Contributing to Raisin Quality. San Joaquin Valley Grape Symposium Proceedings. Univ. of Calif. Cooperative Extension, Fresno County.

Christensen, P., J. F. Thompson, C. D. Clary, and M. Miller. 1988. Feasibility of Tunnel Dehydration Versus Sun-Drying for Raisins. IN: Raisin Research Reports, 1988. California Raisin Advisory Board, Fresno, CA. pp 179-214.

Clary, C. D. and J. Moso. (unpublished) 1983. Survey of Grape Dehydrators. V.E.R.C., California State University, Fresno, CA.

Clary, C. D. and D. Schmidtgall. 1997. Evaluation of EE-Muls-Oyle for Production of Low Sulfur Golden Seedless Raisins. Viticulture and Enology Research Center, California State University, Fresno, CA 93740-8003.

Clary, C. D. 1996. Use of Liquid Media for Dehydration of Seedless Grapes. CATI Publication 960903. Viticulture and Enology Research Center, California State University, Fresno, CA 93740-8003.

Clary, C. D., and V. E. Petrucci. 1991. Method and Apparatus for Dehydrating Matter. U.S. Patent 5,380,189 10 January.

Cruess, W. V., and A. W. Christie. November, 1921. Some Factors of Dehydrator Efficiency. Calif. Agr. Expt. Sta. Bull. 337.

Ede, A. J., and K. C. Hales. 1948. The Physics of Drying in Heated Air with Particular Reference to Fruit and Vegetables. Dept. of Scientific Research and Industrial Research - Food Investigation Report No. 53. His Majesty's Stationery Office, London.

Eissen, W. and W. Muhlbauer. 1983. Development of Low-Cost Solar Grape Dryers. Proc. on Int. Workshop on Solar and Rural Development. May, 1983. Bordeaux, France.

Erke, K. H., N. K. Dokoozlian. 1982. The Influence of Mold Species and Their Growth Factors on the Deterioration of Raisin Trays and Associated Fruit. IN: Raisin Research, California Raisin Advisory Board, Fresno, CA pp 227-233.

Fisher, C. D. 1958. Outline of California Raisin Advisory Board Research on Removal of Damaged Raisin Berries. IN: Twenty Years of Raisin Research, California Raisin Advisory Board, Fresno, CA pp 473-480.

Jacob, H. E. January, 1944. Factors Influencing the Yield, Composition, and Quality of Raisins. Calif. Agr. Expt. Sta. Bull. 683.

Kasimatis, A. N., E. P. Vilas, F. H. Swanson, and P. Baranek. 1977. Relationship of Soluble Solids and Berry Weight to Airstream Grades of Natural Thompson Seedless Raisins. Amer. J. Enol. and Vitic. 28:1.

Nichols, P. F., and A. W. Christie. October, 1930. Dehydration of Grapes. Calif. Agr. Expt. Sta. Bull. 500.

Nury, F. S., J. E. Brekke, and H. R. Bolin. 1973. Fruits (IN:) Food *Dehydration*. (W. B. Van Arsdel, M. J. Copley, A.I. Morgan, eds.) AVI Publishing, Wesport, CN 2:158-198.

Ponting, J. D., and D. M. McBean. 1970. Temperature and Dipping Effects on Drying Times of Grapes, Prunes and Other Waxy Fruits. Food Technology, Vol 24 No. 12 pp 85-88.

Petrucci, V. E., C. D. Clary, and S. A. McIntyre. 1981. Use of Methyle Oleate and Potassium Carbonate for Production of Golden Seedless Raisins. Bulletin 880302. California Agricultural Technology Institute, California State University, Fresno, CA 93740-8003.

Petrucci, V. E., C. D. Clary, and M. O'Brien. 1983. Grape Harvesting Systems. (IN:) *Principles and Practices for Harvesting and Handling Fruits and Nuts* (M. O'Brien, B. F.

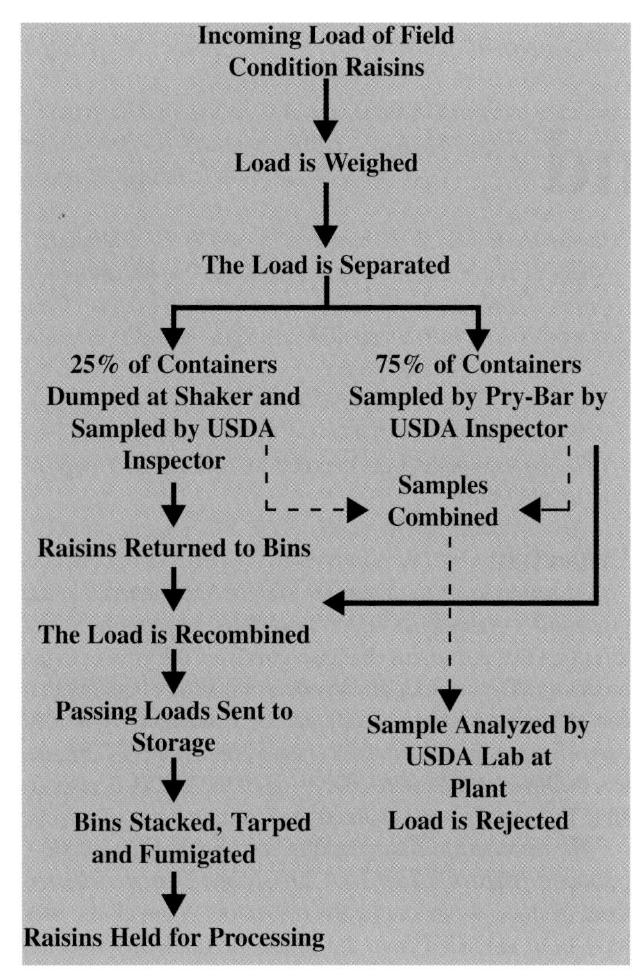

Fig. 9.1: Flow Diagram of Raisins Delivered to the Processing Plant.

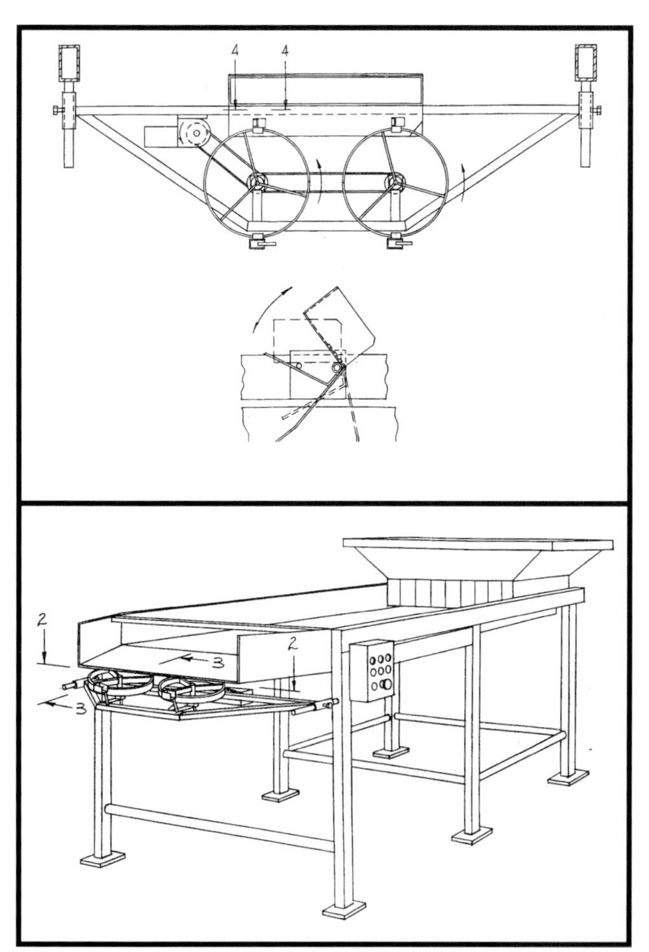

Fig. 9.2: Mechanical sampling device to collect a random sample at the end of the inspection shaker. *Top:* Sampler mounted on shaker. *Bottom:* close-up view of sampler.

depending on the reason for rejection.

The raisin sample collected at incoming inspection is evaluated for moisture content. The raisin sample must be below a moisture content defined by the RAC to be accepted. The maximum moisture content is established by the RAC for each type of raisin (e.g. natural, dipped, etc.) and measured by USDA inspectors. The range of maximum incoming moisture is about 14 to 16% (wet basis). Moisture is critical because it influences the stability of the field condition raisins during extended storage. Raisin moisture content also affects the performance of the raisin processing equipment.

Raisin quality is determined by airstream sorter and any defect identified by the inspector is quantified using specific protocols. Van Diest describes airstream grade and determination of defects in detail in **Chapter 12**. Depending on the season, there may be minimum standards for quality and, in some years, the processor pays premiums for exceptional quality. Although the airstream sorter is often suspect in accuracy, it is a rapid method and gives a very good indication of how the raisins will process, since similar vertical streams of air are used in processing. Incoming inspection in the USDA laboratory at the processing plant is described in detail by Van Diest in **Chapter 12**.

Considerable research has been completed evaluating alternatives to the airstream sorter for determination of the quality of incoming raisins (Gunnerson 1987 and 1988; Clary et al. 1988 and 1989). This research focuses primarily on separation of acceptable and inferior raisins using a liquid specific gravity method. A sample of raisins is deposited into a solution of sugar and water formulated to the sugar content of passing raisins; inferior raisins with sugar content lower than the concentration of sugar in the solution float and the more mature raisins with sugar content greater than the sugar solution sink. Although separation was successful, the range of this method was not as broad as the range obtained using the airstream sorter. Thompson et al. (1991) describe use of a pycnometer and determined it was not useful for measuring quality of natural condition raisins. Huxsoll and Bolin (1991) evaluated grading raisins using Near Infrared Transmittance. This method proved accurate and provides moisture, acidity, sugar content and bulk density. Okamura et al. (1991) tested machine vision to determine quality based on visual features such as wrinkles and shape. Although research indicated potential for some of these alternative methods, the airstream sorter remains the designated method for determining quality of raisins.

Storage

In most years, field condition raisins must have a moisture content below 16% (wet basis) to be accepted at incoming

Fig. 9.3: Sisal Kraft covered storage stacks for field condition raisins.

inspection. This moisture content is determined initially by the inspector by feel and the moisture is verified using a DFA dried fruit moisture tester (AOAC, 1980). At 16% moisture, or less, the field condition raisins can be held in the field bins for an extended time before processing. Industry standards generally limit storage to 12 months. At this moisture content the low water activity of raisins inhibits development of microorganisms mostly because of the high concentration of sugar. However, after extended storage, the raisins can become prone to crystallization of sugars or 'sugaring.' This quality factor is not to be confused with accumulation of sugar during maturation of the fresh grapes and is described later in this chapter.

Storage of field condition raisins consists of stacking the 1,000 lb (450 kg) pallet bins in multiple rows usually six to eight bins high on a well-drained, asphalt or concrete surface, and covering the stack with 8 ft (2.424 km) wide rolls of sisal kraft paper (**Figure 9.3**). The sisal kraft is supported by a framework on the top of the stack to create a pitched roof for rain runoff and the sisal kraft is held in place using wood lath and nails. The edges of adjacent rolls of sisal are folded together and the ends are sealed to the ground using oiled sand. Sisal kraft paper is a lamination of a white polyethylene outer layer, kraft paper, tar paper and nylon strand reinforcement. The material is impervious to moisture and most gases and, in effect, creates a large airtight package for storage of the raisins until they are scheduled for processing.

Although there is some variation in handling inventory, raisins of similar incoming quality and moisture are stored together. The performance of the raisin processing equipment is dependent on moisture content and other physical properties of the field condition raisins. Therefore, it is prudent to maintain a consistent supply of raisins based on airstream grade and moisture.

Fumigation

Field condition raisins are fumigated to prevent infestation. A full description of raisin pests is provided in **Chapter 10**. Raisins stored in the stacks are fumigated with aluminum phosphide, which is available under various commercial labels such as Phostoxin. Aluminum phosphide is applied in tablet form. The number of tablets for a dose is determined based on the amount of raisins in the stack. The dose is placed in the stack by cutting slits at several locations and placing the fumigation material inside the stack. The slits are sealed with tape. Aluminum phosphide reacts with the humidity in the air to form and releases hydrogen phosphide. It is an effective insecticidal fumigant (Farm Chemicals Handbook). Although there are other methods of fumigation, aluminum phosphide is relatively simple to apply because the tablets take some time before they start releasing the gas. Fumigation with aluminum phosphide therefore requires 24 to 48 hours to complete and is repeated as often as every 30 days. When a stack of raisins is scheduled for processing, some of the sisal kraft is removed to vent the stack and ensure there is no residual fumigant present. The sisal is removed, the bins are un-stacked and moved to the processing area.

Methyl bromide is another material that has been used in the fumigation of raisins. This material is stored as a compressed liquid and when released forms a penetrating and highly toxic gas. Methyl bromide is usually applied in special fumigating rooms that are sealed to protect the surrounding area. This material is highly effective and fumigation can be completed in hours. However, special training and handling is required. Generally, methyl bromide fumigation is used after the bins have been removed from the stacks just prior to processing. The fumigation room is loaded with bins full of raisins, fumigated and then vented. Methyl bromide is being reviewed for use as a fumigant and it is likely it will no longer be labeled for this purpose in the near future.

Since both aluminum phosphide and methyl bromide are odorless and highly toxic, standard operating procedures and training are required to handle and apply these materials. The materials are formulated with chloropicrin as an indicator since it is easily detected at low concentration because of its odor.

Storage Containers

A combination of factors in the early 1980s, including surplus raisins being held in storage and rising costs for wood pallet bins, prompted evaluation alternative storage containers for unprocessed raisins. Clary and Thompson (1985) conducted an evaluation of fiber board containers as an immediate low cost replacement for wood bins. The 2 ft (60 cm) fiber board bin was found to be competitive and performed as well as wood in most respects. In 2000, a plastic bin was introduced for use with raisins.

Processing

As of the 1999-2000 season there were 21 raisin processors in California (RAC, 1999). With the exception of one processor, all are located within about a 40-mile (64 km) radius of the city of Fresno, CA. A list of processors and other information about raisins can be found at http://www.raisins.org or in **Appendix B**.

Since raisin packers maintain some degree of privacy in their methods of processing, this text will describe the general sequence of processing raisins based on equipment manufac-

Fig. 9.4a: Raisin fractions from processing: C-Grade Raisins.

Fig. 9.4b: Raisin fractions from processing: Stems and substandard raisins.

Fig. 9.4c: Raisin fractions from processing: Scalped cluster rachis'.

tured by Valley Welding and Machine Works, Fresno, CA. The two major engineering companies that design and manufacture processing lines for the California raisin processors are Valley Welding and Machine Works and Commercial Manufacturing. Paste grinders, packaging machines, case sealers and glue machines are available from Elliott Manufacturing Inc.

Raisin processing consists of a series of mechanical operations to prepare raisins for consumption by removing stems and off-grade raisins (**Figures 9.4a, b** and **c**). Generally, processing systems are divided into dry and wet stages

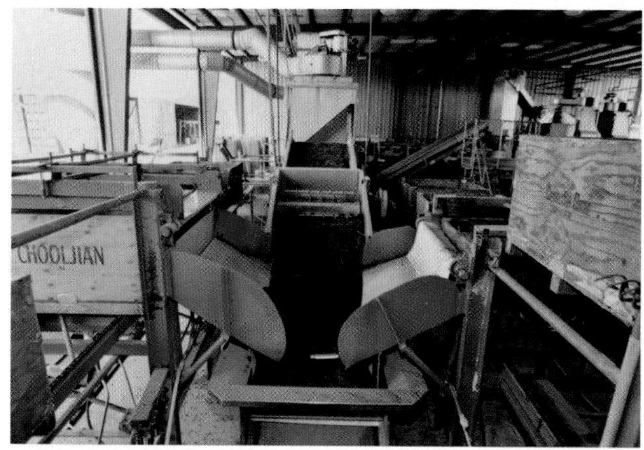

Fig. 9.5: Bin dumpers and metering wheel to control the flow of raisins into the dry line.

Fig. 9.6: Scalping shaker for removal of cluster stems. Note: Raisins attached to the cluster stems are recovered and returned to the process flow.

referred to the dry line and the wet line. In newer dry lines, the incoming bins are dumped directly into a receiving hopper equipped with an incline conveyer. It is important that the raisins have low moisture content so they are dry enough to flow freely. At this low moisture content, the cluster and berry stems are dry and brittle so they can be removed using mechanical agitation. The raisins dumped into the dry line pass under a metering wheel to control the flow of product into the processing system (**Figure 9.5**).

Pre-Scalping

Newer dry line configurations use a scalper immediately after the metering wheel to separate cluster stems from the process flow of single raisins (**Figure 9.6**). The cluster stems have some raisins still attached so they are run through a declustering device or a dry capper. The reason for using this configuration is if the dry capper is used first, it breaks cluster stems into stem fragments about 2 to 4 inches (5 to 10 cm) long that are very difficult to separate from the raisins. The newer configuration leaves the stems mostly intact so they can be separated easily by a scalper.

In some older lines, the dry capper is used to break up the clusters and remove the capstems. The capped raisins continue across a slotted shaker to separate the capstems from the flow, followed by a scalper to remove cluster stems from the

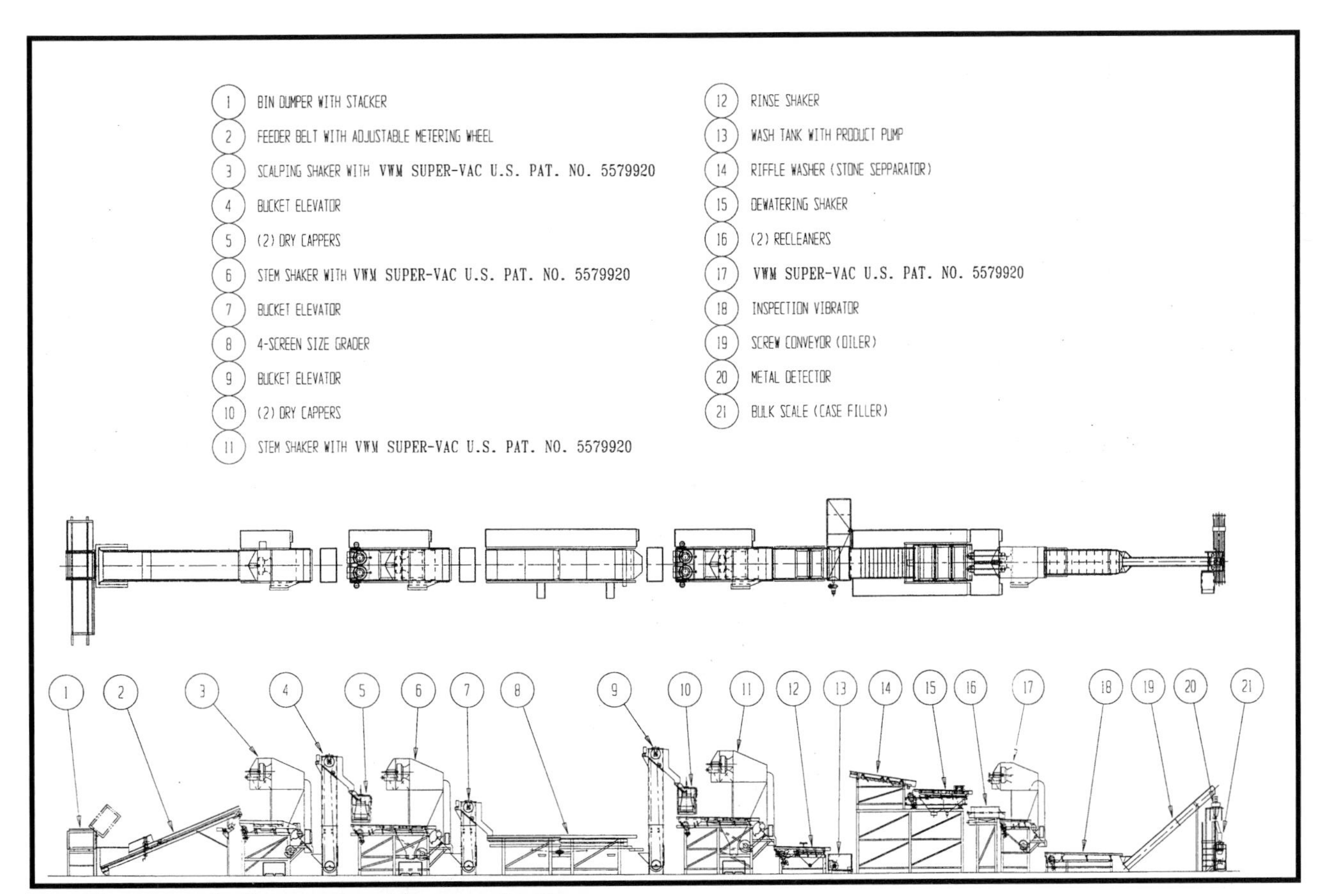

Fig. 9.7a: Schematic diagram of a raisin processing line.

Fig. 9.7b: Photograph of a raisin processing dry line.

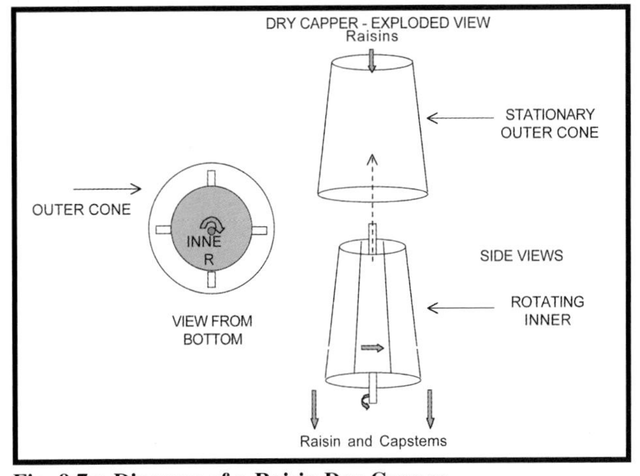

Fig. 9.7c: Diagram of a Raisin Dry Capper.

flow. This configuration is no longer used by California processors.

Dry Capper

The dry capper, also referred to as a capstemer, is the most important component used in raisin processing for two reasons: the dry capper removes most of the capstems so the processed raisins are acceptable for consumption, and the dry capper can be a significant cause of damage to raisins which affects final quality and overall performance of the processing line.

A dry capper consists of a set of vertical rubber impellers rotating inside a vertically oriented cone shaped housing (**Figure 9.7c** and **d**). The interior of the housing is lined with rubber. The raisins are fed into the top of the dry capper housing and they free fall through the dry capper in a spiral pattern induced by the spinning impeller. The raisins are exposed to physical contact with the spinning impellers, the interior of the dry capper housing and other raisins. This contact occurs at high speed and results in the separation of raisins from the cluster stems and separation of most of the capstems from the raisins.

The capstems are removed from the process stream using a slotted screen shaker (**Figure 9.7e**). Larger stems and stem fragments are removed using a vacuum separation system.

Placement of dry cappers in a processing line depends on the age and configuration of the line. Traditional line configurations use a scalper and de-clustering device prior to feeding the dry capper.

Dry cappers are available in two design configurations; the bolted design and a hinged cast outer jacket design. The bolted dry cappers are more difficult to clean and service because they must be disassembled. The hinged cast outer

Fig. 9.7d: Two dry cappers.

Fig. 9.7e: Capstems removed by the dry capper.

Fig. 9.8a: Second and third vacuums in dry line.

Fig. 9.8b: Intake to a vacuum unit. Note the air must pass upward through a eight cm depth of raisins.

Fig. 9.9: Separation area in a SuperVAC. Note the raisins are separated by forced air to improve removal of stems.

jacket design is configured in two halves that separate easily for cleaning and service, as well as provides more accurate paddle clearance specifications. Another improvement is the addition of variable frequency motor drives which electronically control motor RPM in place of adjusting split shives. These devices are useful at the dry cappers as well as other machinery in the processing line.

Vacuum Separation

Traditionally, vacuums have been used to draw air vertically through the flow of raisins (**Figure 9.8**). Light materials including stem fragments, leaves and even some raisins are drawn into the vacuum system and separated as waste using an air leg. The velocity of the air stream dictates the amount of material removed. In a few older lines, a positive pressure air stream is produced using fans instead of vacuums but the principle of separation is the same. Vacuums are adjusted by either adjusting vents on the air leg to increase or decrease suction or use of variable frequency drives to control vacuum fan motor speed. There is also an adjustment at the intake of the vacuum to adjust for the depth of the flow of raisins. The traditional vacuum separation devices are used at several locations in the processing line. Since it is not possible to pull clusters stems in the bottom of the process flow through the flow without removing good raisins, multiple vacuums provide repeated exposure to the vacuums to pull cluster stems and inferior raisins from the process flow of raisins. Generally, the first vacuum is set to remove as many stems as possible and substandard raisins. The second, third and fourth vacuums are set to remove the remainder of the stems and C-grade raisins.

The traditional vacuums are being replaced by a forced

Fig. 9.10: Sizing shaker for separation of midget and select raisins.

Fig. 9.11: Riffle wash for cleaning and rehydration of raisins. Stones are trapped in the riffle wash.

Fig. 9.12: Two recleaners located above a shaker.

Fig. 9.13: Laser sorter for removal of defect raisins.

air, vacuum combination. One configuration, patented by Valley Welding and Machine Works as SuperVAC, uses high velocity air injected into an upward air flow created by a vacuum fan. The raisin process flow feeds into the SuperVAC through an infeed airlock. The raisins and debris fall in front of the injected air which disperses the process flow leaving sufficient space for the upward draft of the vacuum to draw light materials upward into the air leg. Traditional separation takes place through a layer of raisins about 2 to 2.8 inches (5 to 7 cm) deep. The dispersion of raisins in the SuperVAC scatters the layer of raisins into a configuration that appears that the raisins are swarming bees (**Figure 9.9**). This type of separation system virtually eliminates stems being trapped underneath the flow of raisins.

Sizing Shaker

This shaker in **Figure 9.10** separates raisins by size using perforated screens. Although of equal quality, the smaller midget raisins are useful for industrial applications such a baking and breakfast cereals. By virtue of their size, there are more raisins per unit volume. Since raisins for industrial applications are purchased based on the number of raisins per pound, smaller raisins provide a more appealing character when added to food products. The larger raisins move on through the process and the small midget raisins are collected for further processing later.

Washing and Rehydration

The raisin stream leaves the dry processing stage and is discharged into a water tank. The water is used to wash the raisins and during this step the raisins absorb water. The intent is to increase the moisture content from field condition to a moisture content of not more than 18% (wet basis). This softens the texture of the raisins for consumption and also aids in eliminating any raisins that have been damaged by dissolving the sugar in the tissue of the damaged raisins.

The raisins are transported through the water tank using pumps. In 50% of the California lines, augers are used for additional cleaning and moisture control. The water carrying the raisins is pumped to a riffle washer (**Figure 9.11**), consisting a series of gravity fed troughs are used to agitate and mix the raisins and water. Heavier objects such as rocks are trapped in the riffle washer. Residence time in the water is about one to five minutes. The raisins are separated on a de-watering shaker and the water is re-circulated. The raisins are transported by the de-watering shaker to the recleaner.

Recleaner

The device in **Figure 9.12** is similar to the dry capper from the standpoint of mechanically agitating the raisins . However, it is oriented horizontally and uses screens in place of the cone shaped housing. The metal impellers attached to a shaft running through the center of the screen housing are oriented to convey the raisins through the screen housing in a spiral pattern that tumbles the raisins.

Although the recleaner demonstrates the potential of mechanically damaging the raisins of lower moisture content, the softer texture of the raisins and the water used in this step washes the raisins and removes any residual capstems. The recleaner is also effective in centrifuging surface water from the raisins and removing nubins. The raisins discharged from the recleaner pass through another vacuum unit and onto the inspection belt.

Visual and Electronic Separation

Belts for visual inspection convey the processed raisins under bright lights where workers remove any defects. Visual inspection has been augmented and in many cases has been replaced by electronic separation devices that use laser technology to separate based on size, color, presence of capstems and other physical characteristics (**Figure 9.13**). The process flow of raisins is typically 2 to 4 inches (5 to 10 cm) deep. In order for the laser sorter to obtain a spectral image of each individual raisin, the process flow is fed onto a high speed belt. This separates the raisins so each one can be evaluated by the lasers. Based on programming, the laser sorter is capable of detecting and separating raisins with embedded capstems, damage, and other defects.

The processed raisin product is sampled periodically by a USDA inspector and in-plant quality control personnel. Inspection by USDA ensures the processed raisins meet industry standards for quality and defects as determined by the Raisin Administrative Committee. Individual processors may have their own quality specifications and many food companies insist on even higher standards. As in incoming inspection, the USDA inspector randomly samples the processing flow and evaluates the processed raisins for moisture content, quality and defects. Since the processed raisins are not free-flowing, the quality is determined visually in lieu of the airstream sorter. Specific requirements for processed raisins is described in **Chapter 12** and the regulations can be found at http:///www.ams.usda.gov/standards/dried.htm

Optimization of Processing

There are several factors that affect the performance of the raisin processing line. Initial product condition is critical. The quality of the raisins made in the field is controlled by the quality of the fresh fruit. Since most of the water is removed during drying, the remaining constituent is primarily sugar. If the sugar content of the fresh fruit is low, the quality of the raisins made from this fruit will be inferior. Berg (**Chapter 5**) discusses the effects of fresh fruit maturity on raisin quality.

Field drying practices of grapes can affect the level of embedded cluster and capstems (Patterson and Clary, 1986). Removal of embedded stems with conventional processing equipment is very difficult since the stems are not accessible to mechanical removal. The research showed rolling raisin trays too early when the raisins are still soft significantly increased stems and capstems embedded into the tissue of the raisin (**Table 9.1**). Trays are commonly rolled early to equalize moisture in place of turning the trays. This research indicated the importance of turning trays to equalize moisture prior to rolling the trays.

Gibberellic acid is used by some raisin growers to adjust the crop and enhance quality. The practice is timed at bloom and berry set. Research has shown (Patterson and Clary, 1986) that application of gibberellic acid treatments at bloom had no effect on the capstem removal however, vineyards treated with this material at berry set, produced raisins that were significantly more difficult to capstem resulting in a higher capstem count in the processes raisins (**Table 9.2**). This is due to the increased growth of the cluster stem and capstem making the attachment to the grape (and raisin) tougher and more difficult to remove.

The raisins fed into the processing line must be as uniform as possible from the standpoint of moisture content and relative quality from one lot to the next. Any changes in moisture content and/or quality during continuous processing will change the way the processing equipment performs and may require adjustments to the equipment.

The feed rate of raisins is important particularly at the dry capper since this device will become more aggressive in removing capstems as the feed rate increases. Variation in feed rate results in either poor capstem removal or damage to

Table 9.1
Effect of Rolling Raisin Trays on Embedded Capstems and Embedded Cluster Stems, 1986.

	Embedded Capstems (#/lb.)	Embedded Cluster Stems (#/lb.)
Trays rolled at 25% moisture	49.0b	6.0c
Trays rolled at 20% moisture	11.0a	2.0b
Trays rolled at 15% moisture	8.0a	1.0a

Numbers followed by the same letter within columns are not significantly different – Duncan's NMR Test.

Table 9.2
Effect of Application of Gibberellic Acid on Raisin Grapes – 1986.
Incoming Airstream

Treatment	Sub-Standard (%)	B or Better (%)	Processed Capstems (#/lb)	Processed Capstems % by wt
Vineyard #1				
5 ppm thinning	2.7a	70.2 bc	6.1	0.6
10 ppm thinning	2.7a	73.2 c	8.9	0.9
10 ppm sizing	6.7 b	61.4a	13.8	1.3
Untreated Control	5.0ab	62.8ab	8.6	0.8
Significance of F	.01	.01	ns	ns
Vineyard #2				
5 ppm thinning	0.4	92.2	5.9a	0.7a
10 ppm thinning	0.3	91.0	2.7a	0.3a
10 ppm sizing	1.4	84.5	21.2 b	2.5 b
Untreated control	1.0	87.0	3.7a	0.4a
Significance of F	ns	ns	.01	.01

Numbers followed by the same letter within columns are not significantly different - Duncan's NMR Test

Fig. 9.14: Bulk filling line with weigh feeder.

Fig. 9.15: External sugar crystallization – Note crystallized sugar on exterior of raisins.

the raisins from increased feed rate. Aggressive capstemming results in chipping or 'barking' of the surface of the raisins. The fragments or chips of raisin tissue accumulate on the interior of the dry capper and some of this material will travel with the flow of raisins through the rest of the line leaving a sticky film of sugar. Barking also results in a very sticky and unstable raisin product. Severe barking causes seepage of sugar from the raisin and contributes to external crystallization of sugars.

Dry capper adjustment is accomplished in three ways: the speed of rotation of the internal impeller, the proximity of the outer cone to the internal impeller, and change in feed rate. A higher rate of rotation, closer proximity of the outer cone and the internal impeller and/or increased feed rate increases the aggressiveness of the dry capper to the point that barking will occur. Checking the dry capper regularly is very important when maintaining optimum performance of the processing line. Dry cappers in most newer processing lines are adjusted mostly by changing the speed of rotation of the internal impeller.

The quality of the incoming raisins is the most significant factor in the yield of the processing line. Generally, yield of processed product is about 70% of the incoming weight. The rest of the incoming load is discarded as inferior quality. About 10 to 15% of the incoming load is discarded as stems and chaff. Typically, this loss is called shrinkage. At any time during processing, the equipment holds about 20% of the incoming raisins as resident volume.

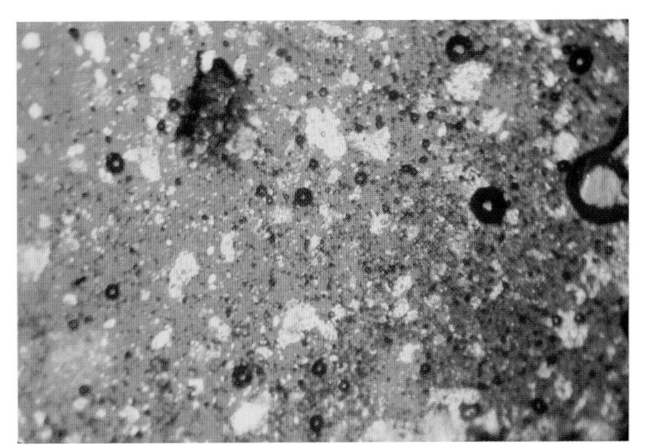

Fig. 9.16: Internal sugar crystallization – Note yellow sugar crystalline structure in cross-section.

It is very important to understand that the mechanical stages of a raisin processing line are effective in removing most of the raisins and debris present in field condition raisins, but not to a level of zero tolerance. Because of this, USDA regulations have permitted tolerances that provide for maximum levels of capstems, stems and other debris in processed raisins. With the increased demand for raisins for the food processing trade, zero tolerance has become a new requirement. Development and use of laser sorters at the end of the processing line have provided the opportunity to remove the remaining debris and to sort out raisins with undesirable characteristics such as color or the presence of embedded capstems. However, the laser sorter will also effectively remove raisins that have been damaged by the dry capper, recleaner or any other components in the processing line. This has the potential of decreasing the yield of the processing line.

Packaging and Shelf Life

Raisins are packaged in containers ranging from 1,000 lb (450 kg) plastic lined pallet bins to 0.5 oz (14 g) pouches. Historically, raisins have been considered extremely stable and therefore did not require packaging techniques related to maintaining moisture for soft texture. Today, raisin packaging is utilizing more advanced packaging materials to maintain quality and preserve moisture content. Quality, as specified by industrial users and the consumer includes, good maturity, minimal defects and good storage stability and shelf life.

An important quality factor related to the shelf life of packaged raisins is the crystallization of sugars present in the tissue of raisins. The highly concentrated sugar solution in raisins has the potential of crystallizing. This is similar to crystallization of unprocessed honey and is referred to as sugaring. External sugaring is the formation of a white to brown sugar crystals on the surface of the raisin (**Figure 9.15**). When sugaring occurs within the tissue of the raisins, the raisins take on a gritty texture (**Figure 9.16**).

Zoecklein et al. (1981) describes factors affecting the

formation of sugar crystals in raisins after storage. Storage method is the most significant factor in sugaring. Processed raisins stored in ambient conditions for extended periods tend to sugar more quickly than raisins stored in controlled temperature environments. Daily variation in temperature and higher moisture content of processed raisins contribute to sugaring. This explains the importance of holding raisins in unprocessed condition until needed.

Processed raisins have a moisture content as high as 18% (wet basis) and are shipped worldwide. Shipment to the Pacific Rim or through the Panama Canal to Europe exposes the processed raisins to high temperature. This factor in conjunction with high moisture content promotes formation of sugar crystals. The result is a sandy or gritty texture and in the case of external sugaring, groups of crystals form on the exterior of the raisins.

The by-products of processing including substandard raisins are often used to produce raisin syrup and paste for use as flavoring in food products. Processed raisins are utilized in two basic segments of the food supply. Raisins are processed and packaged as a single product for consumption and raisins are processed and sold to industrial users as an ingredient in foods.

Photo Credits

Figures 3, 5, 6, 7b, 7d: Chooljian Bros. Packing Company, Del Rey, CA

Figure 7a: Valley Welding and Machine Works, Fresno, CA

References

AOAC Methods. 1980. 220.13 Moisture in Dried Fruits; Official Final Action. pp 361-362.

Clary, C. D. and G. A. Sawyer Ostrom. 1992. Evaluation of a Mechanical Sampler for Raisin Quality Determination; Commercial Installation. 90RAC1. IN: Raisin Research Reports, 1992. California Raisin Advisory Board, Fresno, CA. pp 181-195.

Clary, C. D., G. A. Sawyer Ostrom, E. H. Bowerman and D. J. Pecchenino. 1988. On Site Testing of a Liquid Specific Gravity Device for Quality Determination of Raisins. 88RAB8. IN: Raisin Research, 1985. California Raisin Advisory Board, Fresno, CA. pp 156-170.

Clary, C. D., G. A. Sawyer Ostrom, E. H. Bowerman. 1988. Comparison of Liquid Specific Gravity, Airstream Sorter, and Sight Grading for Determination of Raisin Quality. 88RAB8. IN: Raisin Research, 1985. California Raisin Advisory Board, Fresno, CA. pp 171-178.

Clary, C. D. and J. F. Thompson. 1985. Alternative Containers for Storage and Handling of Unprocessed Raisins. 84RAB24. IN: Raisin Research, 1985. California Raisin Advisory Board, Fresno, CA. pp 212-226.

Gunnerson, R. E. 1987. A Study of Specific Gravity Methodology to Determine the Maturity Grade Level and Substandard Content of Raisin Lot Samples. 87RAB9 IN: Raisin Research, 1987. California Raisin Advisory Board, Fresno, CA. pp 174-181.

Gunnerson, R. E. 1988. Airstream Sorter and LSG Comparison of Maturity Grading Varieties Other than Natural Thompson Seedless. 87RAB9B. IN: Raisin Research, 1985. California Raisin Advisory Board, Fresno, CA. pp 103-143.

Huxsol, C. C. and H. R. Bolin. 1991. Maturity Grading Raisins by Near Infrared Transmittance (NIT). 90RAC3. IN: Raisin Research, 1986. California Raisin Advisory Board, Fresno, CA. pp 183-189.

Okamura, N. K, M. J. Delwiche and J. F Thompson. 1991. Raisin Grading by Machine Vision. 90RAC2. IN: Raisin Research, 1991. California Raisin Advisory Board, Fresno, CA. pp 190-221.

Patterson, W. K. and C. D. Clary. 1986. Effect of Fruit Moisture Content, Sugar Content, Temperature, and Rolling Pressure on Reducing Embedded Capstems and Cluster Stem Fragments. 86RAB4. IN: Raisin Research, 1986. California Raisin Advisory Board, Fresno, CA. pp 174-181.

Patterson, W. K. and C. D. Clary. 1986. The Effects of Gibberellic Acid on Capstemming Raisins. 86RAB2. IN: Raisin Research. California Raisin Advisory Board, Fresno, CA. pp 68-71.

RAC. 1999. Marketing Policy, 1997-98 Marketing Season. Raisin Administrative Committee. 3445 N First St, Fresno, CA 93726; http://www.raisins.org.

Thompson, J. F., M. J. Delwiche and N. R. Raubach. 1991. Raisin Grading Using a Pressure Difference Specific Gravity Technique. 91RAC1. IN: Raisin Research, 1986. California Raisin Advisory Board, Fresno, CA. pp 178-182.

Zoecklein, B. W., S. A. McIntyre, V. E. Petrucci, T. Freeto, H. Bolin and C. D. Clary. 1981. Factors Affecting the Formation of Degradative Crystals in Thompson Seedless Raisins During Prolonged Storage. IN: Raisin Research, Late 1960s-1980, Vol. III. California Raisin Advisory Board. Fresno, CA. pp 1,265-99.

CHAPTER 10.

Insects Affecting Raisins and Their Management

Mark A. Mayse, Ph.D., Z. Jo Harper, B.S., Michael J. Costello, Ph.D., and Kip R. Green, M.S.

Introduction

Goals of Chapter

The primary goal of this chapter is to help the reader gain a better understanding of the economically important arthropods associated with raisin production systems. Herbivorous insects and mites may reach population densities, which cause them to be considered serious pests. On the other hand, a large number of insects and arachnids (e.g., spiders and mites) are recognized for their beneficial activities in vineyards in their ecological roles as predators and parasitoids (i.e., insects which parasitize other insects). Furthermore, those arthropods, which may be detrimental in raisin agro-ecosystems, include species, which attack various parts of the vine (especially fruit clusters) during the growing season, along with those pests primarily attacking raisins in post-harvest/storage situations. Two simple dichotomous keys are included in this chapter which are designed for easy and reliable identification of the larvae (typical damaging stages) of insect pests likely to be encountered in association with raisins, both pre- and post-harvest.

Plant Health Perspective & IPM Philosophy

Integrated Pest Management (IPM) can be viewed as a philosophy which provides farmers and pest control advisors with a rational framework for decision-making regarding pest situations. Rabb (1972) succinctly defined IPM as "Intelligent selection and use of pest-control actions which will ensure favorable economic, ecological, and sociological consequences." Key strategies historically embodied in such pest-control actions include the areas of cultural control, biological control, and chemical control. More recent perspectives in this area have begun to emphasize the value of expanding traditional IPM concepts to the level of agro-ecosystem management. By broadening the conceptual realm of IPM to an agro-ecosystem level, we are thus better able to focus appropriate attention on the subtleties of the growth and development patterns of the actual crop or commodity to be protected from the pests, and also on the myriad interrelationships among non-pest herbivores, generalist predators, parasitoids, and pathogenic organisms as they relate to the basic crop production system.

These new perspectives on IPM have begun to take shape as essentially a "Plant Health" approach to agricultural problem solving. The major emphasis shifts to more clearly understanding and utilizing those factors (biotic and abiotic) which promote the fundamental ability of the crop plant to persist and even thrive in the context of pest populations rather than merely settling for the overly-simplistic goal of reducing pest population densities and overlooking the more subtle aspects of crop/pest/natural enemy interactions. Another way to describe the Plant Health perspective would be seeking to delineate root causes of pest population outbreaks rather than being content with only treating symptoms (i.e. just killing pests) resulting from agro-ecosystem imbalances and shortcomings.

Accurate and well-documented pest population densities which serve as benchmarks for IPM decision-making as to if and when specific tactics should be utilized against the pest to prevent the occurrence of economic damage are essential. The primary benefit of such economic threshold (ET) values of course accrues when the farmer, pest control advisor, or other IPM practitioner uses proper sampling techniques for a given target species and develops a current pest population density estimate which is then compared to the ET benchmark. It is rather difficult to imagine effective and ecologically sound integrated pest management being accomplished without one's embracing this fundamental principle of knowing the level of pest pressure before taking any problem-solving actions.

Part I: Insects Affecting Raisins

Mark A. Mayse, Ph.D. and Kip R. Green, M.S.

Arthopods of Economic Importance in Raisin Production Systems

Arthropod pests in raisin production systems include species, which exert damage in the vineyard during the growing season (i.e. pre-harvest), as well as species that cause problems with raisins during the drying and storage periods (i.e. post-harvest). In addition, pre-harvest pests may either impact fruit directly (i.e. direct pests) or they may feed primarily on leaf tissue or sap and damage the fruit only in an indirect manner (i.e. indirect pests).

Fig. 10.1: Western grape Leafhopper adult.

Fig. 10.2: Variegated Leafhopper adult.

Fig. 10.3: Western Grape Leafhopper nymph.

Fig. 10.4: Variegated Leafhopper nymph.

Pre-Harvest Raisin Arthropod Pests

Leafhoppers (Order Homoptera: Family Cicadellidae) belonging to the genus *Erythroneura* [including *E. elegantula* Osborn, western grape leafhopper (WGLH) and *E. variabilis* Beamer, variegated leafhopper (VLH)] are very commonly found in California raisin vineyards. Adults of these congeneric species are only about one-eighth inch (3 mm) long, with WGLH appearing pale yellow with reddish and dark brown markings (**Figure 10.1**), and VLH are somewhat darker and showing a mottled pattern of brown, green, red, white and yellow (**Figure 10.2**). WGLH nymphs are generally cream-colored (**Figure 10.3**), while VLH nymphs are reddish brown on the side margins with pale yellow more prominent across the dorsal surface (**Figure 10.4**).

Erythroneura leafhoppers can be serious pests in raisin vineyards in several ways. With their stubby piercing-sucking beaks, these mesophyll tissue-feeding pests cause yellow stippled areas in leaves, which reduces the vines' effective photosynthetic area. When sufficiently prevalent, heavy leaf stippling can cause premature leaf drop. Extensive loss of the leaf canopy of course leads to sun burning of developing fruit clusters.

Erythroneura leafhopper nymphs and adults also produce copious amounts of honeydew, a high-sugar waste product that collects on fruit and leaf surfaces and may accumulate dirt or even provide substrate for growth of unsightly black sooty mold. In addition, high densities of *Erythroneura* leafhopper adults (i.e. fliers) may cause severe irritation and annoyance of field workers in vineyards by their presence, particularly during harvest.

A potentially significant new threat to California grape growers is a relatively unusual leafhopper from the sharpshooter group (Tribe Proconiini) known as the glassy-winged sharpshooter (GWSS) (*Homalodisca coagulata* Say) which has recently moved northward from southern California and has now been found in many California counties. Currently, grape growers in the Temecula region are experiencing extremely significant levels of damage and even vineyard destruction due to GWSS activity. GWSS is an extremely large leafhopper, typically reaching one-half inch (12 mm) in length. Adults are very strong fliers, and are readily able to disperse throughout major portions of even large vineyard acreages. The primary threat posed by this new leafhopper species results from its role as an effective vector of *Xylella fastidiosa*, the bacterial agent that causes Pierce's disease in grapes. **Part II** of this chapter provides a detailed description of sharpshooters relating to Pierce's disease.

Until recently, the only serious mealybug pest throughout most of California's grape vineyards has been the grape mealybug *Pseudococcus maritimus* (Ehrhorn) (Homoptera: Pseudococcidae) (**Figure 10.5**, the specimen of the grape mealybug was provided by the Joe and Terrie Antonino vineyard in Fresno, CA). This species has been much more of a problem in table grape vineyards than in raisin and wine grape production systems, perhaps relating to the historically heavy insecticide treatments applied in table grape operations to maintain fruit clean of spotting by *Erythroneura* leafhop-

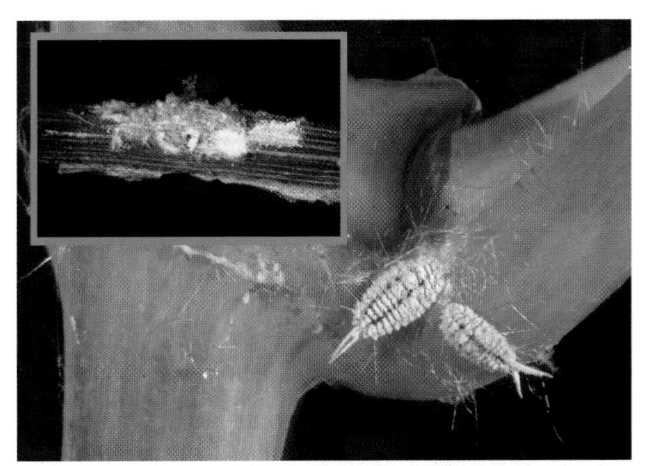

Fig. 10.5: Grape mealybug at base of green petiole. Insert: grape mealybug egg mass.

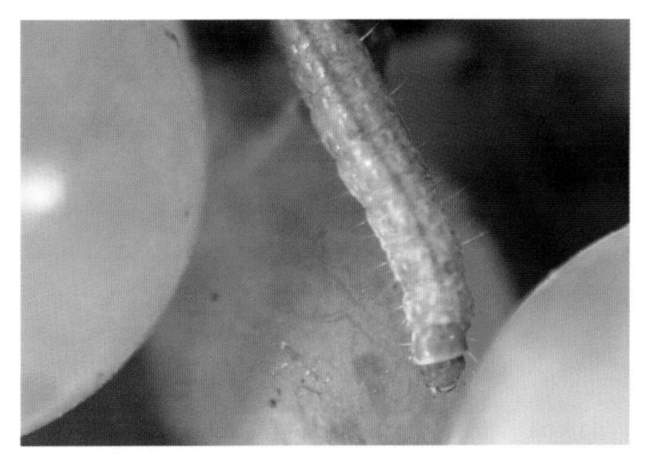

Fig. 10.6: Omnivorous Leafroller larvae.

pers (Flaherty et al., 1992a).

Now the vine mealybug *Planococcus ficus* (Signoret), a new grape pest in the Central Valley of California, has begun to cause major concern even for raisin farmers. Some industry experts suggest that the vine mealybug may be one of the most potentially damaging pests to show up since omnivorous leafroller. Unlike the more familiar grape mealybug, the vine mealybug can transmit grape viruses as it infests leaves, bunches, canes, and roots. Although the presence of ants may indicate infestation by vine or grape mealybug, copious excretion of honeydew which typically resembles candle wax more likely indicates establishment of vine mealybug. Authorities agree that vine mealybug is much easier to exclude from a vineyard than to eradicate.

The two major species of herbivorous spider mites (Class Arachnida: Order Acari) found in raisin vineyards include the Pacific spider mite, *Tetranychus pacificus* McGregor (Acari: Tetranychidae), which can be a major economic pest, along with the Willamette spider mite, *Eotetranychus willamettei* (McGregor), which is generally regarded as a much less significant pest species. University of California scientists suggest that the use of synthetic organic insecticides has upset natural controls of these mites in some situations, even though the reasons have not yet been fully documented (Flaherty et al., 1992b).

Tetranychid mite populations in California are strongly influenced by environmental conditions and soil types (Flaherty et al., 1992b). Light-textured soils (prevalent in the western San Joaquin Valley) typically contribute to hot, dry and dusty vineyard conditions which are favorable for Pacific mite outbreaks. The heavier soils of vineyards located on the east side of the San Joaquin Valley produce more humid and less dusty conditions and, thus, the Pacific mite is generally less favored.

Recent studies indicate that cultural factors such as amount and timing of sulfur play a role in the Pacific mite seasonal abundance, possibly by affecting grapevine chemistry, which translates to significant population impacts via mite reproductive rate and longevity (Costello et al., 2000).

Although feeding damage by the Pacific mite is considered a relatively serious threat to grape farmers, the presence of the Willamette mite in a vineyard may, in fact, have unexpected benefits. Studies by Karban and co-workers (English-Loeb, 1988; Karban and English-Loeb, 1990; Karban et al., 1991) clearly suggest that in many vineyard situations, the Willamette mite feeding actually induces individual grapevines to effectively rally their chemical defenses, which then better prepares those vines to withstand the onslaught of the more serious Pacific mite later in the season. This phenomenon of "induced resistance" has been sufficiently well-documented to prompt the suggestion that some grape farmers might even consider inoculating their vines with Willamette mites early in the season to better prepare for subsequent battles commonly ensuing with the Pacific mite.

Throughout major parts of California's grape-growing regions, one or more of the following caterpillar pests are likely to pose a significant threat of crop damage. Larvae of the omnivorous leafroller (OLR) (*Platynota stultana* Walsingham) (**Figure 10.6**) and the orange tortrix (OT) [*Argyrotaenia citrana* (Fernald), both Lepidoptera: Tortricidae] are commonly found damaging grape clusters by their feeding activities. Historically, OLR has been the primary bunch-damaging lepidopteran in the Central Valley, while OT has filled the same ecological niche in central and north coast vineyards. These pests typically spin protective webbing in and around grape clusters, which is helpful to farmers checking for their presence.

A major secondary problem following OLR and OT feeding involves infection by various bacteria and fungi at the feeding sites as berries progressively ripen. Establishment of these opportunistic microbes produces rotting and fermentation of ripening berries which then become very attractive to such secondary insect pests as vinegar flies and dried fruit beetles (see Post -Harvest Pests below); foliar injury by OLR and OT is generally considered minor (Coviello et al., 1992). The grape leaffolder (GLF) [*Desmia funeralis* (Hubner), Lepidoptera: Pyralidae] is another caterpillar pest which may damage grapes, although GLF infestations vary dramatically from year to year (Jensen et al., 1992a). Contrary to its common name, the GLF larva (**Figure 10.7**) actually rolls the leaves (**Figure 10.7 insert**), reducing photosynthetic capacity of the grapevine but also providing a highly visible aid in

Fig. 10.7: Grape Leaffolder larvae. Insert: Leaf rolled by GLF.

monitoring the pest's densities. Although GLF larvae may feed on fruit at extremely high population levels, economic damage from this pest is generally restricted to only massive, late-season infestations (Jensen et al., 1992a).

For the past 25 years or so, the western grapeleaf skeletonizer (WGLS) *Harrisina brillians* Barnes & McDunnough (Lepidoptera: Zygaenidae) (**Figure 10.8a-d**) has been a frequently encountered Central Valley grape pest. WGLS larvae can be voracious leaf feeders, with heavy second and third seasonal generations at times defoliating entire vineyards (**Figure 10.9**). In addition to leaf loss and occasional fruit damage and resultant increases in bunch rot, WGLS larvae possess long, dark hairs in their fourth and fifth instars, which can be very irritating to field workers whose skin comes in contact with the larvae.

Somewhat curiously, WGLS populations in California have dramatically plummeted during the past four or five years. Although a number of factors undoubtedly account for this fortuitous decrease in WGLS pressure, the growing prevalence and widespread distribution of a particularly virulent strain of granulosis virus is thought by many to have played a major role in reducing WGLS populations.

Fig. 10.8a (above): Western Grapeleaf Skeletonizer larvae colony feeding on grapeleaf.

Fig. 10.8b (left): Western Grapeleaf Skeletonizer adult.

Fig. 10.8 c and d: Western Grapeleaf Skeletonizer larvae feeding on fruit (left) and Western Grapeleaf Skeletonizer egg mass on grapeleaf (right).

Key for Identifying Larvae Commonly Found in Grape Bunches

1. Larva with up to five pairs of fleshy prolegs on abdomen (2). Larva lacks paired prolegs on abdomen (6).
2. Distinct, whitish, slightly convex, and oval tubercles (i.e. pinaculae: enlarged seta-bearing papillae forming flat plates) on upper sides of abdominal segments of larva. Omnivorous leafroller. Larval appearance other than above (3).
3. Pinaculae on abdominal segments are round and not prominent; larva lacks whitish tubercles (in contrast to OLR) orange tortrix. Larval appearance other than above (4).
4. Third instar larva has small black spot on each side of body above second pair of legs), and two black spots near end of abdomen. [First two larval instars lack distinctive markings allowing accurate field identification.] Grape leaffolder. Larval appearance other than above (5).
5. Four to six rows of purple spots along back produce a generally pinkish striped appearance in larva - raisin moth. Crescent-shaped markings above second pair of true legs on each side of larva. - Navel orangeworm.
6. Larva is yellowish or whitish and about one-quarter inch (7 mm) long with a brown head and two large and two small brown tubercles on end of abdomen - dried fruit beetle. Larva is white maggot about 1.4 inch (7 mm) long, lacking a head capsule but with black mouthparts in narrow head region - vinegar fly.

Natural Enemies of Raisin Pests in the Vineyard

A diverse group of generalist insect predators are known to feed on *Erythroneura* leafhopper adults and nymphs in California vineyards. Included among these important natural enemies are *Chrysoperla* spp. (green lacewings) (Neuroptera: Chrysopidae), *Orius* spp. (minute pirate bugs) (Hemiptera: Anthocoridae), *Nabis* spp. (damsel bugs) (Hemiptera: Nabidae), *Hippodamia convergens* Guerin-Meneville (convergent lady beetle) (Coleoptera: Coccinellidae), *Geocoris* spp. (big-eyed bugs) (Hemiptera: Lygaeidae), and *Hemerobius* spp. (brown lacewings) (Neuroptera: Hemerobiidae). In addition, a tiger fly (genus *Coenosia* [Diptera: Muscidae]) is common in San Joaquin Valley vineyards. Adult tiger flies

Fig. 10.9: Vineyard defoliated by Western Grapeleaf Skeletonizer.

Fig. 10.10: Agrarian Sac Spider.

prey upon adult leafhoppers.

Spiders (class Arachnida: Order Araneae) represent an often overlooked group of beneficial arthropods in vineyard agro-ecosystems. Almost exclusively carnivorous, the great diversity of spider species typically inhabiting California vineyards appear to be of considerable significance in reducing populations of certain grape pests. Numerous observations suggest that various spider species (most notably the agrarian sac spider *Cheiracanthium inclusum*, **Figure 10.10**) may influence omnivorous leafroller populations by feeding on larvae in fruit clusters. Laboratory feeding trials have demonstrated that commonly occurring vineyard spiders may consume substantial numbers of *Erythroneura* leafhoppers.

Costello and co-workers (1995) recently developed an outstanding color illustrated field guide to the spiders in San Joaquin Valley grape vineyards. In this four-page field-ready resource, the authors grouped major spider species according to where they are most commonly found in the vineyard: 1) in the grapevine canopy, 2) on the vineyard floor (including ground vegetation), or 3) in both the canopy and on the ground. Besides the excellent color photographs of vineyard spiders, the field guide includes key features allowing accurate identification of species along with notes of significant aspects of their biology and ecological roles in vineyards.

It is generally recognized that the most important predatory mite of Pacific and Willamette mite in California vineyards is the western predatory mite *Metaseiulus* (= *Galandromus*) *occidentalis* (Nesbitt) (Acari: Phytoseiidae). Studies in San Joaquin Valley vineyards indicate that at times *Euseius* nr. *Hibisci* may replace *M. occidentalis* as the dominant predatory mite and thus effectively control the Willamette mite. Another predatory mite, *Amblyseius californicus* (McGregor), has been observed in association with the Willamette mite in some Salinas Valley vineyards (Flaherty et al., 1992b).

Two species of thrips are becoming increasingly recognized as effective natural enemies against herbivorous mites in grape vineyards. Sixspotted thrips, *Scolothrips sexmaculatus* (Pergande) (Thysanoptera: thripidae), although somewhat erratic as predators of spider mites, are known to aid predatory mites through their rasping-sucking feeding activities (Flaherty et al., 1992b). It has been observed that when large numbers of sixspotted thrips are found within an infestation of Pacific spider mite, acaricide applications may not be necessary. Furthermore, results of both vineyard and laboratory studies indicate that the western flower thrips, *Frankliniella occidentalis* (Pergande) (Thysanoptera: Thripidae), preys upon Pacific mite eggs, and that under certain conditions herbivorous mite population densities may be affected (Flaherty et al., 1992b).

Numerous studies clearly indicate that the most important natural enemy of *Erythroneura* leafhoppers in vineyards is the *Anagrus* wasp (Hymenoptera: Mymaridae). Formerly known as *Anagrus epos*, it is now recognized as a complex of at least three different species in the *Anagrus* genus (Triapitsyn and Chiappini, 1994; Triapitsyn, 1995 and 1998). *Anagrus* spp. wasps, which parasitize the tiny eggs of the one-eighth inch (3 mm) leafhoppers, are among the smallest insects one is likely to encounter. A relatively short life cycle leading to nine to 10 generations during a growing season makes *Anagrus* capable of parasitizing 90% or more of all western grape leafhopper eggs deposited after July (Wilson et al., 1992a). Studies further reveal that *Anagrus* biotypes from California, other states and Mexico differ in their ability to locate and parasitize the two species of *Erythroneura* leafhoppers (i.e. WGLH AND VLH) (Wilson et al., 1992b).

A much larger parasitic wasp, *Aphelopus albopictus* Ashmead (Hymenoptera: Dryinidae), attacks *Erythroneura* leafhopper nymphs of the third, fourth, and possibly fifth instar. Parasitism rates (measured as parasitized adults) up to 77% have been reported and adult leafhopper mortality has averaged 10 to 40% (Wilson et al., 1992a).

Researchers have found that parasitic wasps attacking larvae of the omnivorous leafroller seldom account for more than 10% mortality, even when OLR populations are large and damaging. According to Coviello et al. (1992), the most common OLR parasitoids include the following:

Goniozus platynotae Ashmead (Hymenoptera: Bethylidae)

Tricholgramma spp. (Hymenoptera: Trichogrammatidae)

Apanteles spp. (Hymenoptera: Braconidae)

Microgaster phthorimaeae Muesbeck (Hymenoptera: Braconidae)

Macrocentrus ancylivorus Rohwer (Hymenoptera: Bra-

Fig. 10.11: Dried Fruit Beetle - *Top:* Larvae, *Middle:* Pupae, *Insert:* Adult feeding on Raisin, *Bottom:* Adult.

Fig. 10.12: Sawtoothed Grain Beetle - *Top*: Larvae, *Middle*: Pupae, *Bottom*: Adult.

conidae)

Cremastus platynotae Cushman (Hymenoptera: Ichneumonidae)

Elachertus proteoteratis (Howard) (Hymenoptera: Eulophidae)

Spilochalcis spp. (Hymenoptera: Chalcididae)

Erynnia tortricis (Coquillett) (diptera: Tachinidae)

Nemorilla pyste (Walker) (Diptera: Tachinidae).

A recent study found that activity of tachinids, braconids and ichneumonids accounted for roughly 50% parasitism of OLR in a Madera County vineyard (Mayse et al., 1998).

In Central Coast vineyards, the dominant parasitic wasp attacking the orange tortrix is *Exochus nigripalpus subobscurus* Townes (Hymenoptera: Ichneumonidae), accounting for 95% of parasites found (Bettiga and Phillips, 1992). Three additional internal parasitoids of OT larvae include the following: *Apanteles aristoteliae* Viereck (Hymenoptera: Braconidae) (2%), *Nemorilla pyste* Walker (Diptera: Tacinidae) (2%), and *Dibrachys cavus* Walker (Hymenoptera: Chalcidae) (1%) (Bettiga and Phillips, 1992).

As suggested earlier in the discussion of why the western grapeleaf skeletonizer may be dramatically less abundant than in recent years, the most effective biological control agent of WGLS is a granulosis virus which causes mortality of eggs, larvae and pupae. Disease-carrying adult moths typically transmit the virus from one generation to the next, and infected males can transmit the virus to healthy females during mating (Stern et al., 1992).

Studies show that the most abundant natural enemy attacking the grape leaffolder is the external larval parasitoid, *Bracon cushmani* (Muesbeck) (Hymenoptera: braconidae) (Jensen et al., 1992). This tiny parasitic wasp (roughly the size of a gnat but with a relatively fat abdomen) generally reproduces on GLF larvae of third instar and older. Parasitism rates of 30 to 40% and even greater are frequently encountered, and thus the resultant reduction in size of the second and third GLF broods tend to stabilize the pest's population (Jensen et al., 1992).

Post-Harvest Raisin Insect Pests

The primary pests of raisins in post-harvest and storage situations include two types of beetle and two species of caterpillars. Beetle pests (Order Coleoptera) include the dried fruit beetle (DFB) *Carpophilus hemipterus* (L.) (Family Nitidulidae) (**Figure 10.11**) and the sawtoothed grain beetle (STGB) *Oryzaephilus surinamensis* (L.) (Family Cucujidae)

Fig. 10.13: Indian Meal Moth - *Top Left:* Larvae, *Top Right:* Pupae, *Bottom:* Adult.

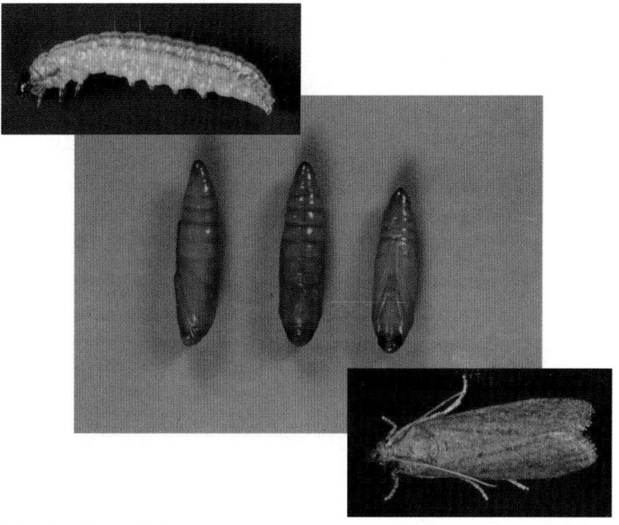

Fig. 10.14: Raisin Moth - *Top:* Larvae, *Middle:* Pupae, *Bottom:* Adult.

(**Figure 10.12**), while the two lepidopteran pests (both in the Family Pyralidae) include the Indianmeal moth (IMM) *Plodia interpunctella* (Hubner) (**Figure 10.13**) and the raisin moth (RM) *Cadra figulilella* (Gregson) (**Figure 10.14**). DFB adults and larvae feed directly on raisins and reduce the quality of dried fruit with their deposits of excreta and cast skins (Lindegren et al., 1992a). STGB adults and larvae feed with equal gusto in the deep folds and ridges of raisins, leaving excreta as tiny yellowish pellets.

Damage by the two primary species of lepidopteran pests of raisins (i.e., Indian meal moth and raisin moth) results mostly from their contamination of dried fruit with excrement, cast skins, webbing, cocoons, and living or dead larvae (Lindegren et al., 1992a). Research conducted by scientists at the USDA laboratory in Fresno, California, showed that during its development, a single RM larva can damage about 20 Thompson Seedless or nine Muscat raisins.

Stored product raisin pests can best be controlled in the packinghouse through use of sanitary cultural practices, along with fumigation when necessary (Lindegren et al., 1992a). Each of several processing steps which raisins typically undergo between removal from storage and packaging for sale progressively reduces infestation by these stored product pests. For example, raisins with significant insect-feeding damage are lighter in weight than undamaged fruit and thus tend to be automatically blown out during normal dried fruit processing. USDA scientists recommend that farmers who cover and make sweatbox stacks or bins on the ranch should implement fumigation programs (Lindegren et al., 1992a). It is further suggested that raisins should be first fumigated when the stack is completed and that the stack should again be fumigated whenever storage exceeds two or three months.

Key for identifying Larvae Commonly Found in Stored Raisins

1. Larva has three pairs of true legs on thorax along with paired fleshy prolegs on the abdomen. Order Lepidoptera (Indianmeal moth and raisin moth).
2. Larva has three pairs of true legs on thorax but lacks prolegs on abdomen. Order Coleoptera (dried fruit beetle and sawtoothed grain beetle).
3. Larva up to one-half inch (12 mm) long and typically white or cream with dark brown head and body sparsely covered by hairs. Indianmeal moth.
4. Larva up to one-half inch (12 mm) long; four to six rows of purple spots visible along its white back produce a pinkish to purple cast raisin moth.
5. Larva up to one-quarter inch (6 mm) long and creamy white with brownish ends and a pair of pointed tubercles on end of abdomen.
6. Larva up to one-eighth inch (3 mm) long and yellowish white; end of abdomen without tubercles.

Natural Enemies of Post-Harvest Raisin Insect Pests

Several different natural enemies have been found associated with dried fruit beetles. Researchers at the USDA Horticultural Field Crops Laboratory in Fresno have discovered that pathogens are abundant in natural nitidulid populations and include the protozoans *Mattesia* sp., *Helicosporidium* sp., *Nosema* sp., *Ophryocystis* sp., and a cephaline eugregarine, as well as a sphaerulariid nematode (Lindegren et al., 1992a). A very common entomophagous fungus, *Beauveria bassiana*, appears to show promise under lab conditions. In addition, a solitary internal parasitoid, *Zeteticontus utilis* (Hymenoptera: Encyrtidae) is undergoing evaluation for control of dried fruit beetle larvae in Hawaii (Lindegren et al., 1992a).

A granulosis virus has been shown experimentally to be an effective biological control agent for Indianmeal moth on stored raisins (Lindegren et al., 1992b). In addition, the well-known bacterium, *Bacillus thuringiensis* Berliner (B.t.), is also commonly observed in lab cultures of the Indianmeal moth (IMM) and is registered for its control on stored grains and soybeans. Several other pathogens have been reported from IMM larvae in California: two microsporidians, *Nosema plodiae* Kellen & Lindegren and *Nosema heterosporum* have been described, along with a protozoan *Mattesia* sp. (Lindegren et al., 1992b).

Among the parasitic wasps attacking raisin moth and other caterpillars infesting stored raisins, most important is the parasitic wasp *Bracon hebetor* Say (Hymenoptera: Braconidae) (Lindegren et al., 1992c). In addition, two ichneumonids, *Devorgilla canescens* (Gravenhorst) and *Mesostenus gracilis* Cresson (Hymenoptera: Ichneumonidae), may also be of importance, particularly early in the growing season. However, since all three of these parasitoid species attack only well-grown larvae that have completed their feeding and damage, thus the major control benefit is a reduction in future populations (Lindegren et al., 1992c).

Part II: Effects of SharpShooters & Pierce's Disease

Z. Jo Harper, B.S. and Michael J. Costello, Ph.D.

Pierce's Disease of grapes (PD) has been a concern for the California grape industry almost as long as grapes have been commercially grown, but it has only come to the forefront of attention since the PD epidemic in Temecula which began in 1997. This came to pass because of the invasion into California of a new insect vector of the disease, the glassy-winged sharpshooter (GWSS). The volatile combination of a lethal grape disease, an insect vector which likes to feed upon and can reproduce on grapes, and the wide variety of alternate plant hosts throughout the state which can support GWSS has the California grape industry concerned about its very survival. This chapter is meant to summarize the biology, ecology and history of PD and its sharpshooter vectors, including the potentially devastating GWSS.

Casual Agent of Pierce's Disease

The cause of Pierce's disease is *Xylella fastidiosa* (*Xf*), a bacterium which lives in the water conductive tissue, or xylem, of plants. *Xf* can infect a wide variety of plants, including crop plants (e.g., almond, grape, stone fruits, alfalfa, and citrus), native plants (e.g., wild blackberry, elderberry, wild grape), landscape plants (e.g., oleander, elm, maple and sycamore), and weeds (e.g., mallow and lamb's quarters). Some pathotypes of *Xf* have wide host ranges; however, in many plant species, few or no symptoms of disease are noted (Freitag, 1951). Tests in the 1940s identified 91 of 116 tested plant species as hosts of *Xf*, then known as "Pierce's disease virus" (Freitag, 1951). The same strain of *Xf* causes PD of grape, alfalfa dwarf (Frazier, 1943) and almond leaf scorch (Davis, 1978). Other strains of *Xf* cause phony peach disease, citrus variegated chlorosis, oleander leaf scorch, and leaf scorch in a wide range of ornamental plants (Purcell et al., 1999). Recently *Xf* diseases have also been reported in coffee (Beretta, 1996). In grapes, bacterial infection is often progressive and lethal.

Xf is nonmotile, lacking flagella or other means of movement, incapable of self-propulsion or movement within the highly turbulent fluid environment in which it lives and reproduces. Xylem tissue is a conduit from the roots to the shoots and leaves of plants, conducting water, minerals and plant hormones, as well as extremely low concentrations of

Fig. 10.15: Pierce's Disease – Leaf symptom on var. Fiesta.

carbohydrates and nitrogen (sugars and amides). Nitrogen concentrations of xylem fluid are typically two orders of magnitude less than that of leaf tissue and 10 times less than phloem concentrations (Mattson, 1980). Because the fluid consists primarily of water, nutrient levels are very dilute and under strong negative pressure. *Xf* has specialized extracellular attachments, which allow them to attach to surfaces in the turbulent habitat of the xylem, and is also specialized to absorb nutrients from a low concentration.

Symptoms of PD on grapevines are typical of water stress and often appear at mid- to late-season when temperatures increase and water becomes limiting. Foliar symptoms include leaf chlorosis followed by necrosis (**Figure 10.15**). Bunches on infected shoots will shrivel, dry and remain on the cane. Infected canes may be stunted and deformed. Often the petiole will remain on the cane long after the leaf blade itself has fallen off (**Figure 10.16**). Infected canes grow in an abnormal zigzag pattern, and wood matures unevenly, leaving "islands of green" (wood that has not matured) against the tan background of mature wood (**Figure 10.17**). On some wine grape varieties leaf scorching is preceded by intense color changes (bright yellows, oranges and reds), but this is not the case with raisin varieties, and the disease symptoms progress from green to dull yellow to brown (**Figure 10.15**). How PD actually kills grapevines has been a matter of considerable debate.

In July, 2000, the *Xylella fastidiosa* Consortium, Sâo Paulo, Brazil, announced the complete genome sequencing of *Xf* clone 9a5c, casual agent of citrus variegated chlorosis. The researchers found that the pathogenicity and virulence of *Xf* involves toxins, antibiotics and ion sequestration systems. In addition, they found that interactions between bacteria and between host and bacterium, are mediated by a group of proteins that have been identified in animal and human pathogens. These proteins from different species arise from genes that are orthologs, that is, genes that have evolved from a common ancestral gene. Orthologs retain the same function in the course of evolution. The presence of these same proteins in *Xf* indicate that the molecular basis for the ability to cause disease caries little from one species to the next and is independent of the host. Understanding the molecular basis for pathogenicity for *Xf* and other plant pathogens may in the future lead to methods of therapies and of improving host resistance to the pathogen (Simpson, 2000).

Fig. 10.16: Pierce's Disease – Petiole absent leaf blade on var. Fiesta.

There is little doubt that plugging of the xylem by the bacteria themselves, gums produced by the vine, and tyloses are a significant cause of vine decline but it has also been proposed that phytotoxin production and plant growth regulator imbalance may also play a role. Typical symptoms on the highly susceptible Fiesta variety shows marginal leaf chlorosis, followed by necrosis (**Figure 10.15**). Leaf blades abscise at the terminal end of the petiole (**Figure 10.16**) which is a unique symptom. The shoots show definite green islands as the remainder of the cane matures into a normal tan color (**Figure 10.17**).

PD is endemic throughout the United States in areas with mild climates, including Southern states along the Atlantic Ocean, Gulf of Mexico and California. Early European settlers attempted to grow varieties of *Vitis vinifera* in Florida and other parts of the Southeast as early as the 1600s. They experienced extensive vine disease and we now believe that PD was responsible for much of this (Crall et al., 1957).

Sharpshooter Vectors of Pierce's Disease

Insects that transmit (vectors) PD are members of the leafhopper family Cicadellidae. Sharpshooters are a highly specialized group of leafhoppers that fall into the subfamily Cicadellinae and the tribe Proconiini. In contrast to leafhoppers, which feed on the mesophyll (non-vascular leaf tissue), such as the Western grape and variegated leafhoppers, sharpshooters inject a feeding tube directly into the plant's xylem. Because of the low concentration of nutrients in the xylem, sharpshooters have extremely high daily feeding rates (100-300 times dry body weight per day), a high efficiency of conversion for all organic compounds, and an ability to adjust feeding rates to diurnal fluctuations in plant xylem fluid chemistry (Brodbeck et al., 1992). On grapes, the feeding rate of GWSS best correlates with the concentration of total amino acids, amides and glutamine (Brodbeck et al., 1992).

Because all sharpshooters are xylem feeders, they actively seek fresh, succulent host plants, and the quality and quantity of plant xylem fluid they consume is directly related to fecundity and longevity. When Brodbeck et al. (1992) subjected host plants to a high level of water stress, the feeding rate of GWSS best correlated with xylem tension.

Fig. 10.17: Pierce's Disease on a mature cane showing alternating green islands of immature tissue between nodes.

Fig. 10.18: Green Sharpshooter adult.

Fig. 10.19: Redheaded Sharpshooter adult.

Fig. 10.20: Blue-green Sharpshooter adult.

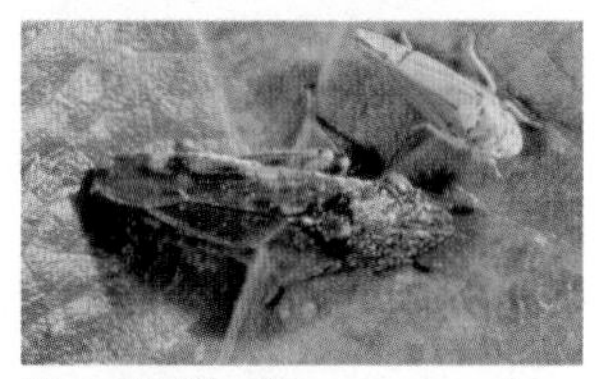

Fig. 10.21: Size comparison of Glassy-winged sharpshooter (bottom) and GHS (top).

This need for well nourished, well watered host plants draws these insects into irrigated farmlands and crops as well as urban and residential landscapes where they can easily spread PD.

Four species of sharpshooter in California are known to transmit PD to cultivated grape, whose habitats, hosts and behaviors are summarized in **Table 10.1**. These are the:

1. Green sharpshooter (GSS, *Draeculacephala minerva*) (**Figure 10.18**).
2. Redheaded sharpshooter (RHSS, *Carneocephala fulgida*) (**Figure 10.19**).
3. Blue-green sharpshooter (BGSS, *Graphocephala atropunctata* (**Figure 10.20**).
4. Glassy-winged sharpshooter (GWSS, *Homolodisca coagulata*) (**Figure 10.21**).

The GSS adult is about one-quarter inch in length with a sharply pointed head (**Figure 10.18**). Males are slightly smaller than females. Adults are typically bright grass-green during the entire year, and in areas of the San Joaquin and Sacramento valleys, most GSS become dark green to brown in fall through early spring. Females have a pale green underneath, whereas males have a black underneath. The GSS is active through the year, and typically has three generations per year in the San Joaquin Valley. GSS occurs in all principal grape-growing areas of California, and is more widely distributed over the state than RHSS. Grasses are

Table 10.1

Summary of Established Pierce's Disease Vectors.

Vector	*Blue-green Sharpshooter*	*Green Sharpshooter*	*Red-headed Sharpshooter*	*Glassy-winged Sharpshooter*
Scientific Name	*Graphocephala atropunctata*	*Draeculacephala minerva*	*Carneocephala fulgida*	*Homalodisca coagulata*
Breeding Habitat	Riparian	Grasses in wet spots	Grasses in wet spots	Many trees and vines; does not reproduce on oleander.
Breeding Host	Woody Perennials	Hedges, nutgrass, water grass	Hedges, nutgrass, water grass.	Prefers citrus; wide range of hosts.
Transmission Efficiency	High	Low	High	High
Occurrence in Breeding Habitat	Frequent	Frequent	Sporadic	Frequent
Movement Into Vineyard	Along riparian edge	Along pastures & ditches. Flight activity primarily at night into vineyard.	Along pastures & ditches. Flight activity primarily at night into vineyard	Strong flier. Moves easily throughout vineyard.
Monitor	Yellow sticky traps	Sweep net	Sweep net	Yellow sticky traps

Fig. 10.22: Female Glassy-winged Sharpshooter (left) and male (right).

Fig. 10.23: Side view of Glassy-winged Sharpshooter adult.

preferred food and breeding plants for GSS and it is primarily found in wet environments, such as marshes and bogs, along streams and ditches, in well irrigated alfalfa fields, lush grass pastures and any wet areas created by irrigation practices. Populations can also be found in unripe grain fields. GSS is an accidental or incidental occupant of vineyards; when pastures or natural grasslands dry up, they will migrate into well irrigated vineyards, especially in the late summer months (Purcell and Frazier, 1984). It is fairly efficient at transmitting alfalfa dwarf to alfalfa and from alfalfa to grape, but it does not reproduce on grapes, and is not an efficient vector of PD to grape from host plants other than alfalfa.

RHSS is paler green in color and smaller than GSS and can be further distinguished from the GSS by its brownish to reddish and rounder head (**Figure 10.19**). Male RHSS are slightly smaller than females and have smoky wing tops. It typically has four generations per year in the San Joaquin Valley. As with GSS, grasses are the favored hosts, but RHSS favors Bermuda grass and sparser, drier conditions. RHSS are more apt to be found in pastures that are not well irrigated and have less dense vegetation (Purcell and Frazier, 1984). RHSS are also commonly found in drier weedy areas adjacent to pastures, irrigated landscapes and ditch banks. Favored weed species of RHSS include puncture vine and cocklebur. As with GSS, RHSS is only an accidental and temporary feeder on grapes, and does not reproduce on grapes. However, it is more efficient at transmitting PD to grape than GSS (Hill and Purcell, 1997).

The BGSS is intermediate in size between the GSS and RHSS and is less robust than both. In northern California it is green or bluish green with an ivory to yellowish rounded head, which bears several characteristic black marks (**Figure 10.20**). It has a yellow underneath with yellow legs. In southern California it is often bright blue. BGSS occurs throughout most of the grape-growing regions of California. It is very common and often abundant in areas of the coastal fog belt. In the San Joaquin Valley, this sharpshooter is sometimes found in large numbers, usually closely confined to stream banks but rarely in adjacent vineyards. BGSS has many hosts, but prefers vines, shrubs, trees, and perennial herbs. Wild grape, blackberry, elderberry, and willow are especially common hosts. BGSS is more important in the spread of PD in north coastal vineyards than either GSS or RHSS, as it may migrate into vineyards from riparian areas in great numbers. It may be present on the vines from the time of early spring growth until growth ceases in the fall, during which time it may reproduce continually.

A New Vector for Pierce's Disease

GWSS is native to the New World, is prevalent in the southeastern United States (Kaloostian et al., 1992), and has been found from Wisconsin and northern Mexico (Young, 1958). It was first detected in California in 1994, but may have been in the state several years prior to that. The GWSS is twice as large as any of the other sharpshooter species, and quite robust (**Figure 10.21**). Female adults are about one-half inch

Fig. 10.24: Glassy-winged Sharpshooter egg mass.

Fig. 10.25: Female Glassy-winged Sharpshooter with egg mass appear as greenish blisters.

Fig. 10.26: Egg parasite (*Gonatocerus ashmeadii*).

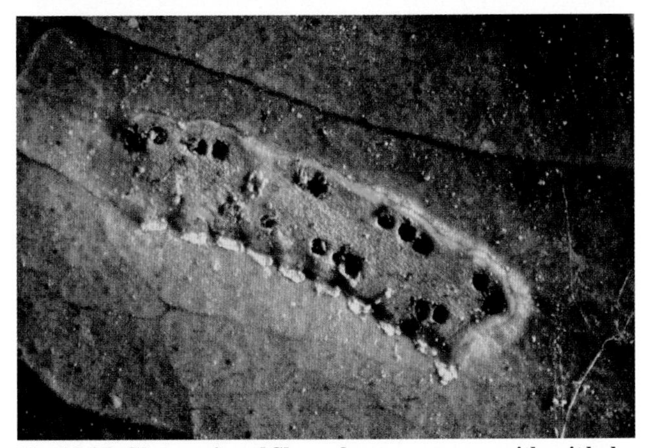

Fig. 10.27: Glassy-winged Sharpshooter egg mass with exit holes of *Gonatocerus ashmeadii*.

long, males are slightly smaller (**Figure 10.22**). The color is generally brown to black, the head and thorax with numerous ivory or yellowish spots. The sturdy piercing-sucking mouthparts combined with well-developed sucking bulbs needed to draw up the plant juices gives the insect a distinctive 'jowly' appearance (**Figure 10.23**). Immature GWSS (nymphs) are gray-brown and look like small adults without wings (or with wing pads). There are four nymphal instars, and GWSS matures in about 10 to 12 weeks. Adults live about two months.

GWSS is a very strong flier and prefers to feed on trees, bushes, and vines and has an extremely broad host range. Adults and older nymphs prefer feeding on stems and twigs rather than leaves of plants. Its preferred host crop plant appears to be citrus and it does well on a variety of ornamentals. Favored herbaceous hosts include sunflower, hollyhock, okra, lambsquarters, cotton, corn, and cowpeas. Favorite woody hosts include oak, ash, silktree, crape myrtle and peach (Turner and Pollard, 1959). GWSS can reproduce on a variety of hosts, including grape, citrus, avocado, crape myrtle, Chinese elm, ash, eucalyptus, sycamore, oak and many more. GWSS will feed on oleander, but does not reproduce on it (A. H. Purcell, personal communication). GWSS is the vector of another *Xylella* disease, oleander leaf scorch, which has caused serious economic losses in the Riverside area of southern California (Purcell et al., 1999). On citrus, GWSS can overwinter in large numbers without any symptoms appearing on the crop (Blua et al., 1999).

GWSS overwinters as an adult (Pollard and Kaloostian, 1961). In the spring, adults gradually migrate to new hosts and eggs are laid. GWSS eggs are laid side by side in groups of about 10, with as many as 27 (**Figure 10.24**). These are laid within host plants inside the epidermis of the lower leaf surface and appear as greenish blisters (**Figure 10.25**). Egg masses may also be laid in the rind of citrus fruits. Two generations of GWSS are reported to occur per year in Southern California (P.A. Phillips, personal communication).

A small egg parasite, *Gonatocerus ashmeadii* (**Figure 10.26**) attacks GWSS egg masses starting in spring. This tiny wasp is in the same family (Hymenoptera: Mymaridae) as the *Anagrus* wasp that attacks the eggs of the Western grape and variegated leafhopper. Parasitized egg masses have tiny holes in them where the parasites have emerged (**Figure 10.27**). Although parasitism rates may reach as high as 80% to 95%, the GWSS numbers are sufficiently high that the population continues to rise from year to year (Blua, Phillips and Redak, 1999).

In California, GWSS was first distinguished in 1994 from specimens collected in Orange and Ventura Counties by Santa Barbara County entomologist Jerry Davidson (Gill, 1994). The specimens were collected by IPM Advisor Phil Phillips on Eucalyptus. At that time, GWSS was found to be established in Fontana (San Bernardino County) and in Anaheim (Orange County), where it is believed to have existed since about 1990. The pest was likely introduced into California before 1990, most likely from the southeastern United

States as eggs on nursery stock. During the ensuing years, GWSS became established in the counties of San Diego, Orange, Riverside, San Bernardino, Los Angeles, Ventura, and Santa Barbara (Blua, Phillips and Redak, 1999). In the summer of 1998, GWSS was noted in citrus orchards in southern Kern County (A. H. Purcell, personal communication).

History of Pierce's Disease

Pierce's disease is named for Newton B. Pierce, the first professionally trained plant pathologist to work in California. In 1941, the American Phytopathological Society formally named the Tulare County grape disease Pierce's disease, and linked the 1935-46 epidemic with the "Anaheim Disease" and the "California Vine Disease" epidemics of the 1890s. During the primary epidemic in southern California, Pierce explored most aspects of the California Vine Disease. He demonstrated it was not caused by fungi or by bacteria that could be cultured from disease vines by common method. He showed that it was not a nutritional disorder and that it was unique, unlike any other disease of grapevine previously described in Europe or the United States.

Pierce investigated the heavy loss of vineyard acreage in Southern California during the 1880s and described the affliction as "California Disease" or "Anaheim Disease" It was first observed near Anaheim during the year 1884 and apparently extended throughout the entire Santa Ana Valley, and eastward into the San Gabriel Valley through Azusa and southerly from Anaheim into San Diego County. It also occurred in Los Angeles throughout the San Fernando Valley and even parts of the Antelope Valley. By 1895, *Xf* was responsible for the destruction of 35,000 acres (86,450 ha) of productive Southern California vineyards. During 1887, Pierce's disease was reported in vineyards of Napa Valley and Sacramento Valley, but did not become epidemic in these northern valleys (Hewitt et al., 1942a).

PD first surfaced in the San Joaquin Valley in the spring of 1920, with the report of a mysterious death of Emperor grapes in Tulare County (California Plant Pathologist, March 1921, cited by Hewitt et al., 1942a). Around 1930 a PD epidemic began in Tulare County (Hewitt, 1939), and by 1934 economic losses in many surrounding counties were alarming (Gardner and Hewitt, 1974). Northern counties such as Alameda, Mendocino, Napa, Sacramento and Sonoma also reported a high incidence of PD during this time. The disease reached a peak in 1941, when a survey conducted by the California State Department of Agriculture found PD in 28 of 45 grape growing counties, with 21% of the vineyards positive for the disease. However, PD incidence declined rapidly thereafter, and 1946 only one-half percent disease incidence was found in the mapped vineyards (Gardner and Hewitt, 1974). By 1948 the disease had all but disappeared in the San Joaquin Valley. However, in Napa and Sonoma counties the disease remained endemic, especially north of St. Helena and in some vineyards near the Napa River.

In the 1990s, considerable economic losses due to PD have taken place in Napa and Sonoma Counties, with some damage occurring in 'hotspots' in the San Joaquin Valley (A. H. Purcell, personal communication). In the North Coast areas of California, PD is generally limited to an "edge effect" pattern" of areas near riparian areas bordering rivers and streams, and vine to vine transmission is uncommon. Part of the explanation for the high incidence of PD on the north coast in the 1990s is the replanting of phylloxerated vineyards (young vines are more susceptible to PD) and expansion of acreage into natural areas (closer proximity to wild reservoirs). A survey of the Napa Valley Pierce's Disease Task Force identified PD as a problem on 10,200 acres (25,194 ha) of Napa County vineyards, with about 2,000 acres (4,940 ha) of vines identified with active PD infections. Reports estimated 517 acres (1,276.99 ha) as having to be replanted, and 60 acres (478.2 ha) taken out of production from 1995-2000, with losses estimated at $17.7 million (NVPDTF, 1999).

Beginning in 1997, an epidemic of PD has occurred in the Temecula Valley of Southern California (Redak and Blua, 1999), with the glassy-winged sharpshooter (GWSS) the vector of PD. Since its discovery in Temecula, the disease has destroyed approximately 180 acres (444.6 ha) with a production value of $5 million.

Epidemiology of Pierce's Disease

In classic theory of plant and animal disease epidemics, three factors are required for disease expression and the development of epidemic disease (Vanderplank, 1963):

1. A susceptible host; all varieties of *Vitis vinifera* are susceptible to some degree;
2. A pathogen of sufficient virulence and quantity to cause disease;
3. An environment conductive to disease.

Whenever these factors do not interact in a complementary way, disease is not likely to develop into epidemics or pandemics. Individual plants may be exposed and may die, but the momentum for widespread disease over a wide geographic location requires all three factors. Prediction and control measures to prevent outbreaks must consider all three factors in order to break the disease cycle and to contain disease spread.

The role of the insect vector is an especially important consideration not only the population density, but also the feeding preference and feeding behavior determine the potential for PD outbreaks. Where the reservoir of infected host plants is not concentrated near grapevines and populations of sharpshooters remain low. PD remains a localized problem, as in the San Joaquin Valley through the past half-century. In Kern County, there had never been a reported case of PD until August 2000, even though GSS and RHSS are endemic to the area (A. H. Purcell, personal communication). It is likely that the introduction of GWSS into the county in about 1997 is responsible for the PD now being observed.

When a host plant is infected with *Xf* but shows no symptoms, it increases the level of potential disease inoculum, which may be transmitted to a susceptible host such as grapevine. Purcell and Saunders (1999) have described three categories of reservoir hosts for *Xf*:

1. Propagative host: *Xf* reproduces within the host;
2. Systemic host: *Xf* moves from one xylem element to another;
3. Pathological host: Measurable distress or symptoms are caused by infection.

Environmental factors such as temperature can influence the bacterium's rate of multiplication, intraplant movement (systemic spread), and winter survival. Propagative and systemic hosts have the greatest potential to serve as sources for vector acquisition of *Xf.*

The pathogen, *Xylella fastidiosa*, is genetically variable (Chen et al., 1992). There are different strains of the bacterium that cause disease in specific plants but not in others. DNA analysis (Machado, 1999) suggests the existence of five groups of *Xf* as of 1997:

1. Citrus group
2. Plum-elm group – includes oak leaf scorch (OLS)
3. Grape-ragweed group
4. Almond group
5. Mulberry group

In 1999, another strain of the bacterium was isolated and identified causing disease in oleander (Purcell et al., 1999). This newly isolated strain is genetically distinct and uniform but does not cause disease in grape.

There is controversy about how similar PD strains may be to each other in their genetic makeup as well as their ability to cause disease in different hosts. Chen et al. (1995), working with Florida PD strains, found a high degree of genetic similarity among the seven PD strains and the seven oak leaf scorch (OLS) strains they worked with even though the strains were pathologically distinct. Hill and Purcell (1997) have speculated that the ability to cause systemic infection in Bermuda grass, a critical breeding host for San Joaquin Valley sharpshooters, may have changed over time. In early work done by Freitag (1954) during the 1940s, PD was transmissible by GSS from Bermuda grass to alfalfa. Subsequent attempts by Hill and Purcell to inoculate and recover *Xf* from Bermuda grass have failed.

Pierce's Disease in the San Joaquin Valley

Thompson Seedless is among the most PD resistant of grape varieties (Hewitt et al., 1942a), while the Fiesta variety is extremely susceptible (A. H. Purcell, personal communication, 1996; Christensen, 2000). No rootstock has been found which will impart PD resistance to the scion. The only satisfactory control measure for PD has been to eliminate diseased vines and to replant using more resistant varieties.

The pattern of *Xf* diseases may be changing in the San Joaquin Valley. Almond leaf scorch is for the first time a concern in Madera County. In addition, new varieties of grapes, including those grown for DOV, may not be as resistant as Thompson Seedless. In addition, the introduction of the GWSS into the San Joaquin Valley increases the potential for the spread of PD and other *Xf* diseases. This increases the inoculum level for all vectors of PD, including the established ones (primarily RHSS and GSS) already here. There are no reports of BGSS populations greatly increasing in the San Joaquin Valley; however, this sharpshooter is a highly efficient vector of PD. Its role in the San Joaquin Valley disease is not documented.

Photo Credits

Figures 1, 2, 3, 4, 6, 7, 9, 10, 15, 16, 17 : Michael Costello.

Figure 5: Randy Vaughn-Dotta.

Figure 5 Insert, 8: V. E. Petrucci.

Figures 11, 12, 13, 14: Charles Burks.

Figures 18, 19, 20, 21, 22, 23, 24, 25, 26, 27, 28: P. A. Phillips.

Overview of Available Resources

One of the most significant references addressing the challenges of effective management of pests in vineyards is the outstanding *Grape Pest Management,* edited by Donald Flaherty and five co-workers. Published through the University of California division of Agriculture and Natural Resources, the second edition of this important work became available in 1992. Nearly 70 specialists in fields ranging from viticulture to vertebrates contributed to this comprehensive volume which provides a remarkable combination of practical and technical information for optimal management of vineyard pests.

An additional and ever-expanding and updated resource for Integrated Pest Management information is the UC IPM website, located at http://www.ipm.ucdavis.edu. This website offers an enormous amount of current information of use to IPM practitioners in a very broad array of cropping systems.

References

Anderson, P. C., B. V. Brodbeck, and R.F. Mizell. 1992. Feeding by the leafhopper, Homalodisca coagulata, in relation to xylem fluid chemistry and tension. Journal of Insect Physiology. 38:8. 611-622.

Beretta, M. J., G. Harakava, Chagas, C. M. et al. 1996. First report of Xylella fastidiosa in coffee. Plant Disease 80 (7).

Bettiga, L. J. and P. A. Phillips. 1992. Orange Tortrix. In: D. L. Flaherty, L. P. Christensen, W. T. Lanini, J. J. Marois, P. A. Phillips and L. T. Wilson (eds.) *Grape Pest Management*, University of California Division of Agriculture and Natural Resources, Oakland, CA, Publication 3343.

Blua, M. J., P. A. Phillips and R. A. Redak. 1999. A new sharpshooter threatens both crops and ornamentals. California Agriculture. 53:2. 22-25.

Brodbeck, B.V., R.F. Mizell, and P.C. Andersen. 1992. Physiological and behavioral adaptations of three species of leafhoppers in response to the dilute nutrient content of xylem fluid. Journal of Insect Physiology. 39:1. 73-81.

Chen, J., C. J. Chang, R. L. Jarett, and N. Gawel. 1992.

37:136-143.

Young, D. A., Jr. 1958. A synopsis of the species of *Homalodisca* in the United States (Homoptera, Cicadellidae). Brooklyn Entomological Society Bulletin. 53. 7-13.

CHAPTER 11.

Diseases Affecting Raisins

Julian Whaley, Ph.D.

Introduction

Diseases reduce the yield and quality of table, wine or raisin grapes. Many diseases of grapevines have been well documented and will cause direct or indirect losses in raisin production. In this chapter we will concentrate mostly on diseases directly affecting the grape bunches during the growing season and after harvest as grapes dehydrate on the ground. In addition we will highlight a few major grape diseases in which new information has recently been published relating directly to establishing new vineyards or affecting raisin quality.

To obtain detailed information on all the major grape and raisin diseases, refer to the 1992 UC Grape Pest Management, Second ed. Oakland: University of California Division of Agriculture and Natural Resources publication 3343. Another excellent source is the American Phytopathological Society's Compendium of Grape Diseases (Pearson and Goheen, 1988). For the latest disease control information as it relates to California conditions, the web site http://www.ipm.ucdavis.edu is recommended. This University of California site is updated frequently and gives disease control recommendations on many crops.

Bunch Rots

Grape berries are an excellent source of nutrients for many microbes. Some of these microorganisms simply grow as epiphytes on the surface of plant parts and may cause no harm to the berries. However, in the course of a growing season, injuries releasing grape juice or opening the berry to infection may provide an opportunity for microbes that could not otherwise penetrate a berry. These conditions set the bunches up for bunch rots during the growing season. When the bunches are harvested and placed on the trays to dry, another set of environmental changes are encountered by the fruit and pathogenic microbes have another chance to infect the fruit. If rain occurs while fruit is on the trays, a dramatic increase in microbial populations occurs on the raisins.

Grape vineyards provide a good source of inoculum for microbes capable of growing on berries as the season progresses. In California many varieties of grapes are grown in the same region. Fruit begins to be harvested in the southern San Joaquin Valley in early summer and varieties continue to mature until the last grape is picked in November. Microbes growing on grape berries may multiply rapidly as bunches mature and are harvested. Sugar levels rise and nutrients become readily available as some berries are injured. By the time raisins are harvested and laid on the ground, inoculum levels of fungi in vineyards are very high all over the San Joaquin Valley and spores are deposited on the drying grapes. A shift in favor of the grapes occurs when osmotic pressures rise as the grapes dry and sugar concentration increases. This makes conditions deteriorate for growth of many microbes because moisture becomes limited in the berries. However, the environment once again favors the microbes when fall weather begins.

In the San Joaquin Valley of California the chance of rain increases dramatically in early October. Growers try to complete their raisin harvest by September 15 if they are going to sun dry the fruit. By this time more dews occur in the evenings, humidity rises and chances for rain increase. If early fall rains persist while the grapes are on the ground, the environment changes in favor of the microbes on the surface of the bunches and disastrous losses can occur.

In the Mexican State of Sonora, such as the Caborca area, berries are dried on the ground during the desert rainy season. This certainly shifts the odds in favor of microbes and damage from moldy fruit is common.

It is well documented that bunch rots occur during the growing season before grapes are harvested for drying. Harvey and Pentzer (1960) review the fungal pathogens of table grapes affecting the bunches during the season and continuing into storage. Species of *Alternaria* and *Stemphylium* can occur quite early in the harvest season, even in the absence of rainfall. These fungi enter through the capstem and cause localized tan to dark brown lesions. In table grapes, these pathogens can continue in cold storage. The blue mold fungus (*Penicillium* spp.) can attack raisin berries before harvest or while on trays. *Botryosphaeria ribis* Gross and Dug can cause ripe rot in Muscadine grapes in the field but many of the berries fall to the ground before harvest.

Bunch rots are known by many common names to growers. Some of these are slip skin, smut, gray-mold, sooty mold, green mold, sour rot, summer bunch rot, etc. Technically, scientists refer to the rots with the name of causal organisms associated with the disease. For example we often use the term Botrytis rot to indicate that the fungus *B. cinerea* is mostly responsible for the decay. It is extremely rare if only a single microbe is responsible for any bunch rot in nature. Once a pathogen infects, or berries are injured, secondary microbes are there shortly to fill their niche in the rotting process.

Nelson (1951, a, b, 1956) studied fungi capable of infecting uninjured grapes and found that *B. cinerea*, *Aspergillus niger* van Tiegh and *Alternaria geophila* Deszew (El-Helaly (et al), 1965) could penetrate sound fruit during the growing season. Fortunately, for California, there are several berry pathogens that are not found here because of our dry growing

Table 11.1

A summary of microorganisms found on grapes by Hewitt (1974).

Acetobacter roseus Vaughn and apparently other species also
Aschochyta sp.
Alternaria tenuis Nees
Aspergillus niger v. Thegh, black and brown spore forms
A. wentii Wehmer
A. flavus Link
A. ochraceus Wilhelm
B. cinerea
Candida sp.
Chaetomium elatum Kze
Cladosporium herbarum Pers.: Fr.
Curvularia sp.
Diplodia natalensis P. Evans (syn. *Botryodiplodia theobromae)*
Emericella rugulosa Thom and Raper) Benjamin
Epicoccum sp.
Fusarium moniliforme Sheldon
Fusidium sp.
Helminthosporium sp.
Heterosporium sp.
Monilinia sp.
Nigrospora sp.
Popularia sp.
Penicillium sp.
Phomopsis viticola Sacc.
Pullaria pullulans Berkh.
Rhizopus arrihzus Fisher
R. stolonifer (Ehrenb. Ex Fr) Lind,
Saccharomyces cerevisiae Meyen ex Hansen var *ellipsoideus*
Stemphyllium botryosum Walker
Trichoderma lignorum (Tode) Hartz
Torula sp.

seasons (Hewitt, 1974). Some of these are black rot caused by *Guignardia bidwellii* (Ell) Viala and Ravaz; anthracnose caused by *Elsinoe ampelina* (d. By); and bitter rot caused by *Glomerella cingulata* (Ston.) Spauld and Schrink. One fungal pathogen that can attack bunches and is usually found only in the Eastern United States is the downy mildew fungus *Plasmopara viticola* Berl. and DsT. However, this fungus has been seen causing slight damage in the San Joaquin Valley in the 1990s where late rains persisted into the early summer (personal observations). Downy mildew has caused serious epiphytotics in the Caborca area of Sonora, Mexico during heavy monsoon seasons in the 1990s (personal observations) and damaged raisin, wine, and table grapes.

Hewitt (1974) reports on a study from 1958 to 1974 whereby some 70 species of fungi in 30 genera were found associated with grape berries in the field in California. Some bacteria, including acetic-acid types were also found in grapes from Napa County south into Kern County. The microbes found in Hewitt's study are found in **Table 1**.

Hewitt summarized his survey by noting that *A. tenuis*, *A. niger* (the brown spored form), *C. herbarum*, *R. arrhizus*, *R stolonifer* and a green-spored *Penicillium* sp. were the most dominant of the fungi occurring on grapes during the growing season. Hewitt's work on sour rot showed that a complex of the organisms listed above are involved and that a sour vinegar odor is the result of the multiple colonization of bunches. Also it seems *Diplodia, Acetobacter,* and some other bacteria, fruit flies (*Drosophila* spp.), and dried fruit beetles all contribute to the conversion of sugars to alcohol and subsequently acetic acid. **Figures 11.1**. **11.2** and **11.3** show sour rot with dripping of juice from berries and insect involvement in the bunch rot.

Fig. 11.1: Thompson Seedless clusters with black exudates resulting from sour rot beginning to drip from bunch on the bottom left.

Fig. 11.2: Sour Rot with black exudates dripped on soil beneath the vine.

Fig. 11.3: Sour rot with black, viscous, syrupy exudates dripping from infected bunch.

In addition to bunch rots characterized as sour rots, there is the summer bunch rot disease complex, as defined by Hewitt. It is difficult to separate this stage from sour rot (Hewitt, 1962; Strobel and Hewitt, 1964). The summer bunch rot is started by *D. natalensis* (**Figure 11.7**) (Barb and Hewitt, 1965). Even though a strong vinegar odor is associated with this disease also, Hewitt separates Diplodia summer bunch rot from other sour rots by association with Diplodia cane blight symptoms on the wood. The pathogen was shown to enter the stigma, transverse the style and from there into the young grape. He believed that the fungus becomes latent in the tissue at the base of the style until the berry sugar reaches

Fig. 11.4: Summer bunch rot on Thompson seedless berries infected by *Rhizopus stolonifer* and *Penicillium* sp.

Fig. 11.5: A Ruby Seedless bunch infected with *Penicillium* sp., *Rhizopus stolonifer*, and *Alternaria alternata*.

10 to 14%, at about which time the fungus resumes growth and causes rot of infected grapes. In addition to *Diplodia*, the dominant fungi found in summer bunch rot were species of *Alternaria, Aspergillus, Cladosporium, Penicillium* and *Rhizopus*. He was able to isolate *Diplodia* from the vineyard soils and built a case for this fungus being the initiator of this type of summer bunch rot.

Duncan et al. (1995) studied the population dynamics of bunch rot microbes throughout the season on five wine grape varieties from 1989-92 in the wine growing areas of California. In the southern San Joaquin Valley where many of the raisins are grown, the fungi dominating bunch rots were species of *Cladosporium, Aspergillus, Penicillium, Geotrichum* and *Alternaria*. **Figure 11.4** shows bunch rots on Thompson Seedless grapes. *Rhizopus stolinfer* and *Penicillium* sp. were isolated from this bunch. **Figure 11.5** shows a cluster from which the same two fungi plus *Alternaria alternata* were isolated. *Botrytis cinerea* was found in significant numbers in only one of five trials in the hot, dry San Joaquin Valley where the raisins are grown. They obtained a reduction in bunch rot with leaf removal techniques. This process increases air circulation and decreases humidity in the canopy making the conditions less favorable for fungal growth and subsequent bunch rot.

Botrytis cinerea has been known to be involved in bunch rots around the world for many years (Ciccarone, 1970). It has been called "slip skin" or "gray mold." In some wine grape growing regions of California and Europe, it is encouraged and results in the production of botrytised dessert wines with an extremely sweet flavor. However, in raisin or table grapes it is definitely not wanted because *B. cinerea* causes decay of berries before they can be harvested or in storage, in the case of table grapes. This fungus is the major pathogen involved in table grape decay in storage and is the main reason for fumigating with sulfur dioxide and keeping the grapes at 32 °F (0 °C).

McClellan (1972), in his Ph.D. thesis at UC Davis, theorized that *B. cinerea* infected flowers and then, as the berry developed, became latent inside the fruit until it ripened. At

Fig. 11.6: A late-season infection of bunches by *Botrytis cineria* (Note sporulation on surface of rachis and berries.)

this point the fungus became active and began the rotting process. This theory was generally accepted for several years. It was said to explain why bunch rot occurred in years with little rain during the growing season. He concluded from his thesis studies and work with Hewitt and others that *B. cinerea* infects through the stigmatic end of the grape flower (McClellan et al., 1973; McClellan and Hewitt, 1973). He reported control of the bunch rot disease by bloom sprays with a fungicide. The theory was that protecting the flowers at bloom resulted in less bunch rot at harvest. Shortly after this work in the mid-1970s, it became common practice to apply bloom sprays. However, no one has been able to find the latent pathogen in berries naturally infected at bloom time. Furthermore, the results of bloom sprays with a myriad of fungicides gave inconsistent results. We conducted several fungicide trials at California State University, Fresno in the late 1970s and provided research reports to the companies sponsoring the research (unpublished). Results were very erratic and generally showed little or no control regardless of timing or rate of fungicide. The current thinking about *B. cinerea* infection of the flower is that pieces of infected flowers remain in the bunches and serve as additional inoculum for infection later in the season. This floral "debris" contains inoculum that survives in the bunches until infection occurs. How much significant inoculum this adds to the *B. cinerea* spores in the vineyard remains unclear. See **Figure 11.6** showing *B. cinerea* fruiting on the surface of berries affected by late season bunch rot in the California State University, Fresno vineyards.

Petez and Pont (1986) found no evidence that *B. cinerea* becomes latent in berries after infecting floral parts. They believe pedicels may be involved in a latent *B. cinerea* infection but not the berries. Savage and Sall (1986) were unable to detect *B. cinerea* in immature berries. De Kock and Holz (1991) found no relation between early infection and subsequent disease development or post-harvest decay on table grapes. Rather, post-harvest decay was due to infection during storage by inoculum present in bunches at veraison or during later stages. Obviously, more information on the infection process by *B. cinerea* is needed before bloom applications of fungicides are scientifically justified and work consistently. Pearson and Riegel (1983) have questioned the need for bloom sprays on wine grapes.

Research reports indicate that many microbes and some

Fig. 11.7: A Thompson Seedless bunch with typical summer bunch rot caused by several different species.

insects are involved in the bunch rot complex of grapes in California. No doubt, various pathogens, saprophytes and epiphytes compete for sites and advantages on the surface and in wounds. As environmental changes occur in a season and the berries mature, the advantages probably shift back and forth until a few hardy fungi take over infection sites and typical bunch rot symptoms are seen. It appears that, in the raisin growing regions of the San Joaquin Valley, pure Botrytis bunch rot is rare according to Duncan et al. (1995). **Figure 11.7** shows typical bunch rot on Thompson Seedless grapes in which several species of fungi are involved.

Microbial Breakdown of Raisins

Microbes found on the bunches at harvest time obviously remain on the grapes as they are placed on trays to dry. Technically, the fruit die at some point in the drying process and any microbial growth at that time is no longer parasitic in nature. Parasitic fungi that continue to develop well in the presence of the nonliving raisins are classed as facultative saprophytes. Environmental conditions change drastically when the fruit is put on trays. Dust, carrying spores of soil microbes, insects inhabiting the soil, debris blowing into the drying grapes and rain all have an effect on the final condition of the raisins. In 1976, 1978 and 1982 disastrous losses occurred with raisins in California due to continual tropical rains followed by cloudy days with high humidity (Flaherty et al., 1992). Brief showers, followed by sunny days with some wind causes little damage, but high humidity following rainy periods results in perfect conditions for rotting of berries on the trays. The drying grapes have been splashed with soil particles containing many microbes, osmotic pressures fall, and adequate moisture is present for spores to germinate. This is a recipe for disaster.

Moisture is the one factor that causes a serious increase in microbial populations on the raisins so growers must use care in placing raisins in bins if the moisture level is 18 to 20%. The University of California recommends that raisins above 20% moisture be put in sweatboxes or sparingly in bins (Flaherty et al., 1992). If the bins are too full, rotting will proceed quickly because the fruit cannot dry adequately and conditions for growth and sporulation of fungi are optimized.

Fig. 11.8: Putrid mold (as defined by USDA standards) on raisins.

The USDA Handbook (developed in Fresno, California) for Inspecting and Receiving of Natural Condition Raisins (1990) outlines their description of what they term as "mold" and "bunch rot." We quote the procedure from the USDA Inspection Handbook for mold and bunch rot determination in raisins.

Mold Inspection Procedure

The following method is suited for all natural condition raisin varieties: Pour approximately 2 lbs (0.9 kg) from the drops after the sample has been run through the airstream sorter at the substandard setting, break up any clusters and mix well. The sample is then spread on the clean flat surface and formed into one long ridge. Rake a handful from the opposite side of the ridge and count the raisins.

Continue this procedure, taking a handful of raisins from alternate sides of the pile until a count of 800 raisins is reached. Substandard raisins are not to be included in the count. Neither shall such material as skins (where flesh has been removed) be counted.

Place sample in a strainer and wash. Then put sample in saucepan, cover with water and bring to a boil. Let simmer for 20 minutes or until reconstituted (do not exceed 30 minutes). Pour raisins into strainer, wash with fresh water and place in a large, deep, white pan. Cover raisins with fresh water, examine and remove raisins affected with mold or rot.

The following is a guide for inspectors when sorting moldy units:

- Score the putrid looking type units when one-half or more of the raisin is affected.
- Score the nodular like, usually dark spot, mold on the raisin when the appearance of the raisin (after boiling) is seriously affected. Such raisins are usually seriously affected when an aggregate of one-eighths or more of the surface is affected. There may be an occasional unit that will aggregate a little less than one-eighths of the surface, but yet the appearance is seriously affected. In such cases score the unit as mold.
- Score the black mold in split (or cracked) raisins when one-half or more of the length of the raisin shows black discoloration in the split or cracked portion. This black discoloration is mold.

Units formerly referred to as "partials" (not scoreable under (1), (2), and (3) of this guide) are not to be considered when calculating the total mold percentage. Moldy raisins are also an index of the degree of wholesomeness and soundness of the product. Allow not more than 5%, by count, (as determined by boil mold test) of this defect in lots that are considered as meeting grade and condition standards.

There are three types of mold generally found in raisins. They are as follows:

Putrid Mold is usually amber in color, sticky, dirty, and generally has a white web-like structure on its surface. A lot can be considered seriously affected by putrid mold when the

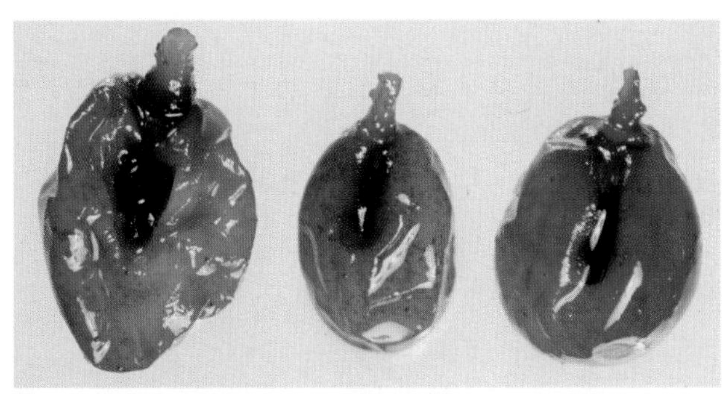

Fig. 11.9: Split mold (as defined by USDA standards) in raisins.

putrid mold is found in clumps of two or three raisins stuck together. Score the putrid type units when one-half more of the raisin is affected. **Figure 11.8** shows putrid mold as defined by USDA.

Split mold is very difficult to detect. As its name implies, split mold is a skin crack containing black mold. Mildew itself is not to be considered as a defect. However, the gray discoloration of raisins and white discoloration of capstems and stems caused by mildew tends to serve as an excellent indicator of split mold presence. Raisins seriously affected by mildew will be held for mold and contamination. Score the black mold in split or cracked raisins when one-half or more of the length of the raisin shows black discoloration in the split or cracked portion (**Figure 11.9** shows split mold).

Nodular mold usually occurs after a heavy rainfall and damp weather. Raisins appear to have a slight greenish appearance and are usually very sticky. The mold actually grows on the raisin skin surface. Score the nodular type mold when an aggregate of one-eighths or more of the surface is affected.

Raisins suspected of containing more than 5% of any of the mold types or a combination of these types must be held for lab analysis in natural condition raisins. Raisins with skin damage or raisins that have been rained on (sticky) are more susceptible to developing an embedded sand condition. When this situation presents itself, the lot shall be held for laboratory analysis regardless of moisture content.

Bunch Rot Inspection

Berries splitting inside of bunches as they enlarge cause bunch rot in grapes. The berries that have split cause the grapes they are in contact with to decay, rot and ferment. Grapes in this condition attract vinegar flies, thereby creating a potential contamination problem. Bunch rot in a container or pieces of bunch rot scattered throughout a container is an indicator of a potential contamination problem of the good raisins in the container. These containers shall be separated and held for microanalysis and a grade breakdown and may require a boil mold test. Bunch rot may be either wet or dry.

Studies of microbial growth on raisins are quite limited compared to those on the different types of bunch rots in the field or rotting of table grapes in cold storage. King et al. (1968) reported a survey of microorganisms on dried fruit. They reported on the survival of *Salmonella* (a human pathogen) on a variety of dried fruit and surveyed the types of microorganisms isolated from the surface of dried fruit products. Most microbial growth ceased as the fruit dried and sugar content went up. Microbial spoilage will not occur in dried fruit having less than 18 to 25% moisture, depending on the fruit and sugar content. High sugar content affects bacterial growth the most, followed by yeasts and fungi. There are, however, a few osmophilic yeasts which can grow in very concentrated sugar solutions. Any pH values below 4.0-4.5 inhibit most bacteria, while yeasts and fungi are not affected as much.

Higher microbial counts are to be expected on dried fruit products such as raisins as they normally receive no heat treatments. Figs are treated with both heat and sorbic acid and have subsequently lower microbial counts. King et al. (1968) took 293 samples of dried fruit products at the processing plants. Counts were then made of bacteria, yeasts and molds. They also isolated for *Salmonella* and coliform bacteria. None were found on any of the dried fruit products, including raisins. Dates, prunes and raisins had the highest microbial counts of all the dried fruit products (apples, dates, figs, prunes, cut fruit and raisins). They believe that this is due to a larger amount of equipment and more hand processing.

Also in this study, they inoculated dried fruit samples with *Salmonella typhimurium* (Loeffler) Castellani and Chalmers and isolated at various times to determine survival of this human pathogen on the fruit. On raisins, survival was less than 50 days on stored product. On dates, it was still detectable after 50 days compared to apples where it survived less than a week.

Tilden et al. (1958) studied mold damage on the 1958 raisin crop following two heavy rains while the crop was on the ground. They examined several hundred damaged and then rehydrated raisins microscopically and cultured from similar samples. In raisins with spotty lesions, the fungal growth was mostly on the skin with little penetration. The infections had little effect on the flavor of raisins. Some of the moldy raisins as classed by the USDA actually were sound berries with only superficial moldy areas. If a deep decomposition of berries occurred, the boil method was successful in identifying this fruit. Fungi identified were species of *Alternaria* in 90% of affected fruit. *Aspergillus* was growing in cracks of 5% of the fruit. *Penicillium*, *Oidium* and yeast species was found in the remaining 5% of the fruit. No *B. cinerea* was found in any of the rained on fruit with damage symptoms.

Erke and Dokoozlian (1982) also studied the microorganisms spoiling raisins after the grapes were laid out for drying. They observed the raisin samples microscopically and cultured them. *Aureobasidium pullulans* (deBary), *Alternaria* sp., and *Cladosporium* sp. had a preference for growing on exposed surfaces while bacteria, yeasts and *Aspergillus niger* caused rot in the lower portions of bunches. Some *Rhizopus* spp. and *Streptomyces* sp. (a common soil actinomycete) also grew on raisin samples. These raisins had been exposed to rains while on the trays during the 1982 season at the California State University, Fresno vineyard.

The first fungi to appear on the raisins grew on the upper surface in the sunlight. The dematiaceous (dark mycelium

Fig. 11.10: Spanish Measles (Esca) leaf symptoms on Thompson Seedless. Note interveinal chlorosis and necrotic areas at the margin and between veins.

Fig. 11.11: Spanish Measles (Esca) showing mild symptoms on foliage and purple streaks and spots in the berries.

and spores) fungi *Cladosporium*, *Alternaria* and *Aureobasidium* colonized this area first. After a few days Cladosporium then dominated the ecosystem. The exposed tray surfaces then became dark with a black growth. *Aureobasidium* was responsible for this colonization and subsequent black color on the trays. Following this sequence, the trays sometimes had growth of *Alternaria* and *Cladosporium* and the fruit developed *Aureobasidium* colonies. The observation that dark spored fungi colonize areas exposed to sunlight had been observed and reported by Diem (1971), Muirhead (1981) and Sussman (1968).

Dokoozlian et al. (1982) also studied the effects of the 1982 rains on raisins placed on several types of trays. They found that fungicide treatments with potassium sorbate, sodium benzoate or potassium metabisulfite on the surface of trays reduced the amount of putrid mold. Fruit dried on paper trays with a plastic coating had lower amounts of nodular mold damage. The only tray in the study to significantly reduce both nodular and putrid mold was the plastic tray when compared to the control (a wet strength paper tray).

Esca and Young Vine Disorders

Until the 1990s we thought of the measles problem in California vineyards as a single disease associated with wood rotting and caused by an unknown pathogen. The symptoms most often seen in California of this common vineyard disease are characterized by interveinal leaf chlorosis and necrosis, shot berries, smaller bunches and purple streaks in light colored berries like Thompson Seedless (**Figures 11.10, 11.11, 11.12, 11.13** and **11.14**). The common names given to this disease in California is Esca, Black Measles or Spanish Measles. References to these measles-type symptoms have been recorded in ancient Greek and Latin literature. The Kitab al-Felahah and the Opus Ruralium Commodorum gave a more detailed description of symptoms of the black measles disease according to Mugnai et al. (1999). These Medieval works were authored by Ibn al-Awam, a Spanish Muslim who lived in Seville at the end of the 12th century and Pietro de Crescenzi, born about 1233 in Bologna.

Often fruiting bodies of *Fomes* sp. (a common wood rotting Basidiomycete) are found on diseased vines with advanced measles symptoms. These conks are usually under the arm of the affected plant. When fruiting bodies appear, the vine has probably been diseased for several years. The word

Fig. 11.12: Spanish Measles (Esca). Some canes produce typical symptoms on foliage and bunches and some canes produce no symptoms.

Fig. 11.13: A comparison of a healthy Thompson Seedless cluster and foliage on the left and a sample from and Esca-infected vine on the right.

Fig. 11.14: Close-up of Spanish Measles (Esca) symptoms at harvest time on Thompson Seedless berries.

esca is from the Latin meaning food, ailment or bait. Grapevine wood infected and rotted by *Fomes* (tinder in Latin), burns slowly and was used to keep fires a glow without a flame (Mugnai et al., 1999). The exact role of *Fomes* sp .in the disease process, if any, is not known. Steinberg (1979) studied the association of interior rotting with typical measles symptoms and found no correlation. He followed the development of measles symptoms in 1431 vines over a four-year period of data collection (1971, 1972, 1977 and 1978) in an old vineyard of Thompson Seedless grapes at California State University, Fresno. He concluded that it was not possible to predict if symptoms would occur on a given vine from year to year. He did find a general trend of increasing severity in the vineyard over time. Steinberg (1979) has also reviewed the measles literature and reported that Rolland (1873) was the first to associate interior decay of the wood with symptoms and Ravaz (1898) found fungal mycelium to be associated with the decay. Eight years later, Ravaz (1906) identified sporophores of the fungus to be *Fomes ignarius* (L. ex. Fr.) Kickx but pathogenicity tests using this organism failed. Vinet (1926) was the first to associate *Stereum hirsutum* Willd. Ex. Fr. with measles. In 1926, Viala published a monograph on the subject and dismissed the possibility of *Fomes* being involved in the disease. He implicated *S. hirsutum* or *S. necator* (a new species) in the disease. In California, Bonnet (1926) correlated the occurrence of trunk rot with

Fig. 11.15: Cross section of a grape rootstock infected with the pathogens involved in causing "black goo" symptoms. Note dark exudates coming from xylem tissue shortly after the cross section was made.

Fig. 11.16: Cross section showing xylem discoloration and dark pith symptoms on the rootstock. Black goo exudates were seeping from the xylem an hour after the cross section was made.

Fig. 11.17: Vascular discoloration, dark pith and orange-colored tyloses in the xylem are evident in dark areas of this rootstock

Fig. 11.18: Cross section of a healthy one year old rootstock. Note clean, white color of the specimen.

Black Measles symptoms and noted that the foliar and fruit symptoms were similar to those described in Europe. Bioletti (1923) compared other vine diseases to Spanish Measles and discussed the differences. Chiarappa (1959) did extensive studies on Spanish Measles. He attempted to correlate wood decay with foliar and berry symptoms. He also showed how a *Cephalosporium* sp. reproduced in vivo some of the symptoms observed in the wood of diseased vines and how *P. igniarius* caused wood decay in vitro. More recently, Larignon and Dubos (1987) began the study of certain imperfect fungi and their role with some Basidiomycetes in causing the measles or esca type diseases.

A more serious symptom can occur in some vines with black measles. These affected vines can wilt and/or defoliate and the berries dry on the vine at mid- to late-season. This phase of the disease is known as apoplexy. Both the measles and apoplexy symptoms may appear and disappear from one season to the next on affected vines. Dormant sprays with sodium arsenite were used for many years to reduce symptoms of this disease. No one understands how this fungicidal compound worked. It has been banned for use in California because of its human toxicity. There is currently no replacement available.

Mugnai et al. (1999) has extensively reviewed the literature on the measles disease complex. Much more research emphasis has been placed on measles globally in the 1990s and we are getting a clearer picture of the role of various fungi in the disease complex. Europe has seen the disease increase dramatically since the banning of sodium arsenite treatments and some areas are severely affected even with the use of this chemical.

We now know that a third related syndrome affecting young grapevines is a part of the measles complex (Mugnai et al., 1999). Wood discoloration and streaking is a common symptom in measles and has been noted for many years. Petri (1912) described these symptoms in southern Italy. The plants showed a general decline similar to symptoms caused by some viruses that we know today (arricciamento or court-noue). At the time the cause was unknown. He reported wood related gummosis and brown to black wood streaking. These streaks were sometimes seen in xylem vessels along with gummy masses that extended deeply into the trunk and sometimes into the roots. Recently in many parts of the world, this streaking and gumming in xylem tissue has been reported in young vineyards. These newly planted vineyards sometimes have a high mortality rate and show extensive stunting of vines.

Fig. 11.19: A first year vineyard in the San Joaquin Valley of California that was planted eight months before this photo was taken. Most of the Cabernet Sauvignon on Freedom vines on the left were infected by the fungi involved in the "black goo" disease. A different nursery provided the same variety of rootstock/scion shown on the right and these vines were not infected. Note the difference in vigor, even though both were planted at the same time. Many of the vines on the left subsequently died of the disease.

Mugnai et al. summarize the associated microorganisms isolated from affected young grapevines in their 1999 paper. Brown wood streaking, slow dieback, black goo (referring to the black gummy exudates coming from diseased xylem tissue when cut in a cross section), *Phaeoacremonium* grapevine decline, wood discoloration of rooted cuttings before planting, and decline of young grapevines all had one thing in common: Species of *Phaeoacremonium* were isolated from all the diseased wood. This fungus is probably identical to *Cephalosporium*. This, most likely, is the fungus implicated in measles by Chiarappa in 1959 when he referred to finding a Cephalosporium-like fungus associated with measles.

Acremonium sp. was also isolated from young rooted cuttings with discolored wood in Italy (Frisullo et al., 1992). Morton (1995) reported on a mystery disease hitting young vines in California, and in 1997 gave an update on what she called the "black goo disease." Young vineyards with black goo began to decline shortly after planting infected rooted cuttings. This was a very good descriptive name for the disease because copious amounts of a black oily substance oozed from xylem shortly after making cross sections of affected plants. Vineyards from all parts of California were afflicted. The author has seen this disease in several young vineyards and a species of *Phaeoacremonium* was isolated from all infected plants we studied in the San Joaquin Valley of California (unpublished data). Extensive losses were incurred in identifying, removing and replanting hundreds of acres in our own experience. **Figures 11.15, 11.16** and **11.17** show cross sections of some of the infected rootstocks that we studied. A cross section of a healthy rootstock is shown in **Figure 11.18**. The "black goo" type symptoms were obvious on infected vines. We determined that certain nurseries were producing infected rootstocks and others provided very healthy plants. **Figure 11.19** shows infected Freedom rootstock with Carbernet Sauvignon scion wood on the left and healthy plants from a different nursery on the right. All cultural practices were similar and the vines were planted on the same date. Scheck et al. (1998) also found this young vineyard disorder to be serious in California. The species of *Phaeoacremonium* isolated from *Vitis vinifera* L. were *aleophilum*, *angustius* and *chlamydosporum* in California, Europe, South Africa and South America. In addition, other species in this genus were isolated from oaks, stone fruit and humans (Mugnai et al., 1999).

In the Mugnai review, she discussed the following related conditions or diseases that exist in the measles-young vineyard disorder complex. It is proposed that the young vineyard disorder (black goo disease) be named "Petri grapevine decline" in honor of L. Petri, who was the first to relate wood brown-streaking and wood gummosis to fungal infection by *Acremonium*-like fungi and saw the close relation between brown wood streaking and esca (measles). The causal fungi for this part of the disease complex are thought to be *Phaeoacremonium chlamydosporum*, *P. aleophilum*, *P. inflatipes* and perhaps *P. angustius*.

The black measles or apoplexy type disease with interveinal chlorosis, purple streaks in berries, wilt of branches and wood discoloration with gummosis is caused by one or more species of *Phaeoacremonium* getting into the plant through wounds during the first years of growth or later. Mugnai proposed calling the disease "brown wood-streaking" in deference to Petri or "young esca" since it may eventually turn into "true esca." The third part of the complex is when mature and old vines are infected by wood rotting fungi by such as *Fomes* species or other wood rotting Basidiomycetes. When white rot and brown wood streaking are present the affected vines may show the entire range of symptoms we know as esca or black measles.

It is obvious that great strides have been made in understanding this disease complex recently and it is equally clear that much more research is needed to understand how all the different parts of the complex interact.

Pierce's Disease

Early settlers in the United States attempted to grow the European grape varieties of *Vitis vinifera* in Florida and other parts of the southeast in the 1600s. They experienced extensive vine disease and we think now that Pierce's disease was responsible for many of these epiphytotics (Crall et al., 1957). A virus was thought to be the causal agent of Pierce's disease and, in 1965, Frazier reported that several small leafhoppers and spittle insects spread the pathogen. Frietag (1951) showed that the host range for the pathogen included shrubs, grasses, and weeds; the pathogen also causes alfalfa dwarf disease. Vines of *V. vinifera* were planted in California in the 1800s and the disease was found in Southern California in the 1880s. It may have been introduced from the Gulf coastal plains area (Flaherty et al., 1992). It was known as Anaheim disease, mysterious disease or California vine disease. Some 35,000 acres of vineyards were destroyed from 1883-1886 (Winkler, 1974). The disease was first studied extensively by Newton B. Pierce in California and was subsequently named in his honor. In the San Joaquin Valley, it first was reported in 1917 and major outbreaks killed many vines in the 1930s and 1940s. Since that time, serious epiphytotics have occurred sporadically in the grape growing regions of California. These epiphytotics are localized but can be severe in individual vineyards (Flaherty et al., 1992). Oltman (1998) estimated that Pierce's disease caused over $33 million in losses in the North Coast wine-growing region from 1993-1998. Overall, yields of grapes in California are not currently reduced drastically by this disease but this is not comforting to individual growers who suffer serious economic losses to their individual vineyards.

Research by Goheen et al. (1973) and Hopkins and Mollenhauer (1973) showed a "Rickettsia-like organism" to be associated with vines affected by Pierce's disease. This type of microorganism was known to be the causal agent of phony peach disease and alfalfa dwarf also. Since that time these bacteria are known to cause other diseases by growing in the xylem and causing symptoms typical of wilt diseases where water uptake is inhibited. Davis (1978) grew the first of these xylem-limited bacteria in axenic culture. Bacterial taxonomists proposed the name *Xylella fastidiosa* to include many pathotypes and possibly different subspecies of this fastidious, Gram negative, xylem-limited bacterium (Wells, 1987). Each pathotype has its own host range. For further detailed information on Pierce's disease please refer to **Chapter 10, Part II** – Effects of Sharpshooters and Pierce's disease.

Powdery Mildew

Powdery mildew is a serious disease of grapevines in California. Symptoms include a white to gray powdery surface on green parts of the vine (**Figures 11.20, 11.21, 11.22 and 11.23**). It occurs throughout the raisin producing areas of the state. The causal agent is *Uncinula necator* Burr., a fungus classified as an obligate parasite. This means that it cannot live on dead tissue or be grown in axenic (pure) culture in the laboratory. This pathogen is an Ascomycete and has both a sexual and asexual stage important to disease development. The asexual (Imperfect) stage of the genus causing powdery mildew is named *Oidium*. The fungus grows on the surface of a grape berry, leaf, rachis or pedicel and puts a haustorium (nutrient-absorbing structure) into epidermal cells. Mycelium on the surface of the plant is fed by nutrients absorbed

Fig. 11.20: Powdery mildew mycelium and spores developing on this Thompson Seedless bunch resulted in stunting and cracking of berries.

Fig. 11.21: Heavy powdery mildew infection on a Thompson Seedless bunch.

from epidermal and associated cells through the haustorium. The fungi causing powdery mildew on many crops prefer warm, dry climates. With most fungi, external water or high relative humidity is needed for spores to germinate. The conidia of powdery mildew fungi are capable of moving water from one side of a conidium to the other to provide enough water to half the cell to allow germination (Yarwood, 1926). The grape powdery mildew fungus spores will germinate and infect grapes at temperatures as low as 45 °F (7 °C). The optimum temperature for infection is between 70 and 85 °F (21-29 °C). Mildew colonies are killed if exposed to a temperature of 104 °F (40 °C) for six hours. In these same studies, the powdery mildew fungus on grapes grew well at 70 to 90 °F (21-32 °C). The conidia germinated well at 0% relative humidity (Delp, 1954).

Thompson Seedless grapes are very susceptible to powdery mildew and, if uncontrolled in a year of extreme disease pressure, it can completely destroy the crop. This disease can now be found routinely throughout the entire San Joaquin Valley where raisins are grown. Until the mid 1960s, powdery mildew of grapes was difficult to find south of Modesto, California (personal observations). The majority of damage in the San Joaquin Valley is to the berries. Even in years of severe powdery mildew epiphytotics, it is rare to see significant damage on foliage. The rachis, pedicel and berry sometimes turn whitish gray with heavy mildew and foliage may remain relatively free of symptoms.

Disease cycles of pathogens can be best understood by taking the disease process one step at a time. We can start at any part of the cycle, but survival of the pathogen is the usual place to begin. In the case of powdery mildew of grapes, survival involves the dormant stages during the winter. Production of primary inoculum involves the type of inoculum that starts the disease off as the season begins. Dissemination of the inoculum must take place next so that it reaches the infection court. In this case the infection court is any green part of the vine. Inoculation is the next step where the inoculum is placed on the infection court. In the case of powdery mildew, spores are placed at random on the host by the wind. Inoculum must penetrate host tissue to continue. This involves germination of spores and the fungus entering the host. Infection is the next part of the process. This is where the fungus establishes a nutrient relationship with the host. There is an incubation period in diseases. This is the time between infection and the first symptoms. After the symptoms begin to show there is production of secondary inoculum. This is sometimes referred to as the repeating stage of the disease cycle and the host is repeatedly infected throughout the season. After the last production of secondary inoculum, the pathogen is ready for the survival or overwintering stage to form and the cycle begins again.

Fig. 11.22: Heavy powdery mildew infection on cluster rachis.

Fig. 11.23: Powdery mildew scarred and cracked Thompson Seedless berries.

Survival of the powdery mildew fungus begins with it overwintering as mycelium in bud scales or as ascospores in cleistothecia on bark on cordons or trunk. The cleistothecia (sexual stage) are not always present in the raisin growing areas. In the spring, ascospores are released and disseminated by splashing rain or wind to new leaves and shoots where the initial infection occurs. As new shoots emerge, they can also be infected by mycelium that has overwintered in the bud scales. These two types of inoculum are considered primary types. Inoculation takes place when ascospores are carried to a green surface of the vine by wind or splashing rain. Furthermore inoculation by mycelium in the infected buds occurs when the fungus begins to grow over the emerging shoots. Penetration of the green host tissue by the primary inoculum is done by mycelium or germ tubes of ascospores forming infection pegs and forcing their way through the cell walls and into the interior of the host epidermal cells.

Infection begins when a nutrient relationship is established with the host. Since the powdery mildews are obligate parasites, they do not benefit by the host cells dying. The fungus is best served when the host cells thrive and produce nutrients that are used for fungal growth. The haustorium does not actually penetrate the cytoplasm of the cell; otherwise the fungus would die out after host cells die. Rather the haustorium positions itself next to the cytoplasmic membrane and begins to cause nutrients to flow away from the host and into the pathogen. A few days after infection, depending on the temperature, the colonization part of the disease cycle begins. Mildew colonizes on the surface of any green tissue.

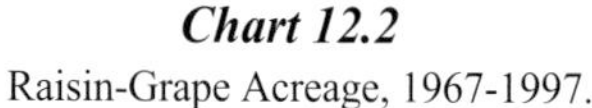

Chart 12.2

Raisin-Grape Acreage, 1967-1997.

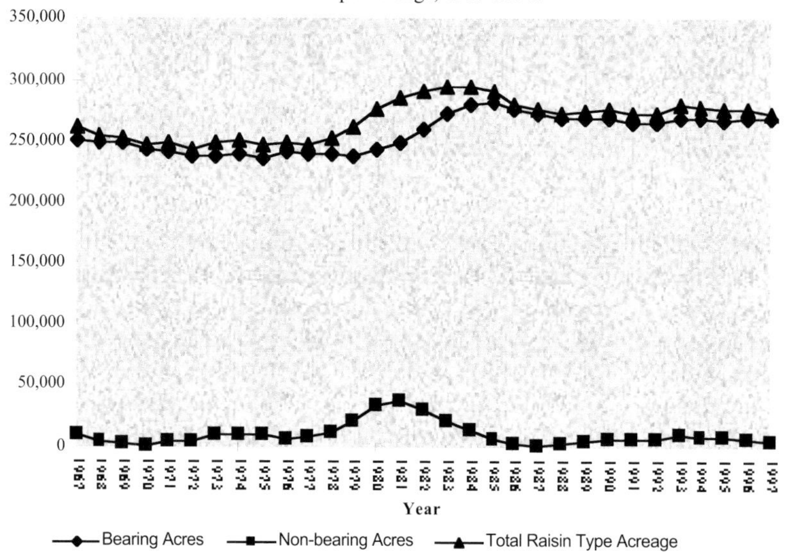

Table 12.2

California Raisin-Grape Acreage by Variety, 1998-99.

Variety	Bearing	Non-bearing[1]	Total Acres
Black Corinth[2]	2,460	19	2,479
Fiesta	4,200	783	4,983
Sultana	166	0	166
Thompson Seedless	262,956	4,274	267,230
Other raisin	61	64	125
Total raisin	**269,843**	**5,141**	**274,984**

[1] Non-bearing includes plantings for 1996, 1997 and 1998.

[2] Black Corinth/Zante Currant.

Source: 1998 California Grape Acreage by California Agriculture Statistics Service, June 10, 1999.

utilization of raisin variety grapes from the 1967-68 crop year through the 1998-99 crop year.

Some grape varieties, such as the Thompson Seedless and Zante Currant, have multiple uses. The Thompson Seedless, for example, is sold into each of the four outlets. Other grape varieties, particularly some wine grapes, are suited for use only in one of the outlets.

The percentage of grapes that are marketed in each outlet varies each year, depending on the market needs for each outlet. Due to the storability of the dried (raisin) outlet, oftentimes the dried outlet is oversupplied when the other three outlets are saturated and do not need the grape tonnage. **Chart 12.3** shows the tonnage of grapes in each outlet and the percentage each outlet is of the total for the 1998-99 crop year.

Yields

Grape yields fluctuate from year to year. One of the lowest yields for raisin grapes was 5.63 tons (5.067 t) of grapes per bearing acre when a spring frost damaged the 1972-73 crop. **Table 12.4** and **Chart 12.4** show the raisin variety grape yields, ranging from 5.63 to 11.8 green tons (5.067 to 10.62 t) of grapes per bearing acre, during the period 1967-68 through 1998-99. The average yield was 8.87 green tons (7.98 t) during this period. With new grape plantings in higher producing areas and improved cultural practices, the trend has been for higher average yields per acre.

Raisin Varietal Types

There are nine raisin varietal types defined by the USDA for marketing purposes. These are somewhat different from raisin varieties as defined ampelographically in **Chapter 3**. The USDA types and a brief definition of each follows:

Natural (sun-dried) Seedless raisins are the most popular raisin varietal type, making up about 90% of California's raisin production. This raisin is produced by sun drying fresh seedless grapes, which have characteristics similar to the Thompson Seedless grape when sun dried. The color varies from dark blue to grayish brown.

Dipped Seedless raisins are produced from seedless grapes, which possess the characteristics similar to Thompson Seedless grapes, that have been dipped in hot water after harvest. This process produces slight breaks in the grape skin, which allows the grapes to dry faster during the dehydration process. These raisins are slightly lighter than, but closely resemble, Natural Seedless raisins in color and their uses closely parallel the Natural Seedless raisins.

Golden Seedless raisins are also produced from seedless grapes and are processed in a manner similar to Dipped Seedless raisins. However, following the dipping process, they are moved into a sealed sulfur house and doused with sulfur dioxide (SO2). They are then dehydrated and the resultant raisins vary in color from predominantly golden yellow to dark brown raisins. They are used mainly in baked goods and are popular in fruitcakes.

Zante Currants, sometimes referred to as black Corinth grapes, are produced by sun drying black or white Corinth grapes. They are dark in color and about one-fourth the size of Natural Seedless raisins. Unlike other raisins, they are tart. They are used mainly in baked goods.

Oleate and related seedless raisins are also produced from seedless grapes. They are either dipped in or sprayed with soda, oil, Ethyl Oleate or Methyl Oleate to check or crack the skin either while on the vine or after they have been removed from the vine.

Muscat raisins are made by sun drying Muscat of Alexandria grapes. They are large and sweet with color varying from bluish black to dark reddish brown. Muscat raisins are primarily seeded and marketed as chocolate-coated Muscat raisins. Some are marketed as layer Muscats. Seeded Muscats have both the stems and seeds removed. Layer Muscats are de-stemmed and left in bunches.

Sultana raisins in California are made by sun drying the seedless Sultana grapes. This should not to be confused with the Australian sultanas that are made from the Thompson Seedless grapes. The California Sultana resembles Natural

Table 12.3

Production & Utilization by Raisin Variety Grapes, green tons, 1967-68 through 1998-99.

Crop Year	Total Production	Dried	Crushed	Canned	Fresh Sales
1967-68	1,635,000	751,000	615,000	54,000	215,000
1968-69	2,135,000	1,110,000	709,000	64,000	252,000
1969-70	2,155,000	1,007,000	845,000	66,300	236,700
1970-71	1,871,000	820,000	851,000	53,700	146,300
1971-72	2,312,000	893,000	1,204,000	58,400	156,600
1972-73	1,344,000	436,000	717,000	50,500	140,500
1973-74	2,376,200	967,000	1,210,200	59,000	140,000
1974-75	1,970,000	1,021,200	754,100	61,200	133,500
1975-76	2,201,000	1,249,600	729,000	52,700	169,700
1976-77	1,957,000	981,000 [1]	755,000	48,000	173,000
1977-78	1,935,000	1,132,000	594,000	54,000	155,000
1978-79	1,670,000	758,000 [2]	702,000	55,000	155,000
1979-80	2,320,000	1,376,000	700,000	60,000	184,000
1980-81	2,692,000	1,612,000	778,000	63,000	239,000
1981-82	1,779,000	1,024,000	509,000	42,000	204,000
1982-83	2,642,000	1,530,000 [3]	774,000	35,000	303,000
1983-84	2,391,000	1,774,000	330,000	35,000	252,000
1984-85	2,227,000	1,387,000	580,000	30,000	230,000
1985-86	2,532,000	1,602,000	559,000	45,000	326,000
1986-87	2,045,000	1,180,000	505,000	40,000	320,000
1987-88	2,170,000	1,430,000	440,000	40,000	260,000
1988-89	2,570,000	1,810,000	415,000	40,000	305,000
1989-90	2,570,000	1,850,000	370,000	40,000	310,000
1990-91	2,345,000	1,730,000	268,000	40,000	307,000
1991-92	2,165,000	1,562,000	284,000	41,000	278,000
1992-93	2,670,000	1,554,000	785,000	46,000	285,000
1993-94	2,354,000	1,642,000	397,000	45,000	270,000
1994-95	2,389,000	1,884,000	197,000	38,000	270,000
1995-96	2,252,000	1,532,000	432,000	35,000	253,000
1996-97	2,192,000	1,308,000	618,000	36,000	230,000
1997-98	2,883,000	1,778,000	786,000	44,000	275,000
1998-99 [4]	1,960,000	1,255,000	449,000	36,000	220,000

[1] Doesn't include 293,000 green tons lost due to rain.
[2] Doesn't include 248,000 green tons lost due to severe weather.
[3] Doesn't include 360,000 green tons harvested but not sold and 60,000 green tons not harvested.
[4] Preliminary.

Source: California Fruit and Nut Statistics, 1965-77, 1973-84, 1985-86, 1986-87. California Resource Directory by California Agricultural Statistics Service.

Seedless raisins, but are less plump and contain more acid and occasionally has semi-hard (undeveloped, vestigial), seeds.

Monukka raisins are produced by sun drying Black Monukka grapes. They are usually large, dark, and plump.

Other Seedless raisins include all raisins produced from Ruby Seedless, Flame Seedless and other seedless grapes not included in any of the other raisin varietal types. The raisins are usually large, dark, and plump.

Marketing

The Buying & Selling of Raisins

Grower Sales

Prior to 1900, raisin sales between growers and packers were negotiated primarily on an individual basis. Various attempts to organize a cooperative raisin packing plant resulted in the formation of the California Associated Raisin Company in 1912. The name was changed to Sun-Maid Raisin Growers of California in 1922.

In 1966, the Raisin Bargaining Association (RBA), was formed by farmers whose desire was to organize a cooperative, to represent them as a single body in negotiating an annual price for their raisins with the industry's packers. The RBA grew over the years to a membership of 2,000 raisin farmers, making the largest bargaining cooperative of its type in the United States. The Raisin Bargaining Association is responsible for negotiating the free tonnage price for its members' raisins.

There is a third group of raisin growers that is not a member of Sun-Maid or RBA. This group is known as independent growers. Each grower group's representation of the industry varies from year to year. As of 1997, Sun-Maid represented almost 30% of the grower tonnage, RBA represented about 40% and the Independents about 30%. **Table 12.5** shows the quantity of all raisins and Natural Seedless raisins produced by crop year, along with RBA announced field price and the seasonal average price for Natural Seedless Raisins (see **Chapter 2**).

Domestic Marketing

Buyers in the domestic raisin market are categorized as retail, industrial-institutional (including mass feeders, cereal makers, bakers, confectioners), and government (military, school lunch programs, etc.). Each customer requires specifications that determine the variety of raisin and the packaging used. For example, many retail chain stores buy a specific grade and use the store's own private label or brand. Industrial-institutional users generally order supplies in large bulk fiber containers.

Some packers maintain their own sales departments, while others sell through food brokers. Brokers are independent sales agents, acting for the packer, and are paid on a commission basis.

Most raisins are moved to the domestic market by either rail or truck. To speed delivery, many packers maintain an inventory of raisins in the eastern part of the country. Each year, the industry pack varies between bulk and consumer packages. **Tables 12.6** shows the breakdown between consumer packages and bulk raisins for both the domestic and export markets from 1967-68 through 1998-99. During the 1998-99 crop year, California raisin shipments comprised 22.8% domestic consumer packages, 37.1% domestic bulk, 1.9% government purchases, 9.2% export consumer packages, and 28.3% export bulk.

Table 12.7 shows the raisin shipments by raisin varietal type from 1967-68 through 1998-99. The most popular rai-

regulate 32 commodities in 33 states. In addition there are some freestanding legislated programs that regulate domestic and imported commodities.

Early History

Hedlund also made the following remarks in the same speech concerning the early history and scope of the Federal Marketing Orders:

"The concept of marketing orders can be traced indirectly to the turn of the century shift from the suburban truck-crop and farm orchard system to specialized production on a large scale in areas far from market. Rising productive capacity for horticulture crops, especially in the South and Far West, highlighted the impact of short-term supply variation on prices and income. Producers, seeking to improve the marketing of their crops, turned first to cooperatives. Nearly 700 cooperative marketing associations for fruits and vegetables were organized between 1910 and 1920.

The various commodity groups, which tried voluntary market rationing, found that orderly marketing—an even flow to market with uniform quality and packaging—generally paid off. They also learned that strictly voluntary marketing programs almost always failed because those outside the programs share in the benefits, but bear none of the costs and have none of the limitations. Consequently, many members dropped out.

Table 12.5

Raisin Production, Field Price and Average Producer Price, 1967-68 through 1999-2000 by short tons, dry weight.

Crop Year	All Raisin Variety Production	RBA Announced		N.S. Season
		Natural Seedless (N.S.) Production	Free Tonnage (N.S.) Field Price	Average Producer Price
1967-68	181,821	161,320	$305	$291
1968-69	263,976	240,949	313	259
1969-70	251,357	227,429	318	261
1970-71	192,937	176,066	320	278
1971-72	194,924	172,347	325	319
1972-73	104,152	91,258	500	500
1973-74	225,568	198,753	700	700
1974-75	242,112	212,390	640	605
1975-76	283,595	253,271	648	607
1976-77	141,924	117,605	1,050	1,050
1977-78	248,942	218,813	840	856
1978-79	99,433	74,410	1,600	1,600
1979-80	303,147	263,108	1,150	1,160
1980-81	313,645	254,657	1,200	1,167
1981-82	257,227	224,463	1,275	1,230
1982-83	254,530	205,700	1,300	1,300
1983-84	387,334	347,943	1,300	573
1984-85	334,506	299,473	700	589
1985-86	407,110	362,657	820	599
1986-87	380,565	346,944	900	761
1987-88	385,218	352,498	945	809
1988-89	415,914	379,053	1,025	917
1989-90	432,550	335,501	1,115	987
1990-91	395,334	357,250	1,115	879
1991-92	387,205	352,659	1,155	962
1992-93	416,080	371,516	1,155	901
1993-94	437,432	387,007	1,155	905
1994-95	421,889	378,427	1,160	928
1995-96	370,758	325,911	1,160	1,007
1996-97	313,822	272,063	1,220	1,049
1997-98	432,629	382,448	1,250	947
1998-99	281,082	240,469	1,290	1,290
1999-00	348,016	299,910	1,425	1,211

Source: RAC Marketing Policies and final payment statements, Raisin Bargaining Association, and Clyde E. Nef (1998).

Because of this basic weakness in a voluntary organization, farmers and their representatives looked for other methods to set up workable marketing programs. The McNary-Haugen Bill of 1926 would have made it possible for agricultural cooperatives or other agencies, in concert with the Federal Government, to work towards promoting orderly marketing and stabilizing markets. This bill was vetoed, but it embodied major principles which were later incorporated in the Agriculture Marketing Agreement Act of 1937, the basic authority for Federal Marketing Orders today."

Scope

Over the years, the Agricultural Marketing Act of 1937 has been amended a number of times in order to provide a variety of regulatory activities from which producers may choose the right combination to suit a given marketing situation. These regulatory activities include maturity and/or size requirements, programs for allocating supplies among markets on a seasonal basis, container and pack regulations, marketing and production research projects, marketing development programs including paid advertising and credit for brand advertising, quality regulations for imported commodities, and reporting requirements. The principal costs of administering marketing orders are financed by assessments against the handlers i.e., packers and shippers, who are regulated.

Within legal limits, producers may configure a program to meet the marketing needs of their particular industry. The USDA expects the producers who propose a marketing order to identify and analyze their marketing problems and to justify those actions best suited to help solve them. The Secretary of Agriculture is responsible for insuring that a proposed order is legal and in the public interest.

Federal Raisin Marketing Order

Background

Clyde E. Nef, retired manager of the RAC and CALRAB describes the setting for the Federal Raisin Marketing Order as follows (Christensen, 2000.):

Table 12.6

Domestic and Export California Raisin Shipments by Types of Packages, packed short tons, 1967-68 through 1998-99.

Crop Year	Domestic shipments				Export shipments		Crop Year Totals
	Cartons/ bags	Bulk	USDA purchases, reserve	Other governemnt purchases	Cartons/ bags	Bulk	
1967-68	66,945	74,147	16,264	1,003	19,015	43,381	**220,755**
1968-69	61,480	74,914	12,119	2,646	19,052	42,967	**213,178**
1969-70	63,202	71,585	20,266	823	17,613	46,720	**220,209**
1970-71	63,253	70,580	2,937	1,613	17,168	40,811	**196,362**
1971-72	67,362	73,708	4,138	612	18,542	48,362	**212,724**
1972-73	46,014	56,539	0	127	8,583	8262	**119,525**
1973-74	76,359	67,846	0	85	16,857	24,173	**185,320**
1974-75	73,003	64,918	0	972	20,921	41,811	**201,625**
1975-76	74,255	72,077	7,627	507	18,352	35,193	**208,011**
1976-77	64,840	65,985	0	0	18,289	22,486	**171,600**
1977-78	69,904	70,635	0	35	22,760	30,500	**193,834**
1978-79	47,156	59,472	2,079	0	13,052	10,659	**132,418**
1979-80	70,429	71,076	7,421	206	20,057	49,958	**219,147**
1980-81	75,037	88,514	4,252	26	22,812	49,917	**240,558**
1981-82	78,027	90,786	7,065	15	24,399	40,384	**240,676**
1982-83	81,238	90,539	3,094	25	18,413	39,640	**232,949**
1983-84	76,554	94,564	4,664	31	20,767	44,845	**241,425**
1984-85	93,128	112,791	7,882	54	22,881	56,910	**293,646**
1985-86	89,873	123,020	4,562	26	26,184	65,913	**309,578**
1986-87	87,460	119,774	4,778	44	28,531	67,061	**307,648**
1987-88	91,691	127,090	3,528	75	33,552	77,191	**333,127**
1988-99	96,948	140,483	8,492	158	33,514	74,235	**353,830**
1989-90	92,823	130,014	3,724	491	32,969	83,593	**343,614**
1990-91	90,978	130,671	12,232	214	39,396	88,005	**361,496**
1991-92	87,528	132,515	12,692	34	33,647	89,829	**356,245**
1992-93	90,250	130,214	10,465	4	32,286	91,615	**354,834**
1993-94	88,938	134,221	16,167	0	32,259	100,109	**371,694**
1994-95	82,809	127,193	5,304	0	32,086	101,438	**348,830**
1995-96	82,570	122,072	5,627	0	32,096	96,910	**339,275**
1996-97	74,389	128,129	219	0	31834	95245	**329,816**
1997-98	79,005	125,282	755	0	29,366	102,954	**337,362**
1998-99	73,158	121,085	6,041	0	29,437	90,452	**320,173**

[1]Crop year: 1967-74 = Sept.1-Aug.31; 1976-present = Aug.1-July 31.

[2]1975 crop year = 11 mo. statistics - Sept. 1, 1975-July 31, 1976.

Source: RAC Final Shipment Report for each crop year.

After World War II, the United States wartime allies, to whom we had been supplying some food commodities, were in economic recovery and unable to purchase other than basic food commodities. The California raisin industry, producing more raisins than the world market could consume, turned to the Agricultural Marketing Agreement Act of 1937, (often referred to as 'the Act') which allowed them to unite for specific purposes and not be subject to antitrust action. The industry felt that a Federal Marketing Order would be an important tool to improve returns to growers, through orderly marketing. The raisin industry was specifically interested in the minimum grade and condition standards and in the volume control provisions of this law which allow the industry to satisfy all the needs of both the domestic and export markets before siphoning off the extra supplies during years of excessive production. The law allows the regulated raisins to be stored to help supply those years when production is down and used to help supply the market demand during the short raisin production years.

Producers held a series of meetings to discuss the needs of the raisin industry. They developed a proposed federal marketing order for raisins and submitted it to the United States Department of Agriculture (USDA). The USDA held a hearing to receive testimony on the proposed Raisin Marketing Order. The raisin industry went through the referendum process, which allowed it to vote on the proposed Raisin Marketing Order and Agreement. Simultaneously, producers were allowed to vote on the Federal Marketing Order and handlers were allowed to vote on the Federal Marketing Agreement. The Federal Marketing Order and Agreement Regulating the Handling of Raisins Produced from Grapes Grown in California were made effective in August 1949.

To establish or amend a Federal Marketing Order, a proposal must be approved by a two-thirds vote of producers by number and/or volume of those producers who voted in such referendum. To establish or amend a Federal Marketing Agreement, the proposed provisions must be approved by the majority of the volume of the raisin tonnage in the industry. Since the Raisin Marketing Order was established, it has been amended fourteen times through amendatory hearings and referenda.

Marketing Order Administration

The Act provides the legislative authority under which an industry, through an Administrative Committee or Board, can develop regulations to fit its own situation and solve its marketing problems. This system permits the Committee or Board members (people who know the industry and its problems) to design and recommend regulations that will best serve the industry and work towards orderly marketing.

Either an Administrative Committee or Board operates each of the 36 Federal Marketing Programs for fruits, vegetables and specialty crops. The Committee members and alternates are nominated by the industry and subsequently are selected by the Secretary of Agriculture. These members and alternates represent all the producers and handlers within the industry. In addition, many of these committees and Boards have consumer representation through a public member.

Raisin Administrative Committee

The Raisin Administrative Committee (RAC) administers the Raisin Marketing Order. The RAC is the largest of the 36 Federal Committees/Boards. For a number of years, the RAC consisted of 47 members. These 47 positions are comprised of 35 grower representatives, 10 handler representatives, one public representative, and one representative of the Raisin Bargaining Association. Each position has an alternate member. The Committee employs a management staff to carry out their actions and recommendations as well as conduct day to day business, including operations, account-is divided into each respective trade demand to calculate the final free and reserve percentages. In the event that the free tonnage pricing has been established for the year between handlers and growers or that additional information has become known about the crop size, the RAC can meet prior to February 15 to compute and announce interim free and reserve percentages. **Table 12.11** shows the preliminary free and reserve percentages for each raisin varietal type during the 1999-2000 crop year.

Table 12.12 shows the preliminary percentages and the final percentages that the Secretary of Agriculture established for the Natural Seedless raisins from the 1981-82 through 1999-2000 crop years.

Table 12.9

Raisins and Zante Currant Exports from United States, packed short tons, 1995-96 through 1998-99, crop year August 1 - July 31.

Country of Destination	1995-96		1996-97		1997-98		1998-99	
	Natural Seedless	Other*	Natural Seedless	Other*	Natural Seedless	Other*	Natural Seedless	Other*
Austria	80	3			335	36	224	23
Belgium	1,038	322	1,265		511	384	426	355
Denmark	6,139	198	5,853		5,453	0	5467	0
Ireland	87		203		142	0	117	0
Finland	2,894	9	2,483		2,514	7	2589	6
France	340	54	405	2	1,423	147	701	91
Germany	10,543		8,805		10,851	23	7052	14
Israel	157	612	222	59	126	926	125	1081
Italy	15	34	35		84	12	100	0
Netherlands	2,997	836	2,793		3,862	621	3032	229
Norway	2,413		2,608		2,199	17	2435	0
Spain	585	8	588		813	7	650	0
Sweden	5,224	55	5,423		5,825	34	5695	0
Switzerland	421		469		59	0	88	0
U.K.	29,599	1,650	27,688		30,974	674	27393	415
Total Europe	**62,532**	**3,781**	**58,840**	**61**	**65,171**	**2,888**	**56092**	**2228**
Brazil	466	91	642		1,493	252	534	30
Columbia	90		155		408	1	111	0
Costa Rica	56	2	88		220	1	81	1
Dominican Rep.	454	1	470		610	1	500	1
Ecuador	30	1	32		49	3	47	4
Mexico	473		658		504	2	22	0
Panama	403	4	448		589	3	481	1
Puerto Rico					0	0	390	1
Venzuela	377	12	379		518	22	353	9
Others	1,464	26	1,174		328	38	437	6
Total Latin America	**3,813**	**137**	**4,046**	**0**	**4,719**	**323**	**2958**	**53**
Hong Kong	1,312	3,567	1,445		2,065	3,343	984	1203
Iceland	262	1	221		221	1	155	2
Japan	26,404	323	28,822	163	27,027	587	32684	876
Korea	2,527	65	2,814		2,410	18	2522	0
Malaysia	656	522	749		789	651	517	475
New Zealand	1,499	39	1,760		1,548	34	1,459	45
NIS (USSR)	534	6	893		1,706	21	360	0
Phillippines	723	20	1,152		1,441	3	1,381	1
Singapore	2,743	1,770	2,594		2,092	999	2,233	1,060
Taiwan	3,691	123	3,918		4,570	557	3,889	386
Others	2,376	1,432	2,461		3,377	1,356	2,396	1,229
Total other countries	**42,727**	**7,868**	**46,829**	**163**	**47,246**	**7,570**	**48,579**	**5,277**
Canada	6,839	1,309	5,628	183	3,707	696	4,189	528
Grand total	**115,911**	**13,095**	**115,343**	**407**	**120,843**	**11,477**	**111,817**	**8,086**

Varieties include: Golden Seedless, Zante Currants, Dipped Seedless, Oleate and related seedless, Muscats, Monukkas and other seedless.

Source: Raisin Administrative Committee.

Managing the Reserve Pool

As authorized by the marketing order, Reserve pool raisins are disposed of through offers developed and announced by the RAC. Reserve tonnage raisins may be offered for free tonnage use in domestic or export markets through the 10 + 10 offers, government feeding programs; direct or blended export programs, Raisin Diversion Program (RDP); by gift; or for disposal in non-normal outlets. Of these disposition possibilities, the three that have the greatest historical use are the 10+10 offers, the Raisin Diversion Program and the blended export programs.

10 + 10 Offers

When the trade demand formula was implemented in the marketing order, the base tonnage of natural condition raisins which is needed to supply the packed tonnage marketed is reduced by 10% so that the market won't be oversupplied early in the crop year. In years where there is pooling, the committee must make two offers, known as the 10+10 offers. These offers have been made on or before November 15 however, during 1997 a change was made to the Raisin Marketing Order that would allow the offers to be made at a later date during the crop year. In the trade demand formula, the difference in tonnage between the sweatbox base tonnage and the adjusted base tonnage becomes the first 10% offer in the 10+10 offers. For the 1998-99 crop, this tonnage was 31,130 tons (28,017 t) for Natural Seedless raisins (see **Table 12.10**). This offer is allocated to handlers on the basis of their prior year's acquisitions. The objective of this offer is to make available to the industry sufficient tonnage of raisins to meet the same shipment volume as in the prior year. Delaying this offer until early to mid-November allows the majority of the raisin deliveries to be received by handlers. The handlers have a better knowledge of the quantity of raisins they expect to receive and the quantity of raisins they need to buy from the 10+10 offers. It also helps handlers in their financial planning by not requiring them to have a larger line of credit earlier in the season.

Table 12.10
1999-2000 Trade Demand, Raisin Administrative Committee.

	Natural	Dipped	Oleate	Golden	Zante Currants	Sultanas	Muscats	Monukkas	Other
Base shipments (packed tons)	277,305	14,554	252	14,815	3,158	72	3	669	3,304
% Shrink factor (5 year average)	0.93874	0.91900	0.79208	0.90405	0.89171	0.70222	-0.52095	0.82929	0.94005
Shrink %	6.126	8.100	20.792	9.595	10.829	29.778	152.095	17.071	5.995
=Base tonnage (sweatbox tons)	295,401	15,837	318	16,387	3,542	103	(5)	807	3,514
x 90% Formula	90%	90%	90%	90%	90%	90%	90%	90%	90%
= Adjusted base	265,861	14,253	286	14,748	3,188	93	(5)	726	3,163
Physical inventory 7/31/99	101,946	1,523	289	4,978	1,906	40	80	79	1,572
- Desirable inventory	73,809	2,685	6	4,381	573	17	0	152	586
= +/- Inventory adjustment	(28,137)	(1,838)	(283)	(597)	(1,333)	(23)	(80)	73	(986)
=Computed Trade demand	237,724	12,415	3	14,151	1,855	70	(85)	799	2,177
Recommended Trade Demand	**254,475**	**14,635**	**500**	**14,151**	**1,855**	**500**	**500**	**799**	**2,177**

* RAC adopted a policy that any trade demand computed at less than 500 tons will be set at 500 tons.
Source: Raisin Administrative Committee

No handler is required to purchase any of the tonnage offered in the first 10% offer. It is based on the established field price plus 3%, plus the estimated costs incurred by the RAC for the equity holders. This encourages handlers to seek their tonnage directly from growers rather than wait and purchase from the reserve pool, thus creating an active market.

The second 10% offer consists of the same quantity of raisins as in the first 10%t offer and is offered based on each handler's prior year's shipments. This offer is intended to provide handlers with tonnage they anticipate needing for growth. The price is the same for both offers unless the second offer is made at a later time and interest is added to the price. Tonnage remaining uncommitted in the initial offers is re-offered to those handlers who accepted 100% of their initial offered tonnage and allocated based on the tonnage accepted in the initial offers. If needed, there can be up to two re-offers and a final offer.

Raisin Diversion Program

In the early 1980s, the U.S. Government implemented a Payment in Kind (PIK) Program for cotton producers. When surplus cotton crops were produced, the government purchased the surplus from growers, resulting in huge cotton inventories in government storage. The PIK program gave cotton already in government storage to producers who agreed to "fallow" land on which they had historically produced cotton. This helped reduce the surplus cotton supplies.

This program was reviewed by the RAC, which led to the development of an industry program known as the Raisin Diversion Program (RDP). The purpose of the RDP is to avoid adding to raisin tonnage already in storage, which is, adequate to meet next year's market needs and opportunities. Under the RDP, certain raisin acreage is diverted into non-production for that season. The RAC utilizes, in substitution, raisins, which are already in the reserve pool and stored on packer's premises.

On or before November 30, the RAC reviews supply and demand statistics and determines if an RDP is warranted. If warranted, the RAC approves and announces the tonnage available for diversion as well as the allowable harvest costs to be applicable to such diversion tonnage.

Raisin growers who want to participate in the program must submit an application to the RAC. Growers must report the size and location of the production unit they want to divert, the weight of the raisins produced on this unit during the most recent crop year, the packer to whom the raisins were delivered, and whether they will remove the vines or divert (abort) the grape crop on this unit. RAC staff reviews each application and if it complies with the announced provisions of the program, the producer is sent an approved copy of his application. In some years, it has included a lottery drawing.

Beginning in June and before a grape crop could be harvested, each production unit is visited by an RAC Compliance Examiner to determine if the vines have been removed or the grape crop diverted (aborted). Diversion (abortion) is generally accomplished by spur pruning cane pruned varieties that have sterile basal buds, and if necessary, removing the few bunches that may develop while still very immature. Each grower is notified of the results of the Examiner's visit. The RAC staff verifies the weight of the raisins produced on each production unit from delivery reports submitted by the packer. After the subsequent raisin crop is produced (usually during early October), the grower is mailed a diversion certificate equal to the weight of raisins produced on the qualified production unit in the prior crop year.

The grower delivers the certificate to the packer of his choice, just as if he were delivering raisins. Since such growers had no harvest cost, the packer deducts the harvest cost on the grower's total tonnage from his payment of the free tonnage to the grower. The packer then submits this certificate together with the amount of the deducted harvest costs to the RAC. The RAC releases the weight equivalent of the certificate from the reserve pool. The packer then handles this tonnage the same as new crop deliveries from other raisin growers, including volume controls if applicable for that crop year.

The Raisin Diversion Program is totally funded by the raisin industry through the reserve pool.

Table 12.13 shows the raisin diversion tonnage for each crop year since the RDP was implemented in 1985-86 through the 1999-2000 crop year.

Table 12.11
Raisin Preliminary Percentages, 1999-2000.

Variety	Trade Demand (short tons)	Estimated Production (short ton)	Preliminary Free	Percentages Reserve
Natural Seedless	254,475	294,519	56%	44% [1]
Dipped Seedless	14,635	19,012	50%	50% [2]
Oleate Seedless	500	505	100%	0%
Golden Seedless	14,151	16,180	100%	0%
Zante Currant	1,855	4,187	29%	71% [3]
Sultana	500	165	100%	0%
Muscat	500	81	100%	0%
Monukka	799	506	100%	0%
Other Seedless	2,177	2,604	100%	0%

*Percentages shown reflect 65% of trade demand.

[1]Interim percentages of 84.75% free, 15.25% reserve for natural seedless raisins was announced on Feburary 14, 2000. The interim final raate establishing final percentages of 85% free, 15% reserve was published on April 10, 2000.

[2]On October 25, 1999, the Dipped Seedless percentages were announced 100% free, 0% reserve.

[3]Interim percentages of 50.75% freem 49.25% reserve for Zanter Currant raisins January 13, 2000. The interim final rule establishing final percentages of 51% free, 49% reserve was published on April 10, 2000.

Source: Raisin Administrative Committee Marketing Policy, 1999-2000.

Government Purchases

The USDA has used California raisins in government feeding programs such as school lunch, hunger prevention and other government institutions. The RAC periodically establishes a price and tonnage of reserve raisins available for the government feeding programs. Invitations to submit bids are announced by the Department and handlers submit their bids pursuant to the terms and conditions of such invitations. When successful bidders are notified, they can either purchase reserve tonnage directly to fill their bid or ship free tonnage and replace it with reserve tonnage. Documents must be submitted to substantiate government shipments. From November 1965 until December 1997, the USDA has purchased over 227,700 packed tons (204,930 t) of surplus/reserve raisins for distribution into government feeding programs.

Export Programs

In the late 1950s and early 1960s, the U.S. wheat industry developed a program to educate Japanese bakers on the use of wheat flour. The raisin industry worked with the wheat industry to develop and promote raisin bread. At that time, virtually all raisins exported to Japan were imported through importers, not users, and they had a well-organized importer association. Working with the Japanese Dried Fruit Importers Association (JDFIA), the RAC implemented incentives for the importers to purchase California Raisins. The first Merchandising Incentive Program (MIP) was implemented in Japan. The raisin industry slowly developed their export markets for California raisins. Today the export market makes up approximately one-third of California's raisin shipments and Japan is one of the largest single export markets.

Prior to the 1976-77 crop year, in order to be competitive in the export markets, the RAC sold raisins to handlers directly out of the reserve pool at lower prices than the free tonnage prices. In 1976, the raisin crop was smaller and the RAC did not implement reserve percentages. California raisins shipped into the export markets during the 1976-77 crop year were from the free tonnage. When the RAC implemented free and reserve percentages due to a large crop in 1977-78 crop year, the trade demand was large because the prior year's free tonnage was shipped both to the domestic and export markets. The RAC needed to remain competitive in the export market, but could no longer sell California raisins in the export markets through this traditional method directly from reserve pool through reserve offers.

The RAC then developed "blended price" programs to handlers. Initially these programs were 100% raisins from the reserve pool. It progressed to 50% reserve tonnage and 50% cash and now it is a 100% cash adjustment program. To someone not familiar with these programs, they may seem quite complicated. However, they are rather basic programs.

Under a 100% cash Adjustment Export Program for Japan, the following would occur when an established free tonnage price of $1290 per ton and a desired export blended price is $940 per ton. A handler would process and export free tonnage raisins and provide the RAC with copies of an endorsed on-board bill of lading substantiating the export of such raisins to Japan. From this documentation, the RAC would identify the type and number of cases of raisins exported and computes the natural condition equivalent for each shipment. For each ton of raisins (natural condition equivalent) documented, the RAC will pay the handler $350 ($1290-940) as cash adjustment from the reserve pool equity. Whether reserve tonnage at a reduced price, 50% reserve tonnage and 50% cash, or when 100% cash adjustment is used for blending export prices, the impact on reserve pool equity is practically the same.

Other Reserve Offers

The Marketing Order allows the RAC to donate reserve raisins as a gift. Over the years, the raisin industry has donated a small volume of reserve raisins, both in the U.S. as well as to their export markets. Most generally such donated raisins go to natural disaster victims. In addition, the RAC has donated small quantities of raisins to the Fresno Farm Bureau for distribution at conferences and conventions.

Other provisions of the Marketing Order allow the RAC to make reserve offers for free tonnage use when there is a national emergency, crop failure, or a change in economic or marketing conditions. During the eleventh month of the crop

Table 12.12

Natural Seedless Raisin Reserve Pool Final Percentages, 1981-82 through 1999-2000.

Crop Year	Preliminary Percentages		Established by Secretary		Date Established
	Free	Reserve	Free	Reserve	
1981-82	71	29	80	20	2/10/82
1982-83	43	57	100	0	11/3/82 [1]
1983-84	35	65	37.5	62.5	4/3/84
1984-85	53	47	61	39	4/11/85
1985-86	51	49	59	41	3/17/86
1986-87	66	34	66	34	5/1/87
1987-88	63	37	67	33	6/1/88
1988-99	59	41	70	30	5/5/89
1989-90	53	47	73	27	6/13/90
1990-91	56	44	69	31	5/24/91
1991-92	55	45	79	21	7/17/92
1992-93	64	36	71	29	10/7/93
1993-94	62	38	74	26	6/27/94
1994-95	62	38	77	23	5/17/95
1995-96	65	35	79	21	2/26/96 [2]
1996-97	73	27	86	14	4/14/97 [3]
1997-98	46	54	66	34	3/18/98 [4]
1998-99	59	41	100	0	1/15/99 [5]
1999-2000	56	44	85	15	4/10/00 [6]

[1] Preliminary percentages were recommended by the Committee and a proposed rule issued for comment. Before the comment period expired, the crop was severely damaged by rain and upon a recommendation of the Committee, the proposed preliminary action was withdrawn and the crop declared 100% free.

[2] Interim final percentages were published on 2/26/96. The final rule establishing percentages as published in the interim final docket was published on 7/15/96.

[3] Interim final percentages were published on 4/14/97. The final rule establishing percentages at 86% free, 14% reserve, was published 6/16/97.

[4] The interim percentages of 61% free, 39% reserve was announced on October 17, 1997. Interim final rule establishing final percentages was published on 3/18/98. The final rule establishing final percentages at 66% free, 34% reserve, was pulblished 6/1/98.

[5] Interim percentages of 85% free, 15% reserve was announced on November 13, 1998. At the January 15, 1999 RAC meeting, RAC revised their crop estimate to 235,000 tons, which is lower than the trade demand of 256,075 tons. RAC unanimously passed motion to make the Natural Seedless free percentages 100% and reserve percentages 0% for the 1998-99 crop year.

[6] Interim percentages of 84.75% free, 15.25% reserve was announced on February 14, 2000. The interim final rule establishing final percentages was published on April 10, 2000.

Table 12.13

Raisin Diversion Program, 1985-86 through 1999-2000.

Crop Year	Natural Condition Tonnage
1985-86	59,361 [1]
1986-87	103,606 [1]
1987-88	29,815 [1]
1988-89	49,447 [1]
1989-90	0
1990-91	0
1991-92	39,833 [1]
1992-93	24,746 [1]
1993-94	49,482 [1]
1994-95	0
1995-96	61,576
1996-97	2,051 [1,4]
1997-98	0
1998-99	127 [2,4,5]
1999-2000	0

[1] Natural Seedless raisins.

[2] Zante Currants.

[3] Natural Seedles raisins 61,098 & Zante Currants 478.

[4] Vine removal only.

[5] Zante Currant raisins from 1997-98 pool.

Source: Table 3 of the 1995-96, 1996-97, 1997-98 and 1999-2000 RAC Market Policy Reports.

year, the RAC can also make a reserve offer of the quantity of free tonnage shipments during the then current crop year that exceeds the shipments of the first 10 months of the prior crop year. These offers allow the RAC to react to market growth or to a sudden change in the market.

Establishing Minimum Quality Regulations

Under the provisions of the Marketing Order, the Raisin Industry has implemented minimum grade and condition standards for both natural condition (unprocessed) and processed raisins.

The Processed Products Inspection Branch of the United States Department of Agriculture (USDA) inspects raisins at least twice as they move from the producer through marketing channels. The first inspection is called incoming inspection. USDA inspectors examine the raisins, container by container, as they are delivered to the packers. All California raisins must meet the incoming grade and condition standards. These standards establish limits or tolerances for moisture and maturity as well as other grade defects, including molds, mechanical damage, sunburn, sugaring, caramelization and uncured berries. Periodically these requirements are modified and changes may be obtained from the USDA or RAC. The maximum allowable moisture for natural condition raisins is 16% for all raisin varieties except 14% for artificially dehydrated raisins (i.e. Goldens and Dipped Seedless) or oleate-treated raisins. The incoming moisture and maturity requirements are shown in **Table 12.14**. These low moisture limits are needed to help ensure that the raisins will not ferment while being stored prior to processing. Also during processing the raisins are washed, picking up moisture content, and they must meet the outgoing 18% moisture limit for all raisins except the 19% moisture requirement for Monukka raisins. Raisin maturity is mostly measured with the air stream sorter, with the exception that the maturity of Flame Seedless, Monukka, Other Seedless, Sultana and Muscat raisins are determined visually due to their widely different physical characteristics. The incoming tolerances are set at 50% B or Better maturity and 5% substandard. Subject to the agreement between the handler and producer, a producer can deliver poorer quality raisins as meeting raisins under a weight dockage system. The weight of the excess immature and excessive substandard raisins is deducted from the net weight of the incoming load. The grade defect limitations including mechanical damage, sunburn, sugaring, caramelization, mold, uncured berries and contamination are shown in **Table 12.15** and **Figures 12.1 – 12.5**.

The purpose of incoming inspection is twofold. First, the USDA inspection service is a neutral independent third party judging the quality of the natural condition raisins for the

Fig. 12.3: The quality of raisins is measured according to "B or better" maturity and the percentage of substandard raisins in a lot. A USDA Lead Aid Phyllis Thraikill pours raisins onto the feed belt of the air stream sorter which separates the substandard raisins from the "C" grade or better raisins. The air stream sorter on the right segregates the "B" or better raisins from the "C" grade and substandard raisins.

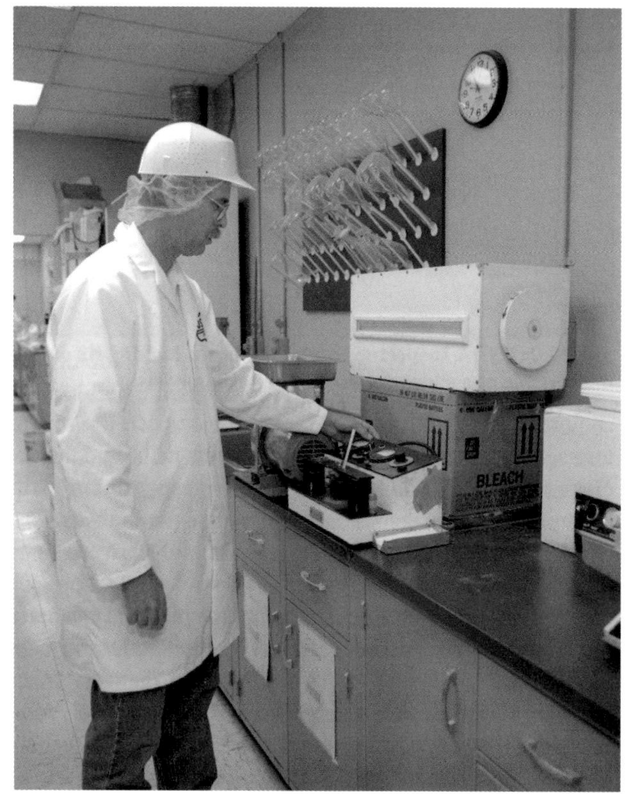

Fig. 12.4: USDA Inspector Erik Palko is shown determining the moisture of raisins using the Dried Fruit Association, DFA, moisture tester.

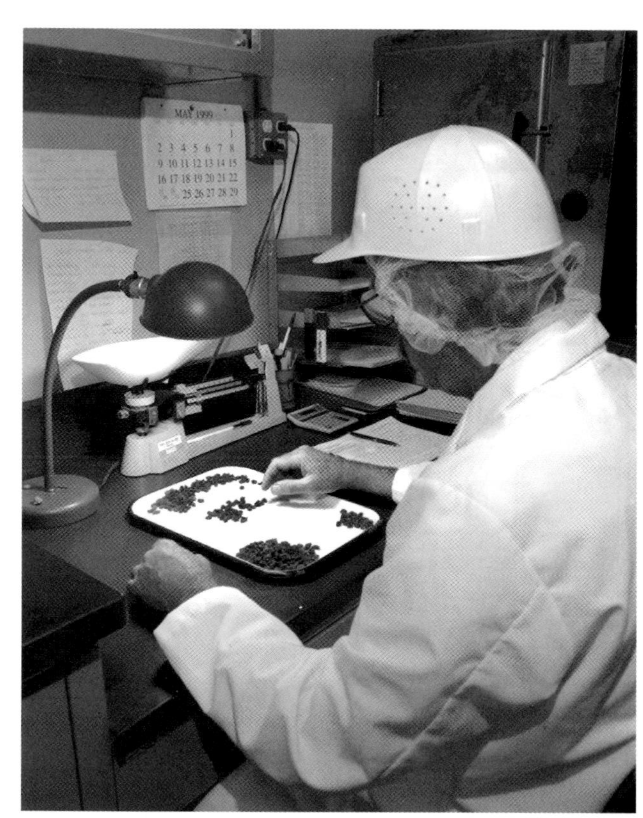

Fig. 12.5: USDA Inspector Erik Palko is determining the moisture of raisins using the DFA moisture tester.

thirds by number of producers of raisins or by their volume of raisin produced from grapes grown in California who voted in a referendum. Once the USDA approved the Raisin Marketing Order, it became Federal Law, enforceable by the U.S. Department of Justice. The USDA is responsible for administrative oversight of this program. Thus, it reviews and approves the recommendations from the RAC. The USDA oversight is to assure compliance with the provisions of the raisin marketing order and the declared policy of the Agriculture Marketing Agreement Act of 1937.

Acknowledgments

Acknowledgment is made to the many people in industry and government whose information and assistance aided in preparing this chapter, especially Norman Engelman, Chairman of the Raisin Administrative Committee (RAC) and President of the Raisin Bargaining Association (RBA), Fresno, California; Clyde E. Nef, retired Manager of the RAC, Fresno, California; Terry W. Stark, General Manager, RAC, Fresno, California; Ronald W. Worthley, Director of Finance, RAC, Fresno, California; Harry Rixman, Export Director, RAC Chief Executive Officer, Fresno, California; Debbie Pilloud, Director of Operations, RAC, Fresno, California; Vaughn Koligian, formerly Chief Executive Officer, RBA, Fresno California; Barry F. Kriebel, President, Sun-Maid Raisin Growers, Kingsburg, California; Pete J. Penner, former Chairman, Sun-Maid Raisin Growers, Kingsburg, California; Floyd F. Hedlund, Retired Director, Fruit and Vegetable Division, Agriculture Marketing Service, USDA, Washington, DC; Joseph C. Perrin, Retired Marketing Specialist, Fruit and Vegetable Division, Agriculture Marketing Service, USDA, Portland, Oregon; Yoshiki "Junior" Kagawa, Processed Products Branch, Agricultural Marketing Service, USDA, Fresno, California; and Glen Yost, Senior Ag Economist, Marketing Branch, California Department of Food and Agriculture, Sacramento, California.

Table 12.16
Summary of Outgoing Raisin Inspection Requirements, 11/15/1985.

	Marketing Order - Grade														
	Grade A					Grade B					Thompson Modified C Grade				
	Thomp.	S. Mus.	Lo. Mus.	Sult.	Curr.	Thomp.	S. Mus.	Lo. Mus.	Sult.	Curr.	Sel/Mix	Midget	S. Mus.	Lo. Mus.	Sult.
Stems	1/96	1/32	3/96	3/96	4/96	2/96	2/32	6/96	6/96	6/96	2/96	4/96	3/32	9/96	9/96
Cap stems	15	10	10	25	1-1/2	25	15	15	45	22	35	35	20	20	65
Discolored, damaged or moldy raisins	4	5	5	4	5	6	7	7	6	7	9	9	9	9	9
Not to exceed damage	2	3	3	2	2	3	4	4	3	3	5	5	5	5	5
Not to exceed mold	2	2	2	2	3	3	3	3	3	4	4	4	4	4	4
Sugared	5	5	5	5	5	10	10	10	10	10	15	15	15	15	15
Seeds		12					15						20		
Well-matured and reason, well-matured	80	80	80	80	75	70	70	70	70	0	70	70	0	0	0
Moisture (All Grades)	18.0%	19.0%	18.0%	18.0%	20.0%	Midget Size: 70.0% thru 22/64, 95.0% thru 24/64. Select Size: Limited to 10.0% thru 20/64, 60.0% thru 22/64.									
Monukkah Type	19.0	Thompsons				Thompsons					Thompsons				
Substandard and Undeveloped	Selec - Sub. and Und. 1.0% Mixed - Sub. and Und. 1.0% Midget - Sub. and Und. 2.0%					Select - Sub. and Und. 1.5% Mixed - Sub. and Und. 2.0% Midget - Sub. and Und. 3.0%					Select - Sub. and Und. 1.5% Mixed - Sub. and Und. 2.0% Midget - Sub. and Und. 5.0%				
	Currants - Muscats - Sultanas Sub. and Und. 2.0%					Currants - Muscats - Sultanas Sub. and Und. 5.0%					Muscats - Sultanas Sub. and Und. 8.0%				

Photo Credits

Figures 1-5: Randy Vaughn-Dotta

References

Christensen, Peter. 2000. Raisin Production Manual. University of California, Agriculture and Natural Resources, Communication Services. Publication #3393.

Nef, Clyde E. 1998. The Fruits of Their Labors.... Malcolm Media Press, Fresno, CA.

University of California Division of Agriculture Sciences. Aug. 1975. Marketing Orders in California: A Description. Leaflet 2719.

Raisin Administrative Committee Reports:

Weekly Report of Deliveries to Handlers
Monthly Comparative Raisin Shipment Data (packed tons)
Monthly Shipment Report Designating Countries of Destination by Varietal Type
Monthly Report of Raisin Industry Shipments in Packed Tons
Accumulative Report of Raisin Industry Shipments in Packed Tons
Annual Marketing Policy Reports
Annual Compliance Plans

Raisin Bargaining Association Information:

Bargains for Raisin Field Price since 1967

California Agriculture Statistics Service (formerly California Crop & Livestock Reporting Service) Reports:

California Agricultural Resource Directory (yearly)
California Grape Acreage
California Fruit and Nut Review
Grape Crush Report
Reports no longer published:
Marketing California Dried Fruits
California Fruit and Nut Statistics
California Fruit and Nut Acreage
California Grapes, Raisins and Wine

California Marketing Field Office, Agricultural Marketing Service, Fruit and Vegetable Programs, USDA Information:

Raisin Marketing Order
Raisin Imports
USDA Marketing Bulletin No. 58 – Raisins

Processed Products Branch, Fruit and Vegetable Programs, Agricultural Marketing Service, USDA Reports:

Inspection and Grading Manuals for: Receiving Natural Condition Raisins; Processed Raisins and; Lab Procedures for Natural Condition Raisins.
U.S. Standards for Grades of Processed Raisins

Additional Sources of Information

See Also ***Appendix A***

Raisin Administrative Committee
3445 N. First St.
Fresno, CA 93726
Mailing address:

P.O. Box 5217
Fresno, CA 93755
Phone: (559) 225-0520
FAX: (559) 225-0652
E-mail: terry@raisins.org
Website: http://11www.raisins.org/

Raisin Bargaining Association
3425 N. First St., Suite 209
Fresno, CA 93726
Phone: (559) 221-1925
FAX: (559) 221-0725
E-mail: rba@lightspeed.net

California Marketing Field Office
Agricultural Marketing Service
Fruit and Vegetable Programs
USDA
2202 Monterey St., Suite 102B
Fresno, CA 93721
Phone: (559) 487-5901
FAX: (559) 487-5906

Processed Products Branch
Fruit and Vegetable Programs
Agricultural Marketing Service, USDA
2202 Monterey St., Suite 102A
Fresno, CA 93721
Phone: (559) 487-5210
FAX: (559) 485-5914
Website: http://11ams.usda.gov/standards/dried.htm

California Agriculture Statistics Service (formerly California Crop & Livestock Reporting Service)
U.S. Department of Food and Agriculture
P.O. Box 1258
Sacramento, CA 95812
1220 "N" St., Rm. 243
Sacramento CA 95814
Phone: (916) 498-5161
Website: http://11www.cdfa.ca.gov

Marketing Order Administration Branch
Fruit and Vegetable Programs
Agricultural Marketing Service, USDA
Room 2525-S
P.O. Box 96456
Washington, D.C. 20090-6456
Phone: (202) 720-2491
FAX: (202) 205-6632
Website: http://11www.ams.usda.gov/fv/moab.html and
http://11www/access.gpo.gov/nara/CFR/Index.html

CHAPTER 13.

Marketing

Clyde Nef, M.S.

The Product

All raisins are dried grapes, but not all dried grapes are raisins. **Chapters 5** and **7** have presented the production systems used to dry grapes and the end product produced. Those dried vine fruits are called currants, sultanas and raisins. Consumers, in some areas of the world, view these dried vine fruits as very different commodities, with specific uses, and, in many instances, not interchangeable. Thus, depending on the market, the marketing strategies for these products can, and do, vary widely. This is the challenge to the dried fruit marketer. The major thrust of this chapter will be to discuss the strategies used to market dried vine fruits produced in California which are known worldwide as raisins, as well as the small annual volume of Zante Currants also produced in California.

Currants, known in California as Zante Currants, are in fact not currants at all, but are dried from a grape variety which belongs to the *Vitis vinifera* species named Black Corinth. The grape berries are small and when ripe are very dark. The dried fruit is very small and almost black. The United Kingdom is the single largest user of currants which are used mainly in biscuits and tea cakes to be eaten with tea. Most ingredients in baked products are measured by weight and/or volume. Because of their small size, a pound of currants will give the appearance of a greater volume of fruit than raisins or sultanas would in biscuits or tea cakes. In the United States, currants have been a major ingredient in the production of mincemeat; with the reduced demand for mincemeat, the major market for Zante Currants has disappeared.

Sultanas, as described in **Chapter 7**, are generally a golden to amber color. Because of the production, harvesting and drying methods used to produce sultanas, they are generally softer and sweeter than sun-dried raisins and they do not have the "caramelized" flavor that characterizes sun-dried raisins. Sultanas have traditionally been used by housewives and by commercial bakeries in major sultana markets as an ingredient in baked products. Most recipes developed in these countries specify sultanas as the dried vine fruit to be used. Major Sultana consuming markets are Belgium, France, Germany, Italy, the Netherlands and the United Kingdom.

It is important to emphasize the distinction of sultanas as a type of dried fruit based on the way sultanas are made. The method of producing sultanas is outlined in **Chapter 7** and does not refer to the variety *Vitis vinifera* Sultana which is a seedless grape. Some consumers perceive sultanas and raisins as completely different commodities even though they are essentially the same grape processed by two different methods. The sultana has a different texture, color and flavor. The basic flavor difference is due to less oxidation of the sugars due to less exposure to the sun and the protection afforded them by the water soluble oil used in the dipping solution. Rack drying in Australia and South Africa also protects the drying grapes from direct sun. Marketers of California raisins who have as their objective to get sultana consumers to use California raisins are challenged to change this perception. They must convince sultana users that raisins come from the same source are readily interchangeable with sultanas.

Sun-dried raisins in most markets of the world, including the United States, were originally produced from Muscat grapes. With the discovery that the Thompson Seedless grape could be sun dried without subjecting it to a potassium carbonate and water soluble oil solution, such natural sun-dried seedless grapes became the almost exclusive type of raisin produced today. Natural sun-dried seedless raisins are the "newcomers" in the world of dried vine fruits. California is the major producer of natural sun-dried seedless raisins followed by South Africa, Mexico, Chile and Afghanistan as more recent and much smaller producers. Due to the drying method, natural sun-dried seedless raisins have a dark, thick skin and a distinctive burnt sugar flavor. Such raisins have been given preference for use in cold breakfast cereals, chocolate and yogurt coating and out of hand eating. Fortunately for California raisin producers, consumers in the United States, Scandinavia and most Pacific Rim markets prefer such raisins over sultanas.

For our purposes in this chapter, a raisin marketer is any raisin industry or raisin packer representative with the responsibility of influencing buyers, retailers, manufacturers or consumers to use raisins. The target market may be as broad as the world or as narrow as a single consumer. The message is always "use raisins," but the vehicle, strategy, timing, etc., are limited only by the creativity of the marketer. It should be assumed that the objective of industry marketers is to expand and increase the use of raisins, while the objective of individual packer marketers is to expand and increase the sales of their own brand of raisins.

The Competition

One way to clearly identify the challenge to marketers of dried vine fruits and more specifically, of California raisins, is to consider their job as a competition for the consumer's food choices. And in the area of food choices, marketers of California raisins must compete with other dried fruits such as sultanas, currants, prunes, dates and figs. This competition must be geared to three broad areas: where the consumer lives, his/her income and his/her life-style. Unfortunately, in those areas of the world which have the greatest need for food, the consumers do not have the income to purchase other than life-sustaining commodities. This area may be the majority of some countries or isolated areas of countries not as economically strong as the United States. Little, if any, dried fruit marketing activity is directed to these areas. Sometimes all three factors of area, income and life-style come into play at once. The author recalls sitting in the office of the Deputy Minister of Food in New Delhi, India and witnessing a United States delegation being told by the Deputy Minister, "If you deliver your raisins free of charge to Bombay or Calcutta, I cannot deliver them to the hungry people of India." He further stated that raisins would probably cause upset stomachs since those people were not used to eating such foods.

In many areas of the world, consumers' diets are strongly influenced by the foods indigenous to the area. Isolation, transportation, storage and traditions limit marketing opportunities in such areas. Europe, the Scandinavian countries, the United Kingdom and areas of the United States and Canada settled by immigrants from these countries have what one could call "meat and potatoes" diets. These areas have been excellent markets for the raisin industry and continue to offer opportunities for dried vine fruit marketers. In breakfast rolls, breads, puddings and hot and cold cereals, these consumers use significant amounts of dried vine fruits.

The Pacific Rim countries have diets based on rice, fish and blander foods. Japan has been an interesting experience for the California raisin industry. Even though it is consistently the number one or two export market, in spite of much effort, marketing raisins directly to consumers has been a failure. Few convection ovens are found in Japanese homes and Japanese housewives shop virtually daily for each day's needs. Over 90% of all raisins exported to Japan are used as an ingredient in baking or confectionery industry products Note: the term confectionery in Japan refers to cookies, cakes, pies, etc. and are a completely separate business from bakeries or candy). It isn't unusual for consumers to buy a quarter or half-loaf, or even three or four slices, of bread. In the United States, to be labeled as raisin bread, the weight of the raisins must equal the weight of the dry ingredients used to produce each loaf. Japanese consumers reportedly do not want their children to eat cake, and a loaf of bread with that many raisins is perceived as cake. Thus the Japanese breads, as good as they are, do not exceed 50% raisins to dry ingredient weight.

The Challenge

As we begin the 21st century, the year round availability of fresh fruits and vegetables continues to increase. There has been an explosion of foods that are prepackaged, precooked, pretrimmed, frozen and generally ready to use. Single parent homes, working mothers, food shopping fathers, and fast-paced life-styles typify today's world. As a result, home food preparation has been and continues to be greatly reduced. More meals are being eaten outside the home. Frozen dinners, microwave ready, can be heated and served together with a prepackaged salad in under 15 minutes. One of the challenges facing dried fruit marketers is fitting their commodity into this life-style.

Having now defined the size of the battlefield as being the consumer's preferences, the marketers of California raisins must begin to develop a strategy for becoming a preference. The strategies to accomplish involve who, what, how, where and when.

The Strategy

The Who

In the vernacular of marketing, the "who" is the target audience. In the case of raisins, as with most foods, research has been designed and conducted to identify the food purchasing decision makers with the greatest long-term potential. Information obtained from consumers by raisin consumption researchers reveals that senior citizens are the heaviest per capita consumers of raisins. Questions then arise. Do we target non-raisin consuming senior citizens to convert them into raisin consumers? Do we target raisin consuming seniors to get them to increase their consumption of raisins? Or should seniors even be considered as a target audience? Coldhearted as it may sound, the return on marketing dollars invested for senior citizens as the target audience is not economically attractive.

Market research has almost universally identified young housewives ages 25 to 44 with children in the home as the largest food purchasing audience with the greatest potential long-term investment return. This is the prime target audience. Research also shows that more men are making food purchasing decisions. Even though this segment of food purchasers isn't large, it is growing and food marketers are including them when formulating food marketing strategies.

The What

Food marketers have learned from experience that there is no effective universal message (the "what") when it comes to marketing. Consumers in different areas of the world and even in different areas of the United States are very different in their eating habits. They also differ in how they relate to and understand the message. Market messages are influenced by the age of the target audience, the competition, the area targeted and even the time of year. Past messages for California raisins included "Eat California raisins, nature's candy." This was dropped because of consumer concern about sugar and candy consumption. "Add a handful of raisins to your favorite food," a non-structured message, turned out to be hard to deal with by recipe-oriented consumers. The objective was to have consumer's add raisins to tossed salads, fruit salads, hot cereals, etc. Marketers found themselves fielding calls asking how large a handful of raisins to add. Since the message was designed to be non-structured, the obvious response was, "How big is your hand?"

The "vogue" food message for the United States consumer in the 1980s and 1990s was low or no fat, no cholesterol, natural, versatile, and no additives or preservatives. Fortunately for the California raisin marketer, this was a perfect fit.

In Europe, the Scandinavian countries and the United

Kingdom, the low or no fat, no cholesterol, natural, versatile and no additive or preservatives message did not have the same degree of interest as in the United States. There is already a high degree of awareness and acceptance of dried vine fruit. However, the consumers were not aware of the differences in production methods between natural sun-dried seedless raisins and sultanas. The result was a message that both natural sun-dried raisins and sultanas were produced from the same grape, and that while they were different in color, flavor and texture, natural sun-dried raisins could be readily substituted for sultanas.

World travel continues to reduce the size of the world. Consumers are being introduced to and experiencing new foods. The opportunity for consumers to try new foods and potentially modify their eating habits continues to grow. Consumption of cold breakfast cereals in Europe, the Scandinavian countries and the United Kingdom is in its infancy. Marketers of raisins need to consider a message to introduce and encourage consumers in these countries to eat cold breakfast cereals and to add raisins to their hot cereals.

A marketing message for Pacific Rim markets is usually totally different from those meant for the aforementioned areas. As stated previously, few ovens exist in homes in Japan and in most other Pacific markets. Consumption of cold or hot breakfast cereals is virtually nonexistent. Home baking and the addition of raisins (or for that matter anything else) to traditional rice dishes is not a common practice. In the late 1950s and early 1960s, the Western Wheat Association introduced a program to Japanese bakers to get them to use wheat flour instead of rice to greatly expand the baking industry. The California raisin industry learned of this program and was able to convince Western Wheat to let them join in their program and include raisins in their baking program. Since Japan has set low tolerances for sulfur dioxide, no Golden Seedless raisins are exported to Japan and this market is virtually a total natural sun-dried seedless raisin market. Almost 40 years later, more than 90% of all raisins exported to Japan are used as an ingredient in bakery and confectionery products. (Thus the message to the Pacific Rim market that is most effective is to encourage consumers to use and increase their use of products which include raisins as an ingredient.

It is worth repeating: there is no universal message. Marketers who invest in consumer research well-designed for each market area and properly analyze will be well-rewarded as they use the information obtained to design their message for each market area. Some old cliches, such as "the early bird gets the worm" and "too soon old and too late smart" have good application here.

The How

The best marketing program is of no value if it doesn't reach the target. The most popular method for delivering the message in the United States market is television. The major negative to this method is cost, with competitive clutter (number of advertisements for other products) in the same time slot, a close second. Program selection, time of day and day of the week are extremely important. Should you place the commercial during children's cartoons, daytime soaps, fringe (early evening) or prime time? Cost and other commercials shown during the same program break can vary immensely. Going "big time" and placing a commercial during the NBA finals, NCAA Playoffs, or the Super Bowl can quickly eat up a budget. One 30-second spot during the 1999 Super Bowl was well over $1,500,000!

Another message method is print. This includes food publications, women's magazines, newspaper food pages and outdoor/sports magazines. Nationwide publications do exist for some magazines; however, many magazines and most newspapers are at best regional and most are local. A further challenge is to develop a message that "leaps off the pages," as the competition in most print methods is very strong.

Finally there is in-store point-of-sale. This includes banners, shelf talkers, or any other printed material that can be placed in the store to attract the consumer's eye to a specific product. The student of marketing in the United States can quickly rate the desirability and effectiveness of the method. Many retailers in the United States place great restrictions on the use of in-store point-of-sale promotional methods. Seldom are marketers able to obtain the use of this method unless they can show the retailer that another method—print or television—will be used in the area. It is of interest that in-store point-of-sale is still quite popular in Europe, the Scandinavian countries, the United Kingdom and most Pacific Rim markets. In the Scandinavian countries, Germany and, to a lesser degree, other European and United Kingdom markets, print continues to be a major marketing message vehicle. This may or may not influence the use of in-store tie-in promotions.

It is important to recognize the consumer is easily saturated or overwhelmed by constant promotions and advertisements. Thus the choice of message vehicle and the message itself are of extreme importance. Marketers must compete for the consumer's mind space as he/she drives, reads, watches television and shops. The message must be attractive, clear, short and memorable if the consumer is to recall it while preparing a shopping list or passing the marketer's product during a shopping visit.

The Where

People all over the world must eat to sustain life. Thus everyone is a potential consumer. For our purposes the world can be divided into five roughly defined areas: Canada, the United States and Mexico; Europe, the Scandinavian countries and the United Kingdom; Japan, China and the Pacific Rim; Eastern Europe, North Africa and the Middle East; and all other countries.

Consumers in Northern Europe, Scandinavia and the United Kingdom are similar to those of Canada and the Northern United States in their eating habits. Southern European consumers relate more closely to those of the Southern United States and Mexico.

Japan, China and the Pacific Rim consumers historically have had diets based on fish, rice and other local foods. What the United States knows as traditional breakfast foods—cereals, bacon and eggs, breakfast rolls, toast and jam—have been and still are relatively foreign to consumers in those areas. To suggest consumers in these areas add raisins to their traditional boiled rice is almost sacrilege. Only in recent years have wheat flour products, including those containing raisins, made inroads into the consumer's diets in these areas.

Eastern Europe, North Africa and the Middle East contain a significant part of the world's consumers. These areas have not been targeted by raisin marketers for several reasons. International politics, accessibility, business terms and payments, and consumer's economics have all influenced the reluctance or inability to develop business in these areas. Some of these areas are vine-dried fruit producers and many of them do import and use sultanas produced in Greece, Iran, Turkey and Afghanistan.

All other countries are very diverse, scattered and less populated. Their eating habits are closely identified with foods that can be grown and harvested locally. The eating habits of Australia, South Africa, South America and New Zealand consumers mirror those of the immigrants who settled those areas. Food trade is mostly from these countries to the Northern Hemisphere. Even though dried vine fruits are produced in months opposite those of the Northern Hemisphere, their periods of heavy marketing are the same.

The When

Raisin consumers have historically purchased raisins during two seasons of the year: just prior to the Thanksgiving/Christmas holiday and the Easter holiday seasons. If the marketer's objective is to take advantage of this predictable market, then promotion activities and distribution of raisins must be scheduled accordingly. Promotion materials and activities should direct consumer's attention to traditional uses for raisins as well as introducing new uses, such as in stuffings, in ham sauce or new desserts.

The other marketer challenge is to encourage consumers to use raisins in the "off seasons." Creative marketers may suggest raisins as a snack, in lunch boxes, in summer salads and, in our "on the run" society, a raisin bagel or Danish during the race to work.

Most of these examples are obviously directed to the United States consumer. About 70% of California's annual raisin shipments are to the United States and Canada markets. The astute international marketer undoubtedly can come up with comparable ideas for the international consumer.

Getting from Packer to Shelf

California raisin packers ship about 60% of their domestic raisins and 70% of their export raisins in bulk containers (2 lbs [0.9 kg] or more). The single most popular bulk container is the 30 lb (13.5 kg) case. For purposes of this discussion, we will assume all of the raisins shipped in bulk containers will be purchased by institutional users (hospitals, prisons, schools, corporate cafeterias, etc.) and industrial users (bakers, cereal manufacturers, chocolate and yogurt coaters, etc.). The California raisin industry, although lacking the statistics to prove it, is of the opinion that the single largest raisin users are manufacturers of cold breakfast cereals.

Fig. 13.1: Retail display of student-produced California State University, Fresno raisins.

The bulk of raisin marketers face a number of challenges. They must provide information to the research and development departments at food companies to convince them to use raisins in their products. They must convince the purchasing departments that raisins are the best value over competing commodities for use in their products. They must convince the operations department that raisins will be available in the quality and quantity needed.

The message of the industry marketers to bulk raisin users could and, probably should, include the promotion of such desirable characteristics of raisins as having low or no fat and no cholesterol, its versatility and naturalness, and its fiber and low sodium. Raisin products (paste and juice concentrate) are excellent fat substitutes in baking and serve as mold inhibitors and humectants and extend the shelf life of breads, pastries and cookies. Packer marketer messages may include competitive price, service, and consistent quality.

Raisin packers ship approximately 40% of their natural sun-dried seedless raisins in consumer packages of less than 2 lbs (0.9 kg). Consumer packages of raisins may be in packers' brands (such as Sun-Maid), private label (store brand) or generic packages. In the California raisin industry, Sun-Maid is generally the recognized packer brand. Other packer brands such as Champion, Airport, Regent and Del Monte have limited distribution and recognition. Most retailers limit shelf space for raisins and in most instances limit their displays to the recognized packer brand and their own private label. The strategy of a packer marketer is to get the retailer to stock and the consumer to purchase his/HER brand of raisins.

Retailer's brands are referred to in the raisin industry as private label. Generally these private labels are the store's name (Vons, Albertson's, Krogor's, etc.) or labels owned by the retailer (Safeway's "Townhouse" label) or by a group of retailers (such as Certified Grocers' "Springfield"). Domestic private labels are all packed by the packer in California from whom they purchase the raisins. Most exported private labels, such as Sainsbury, Asda, Tesko, Marks and Spencers in the United Kingdom, are also packed by the source. However,

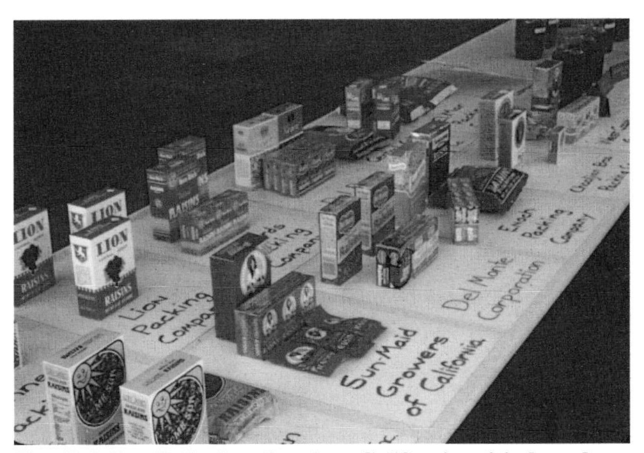

Fig. 13.2: Retail display of various California raisin brands.

some private labels are repacked from bulk containers at the export site.

Generally, retailers display their private label packages side by side on the shelf with packer branded raisins and most often at a price lower than the packer brand. Seldom do retailers promote an individual product (for example Vons raisins). They may promote their label (Vons) but it is a rare instance to see a retailer's brand of any single product promoted.

The importance of private labels in the food business generally, as well as private labels for raisins, has expanded significantly. Mergers, as well as corporate expansion, have resulted in expanded use and awareness of private labels. In 1998 the merger of Albertsons and Associated Food Stores, and that of Smiths and Fred Meyer created two of the largest eight food chains in the United States. Estimates report that 70% or more of all consumer packages of raisins in the United Kingdom and 80% in Germany are private label. In the Scandinavian countries, cooperative buyers dominate the markets and sell under their own labels.

Quality and Perception

All consumers want a high quality product. But what defines quality varies from consumer to consumer. The old cliche that "beauty is in the eye of the beholder" may be modified to state that "raisin quality is in the eye of the consumer/user." Depending on the consumer, quality can be uniform size, freshness (moisture content), absence of foreign matter, sweetness, or it can have no meaning at all.

Quality

Quality to the cereal manufacturer includes uniform size, soft, pliable texture (and yet not too moist) and free flowing. The confectioners sugar generally found on raisins in cold breakfast cereals is not there to add sweetness, but rather is to develop a barrier on the skin of the raisin to reduce the transfer of moisture which would result in dry, hard raisins and soggy flakes. Most cereal manufacturers prefer uniform midsize raisins. In the minds of some manufacturers and consumers, large and small raisins in the same packages signify poor quality. When consumers find something they like, they want another just like the last one.

Quality to many bakers means small (midget) raisins. Ingredients in bakers' formulas generally are measured by weight. Finished bakery products using small raisins give the appearance of lots of raisins even though the weight is the same as it would be if large raisins had been used. Bakers further prefer raisins which can be used directly out of the shipping container and are free flowing. In every case, bakers, as well as other industrial users, want consistency of raisins. Unfortunately, no two raisins are exactly the same in size, shape, or maturity, so users look to suppliers who can consistently provide them with raisins within the parameters of the quality they desire. All the marketing magic in the world will not compensate for the inability to deliver the agreed upon desired product.

Most producers and some consumers include maturity (meatiness) and pliability (softness) as quality factors. Maturity and pliability are often related to moisture content, and some institutional or industrial users do not prefer high moisture raisins. Less mature raisins generally have longer and deeper wrinkles. These are often preferred by chocolate and yogurt coaters as they feel their coatings adhere better to such raisins. Some Japanese bakers have commented that they prefer less mature raisins because they prefer the slight tartness they have over the very sweet taste of more mature raisins.

Perception

Perception is a buzz word of the 1980-1990 marketers. It really doesn't make any difference what the reality is, but rather how it is perceived. The perception is the reality. Raisins have been reportedly perceived as "old" when in reality they are just dry. These same raisins, when soaked in warm water for two to three minutes and drained were then perceived as fresh because they were soft and pliable.

Some consumers perceive sun-dried raisins as natural and other seedless raisins as processed or manufactured. The same grape is used to produce both raisins. Grapes can be exposed to the sun for two to four weeks to produce natural sun-dried seedless raisins which, during this drying process, develop a rather thick skin. The same grape can be dipped in hot water and then dried in tunnels which use natural or propane gas as the heat source to dry them into dipped seedless or golden seedless raisins. Some consumers have commented that tunnel dried raisins are fresher and sweeter.

Summary

For consumer-package consumers, California raisins already are preferred in the United States, Canada, and the Asian and Scandinavian markets. The purpose of the industry marketer's message is to attract nonusers and influence increased use among present consumers. Individual packer marketers should have the same objective. However, in most instances, their objective is the purchase and use of their brand of raisins over another packer's brand or to get the importer to pack the retailer's private label instead of another packer's.

With the message determined, the strategy for transmitting the message must be decided. The cost, area to be covered and promotional methods are major determinants. In most markets, TV is the most popular vehicle; however, it may be prohibitive in cost. Outside the United States and Canada, language and culture become very important factors. Even though we claim the United States and the United Kingdom speak the same language, hard experience has taught us we cannot simply transfer United States advertising and promotion to the United Kingdom with any acceptable level of success. The "Dancing Raisin" did not successfully transfer to any market outside the United States and Canada.

Competition for consumers' preferences continues to increase. Year round fresh fruit availability is challenging the traditional off-season demand for dried fruits. Consumers' life-styles, eating habits, economic status are in constant change. In-home food preparation has decreased. Consumers are increasingly offered new foods, old foods in new forms, more convenient foods, and traditionally seasonal foods all year round.

Failure of marketers to present their commodities in the desired forms, at the correct time, with an acceptable price, and with an attractive message will result in failure. The marketer's message must be clear, attractive, economic, memorable, and sufficiently repetitive to remain in the consumer's mind at food purchasing time. The message's promotional method must sufficiently cut through the competitive clutter to trigger "recall" at the crucial moment when the consumer is making a purchase to encourage that consumer to purchase raisins or products that contain raisins.

A couple of reflections about marketing come to the mind of this author. One is when Kellogg's Europe began using sultanas in their cold breakfast bran flakes cereal. In both their advertising and promotion introduction materials, and in large print on the cereal boxes themselves, it said "Kellogg's Sultana Bran - Now with Sultanas." The other is a past TV commercial where any new food had to be given to Mikey to try first. The results were a favorable expression from Mikey and the comment, "He likes it!" The raisin marketer's challenge doesn't seem to change much in that generally it still comes down to "Try it! You'll like it!"

CHAPTER 14.

Raisins: Versatile Fruits with a 1,001 Uses

Bernadine B. Ferguson and Thomas J. Payne, M.A.

Part I: At The Store

Dried grapes were some of the earliest foods processed by man. Today, food manufacturers everywhere use raisins in products from soups to desserts.

The Historic Raisin

The raisin may well have been the beginning of food ingredient technology. Thousands of years ago, nomadic hunters and gatherers probably discovered raisins when they happened upon grapes drying on the vine. The fruits were portable and could be easily stored.

Between 120 and 900 B.C., the first organized vineyards were developed in the Middle East. Europeans began to cultivate raisins in the Mediterranean region and raisins eventually became a traded commodity throughout the Old World. Some of the earliest baked goods and desserts of Mediterranean, Germanic, French and English traditions contained dried grapes or what have come to be called raisins. Today raisins are used in recipes and food products around the world. Interesting enough, they are a tradition even in areas where raisins have never been produced.

Raisin Usage in the USA

Raisins are a part of North American cuisine and culture. Some of the earliest American cookbooks emulated English and Germanic food traditions. On the West Coast of what was then New Spain and Baja California, Spanish missionaries transplanted vines from the Old World to California mission settlements. Within decades, a grape industry emerged. In 1875 English immigrant, William Thompson, grew a grape variety that was thin-skinned, seedless, sweet and tasty. Today 95% of California raisins are made from Thompson Seedless grapes.

Certainly the first use of raisins in the New World was an extension of our Hispanic heritage. Raisins were used as a preservative in spicy dishes of Early California. These included preserved meats and stews.

On America's East Coast, followed by expansions into the Midwest, raisins were used extensively in German-style breads, other baked goods, and mainstream cooking. As early as the 1900s, dried vine fruits were actually imported from Europe for use in American cooking. With the development of the California raisin industry on the West Coast, new nationwide uses of raisins began to develop serving the ethnic traditions of our nation of immigrants and also the new American culinary identity.

The center of America's raisin industry is in the Central Valley of California, the 'Golden State.' Raisin production is concentrated in a 70-mile (112.7 km) radius around the city of Fresno. Today, 5,000 growers produce an annual yield of about 396,000 tons (356,400 t) on 274,664 acres (111,157 ha). California's growers and processors have developed state-of-the-art growing and processing methods, drawing upon the roots of ancient practices, as well as new food technology. The result is the finest raisins in the world.

See **Appendix H** for a special selection of raisin recipes from the California Raisin Marketing Board.

Raisin Production and Processing Benefits

Raisin vineyards in California are dedicated to grape production. The clean, widely spaced rows ensure that sunlight reaches the raisin grapes. Each fall, grape bunches are hand clipped and placed on clean paper trays where they are dehydrated by nature itself, the sunlight. It takes about three weeks to properly dry a raisin. The bundle of grapes is expertly turned to ensure proper dehydration. Care is taken to produce a sound and well-dried raisin. Field grade raisins at about 15% moisture are placed in wooden boxes and then transported to the raisin processing plant.

California raisins are processed in state-of-the-art raisin processing plants. Raisins are washed with clean warm water, then sorted and inspected. Because of the firm skins of the raisins from the pampering in the vineyard, the California raisin industry is able to rigorously wash the raisin to produce a product which is ready for direct integration into the food product. High tech optical and laser sorters are often used.

Care taken by California's growers and packers in each stage of raisin production produces key technical advantages. California raisins are clean and well dried. A study at an Irish baking plant, for example, revealed that California raisins eliminated the need for additional costly washing and hand sorting. The overall benefit was significant with a savings of more than 15% in the baking cost.

When California raisins are viewed under an electron microscope, scientists note that the skin is mainly unruptured. Sheer skin tests on California raisins show that the skin is actually firm and stable, which allows it to stand up in a manufacturing process without sticking.

For example, baking technicians experimenting with California raisins in Italian panetone formulas discovered that the California raisins plumped more than other dried vine fruits in finished products. Also, good raisin identity was visible in the end loaf.

Health Benefits

Throughout time, raisins have been considered a healthy and nutritious snack. Raisins contain dietary fiber (including soluble fiber) with almost zero fat and offer plenty of nutritional characteristics to manufacturers in every product category. For example, 3.5 oz (100 g) of California raisins contain 0.1855 oz (5.3 g) of fiber, 59% of which is water-soluble fiber. Research indicates that when raisins are consumed at a level of 2.94 oz (84 g) per day, favorable lipoprotein changes occur in the body. Fiber is integral to the prevention and treatment of cardiovascular disease due to its effect on blood cholesterol levels. Soluble fiber helps lower blood cholesterol levels.

Raisins contain many phenolic compounds (or phenols), several of which function as antioxidants and have been shown to slow the potentially damaging cell oxidation process. Recent research has begun to indicate the potential role of phenolic compounds in reducing the risk for both heart disease and cancer.

Because phenolic compounds are found in high concentrations in grape skins, raisins seem to be a particularly rich source for antioxidant nutrients. Three and one-half ounces (100 g) of raisins provides more than 2800 ORAC units (Oxygen Radical Absorbance Capacity) . Based on evidence to date, researchers suggest that daily intake of antioxidants be set at 3,000 to 5,000 ORAC units to have a significant preventative effect.

Raisins also have a carbohydrate distribution, which is of special interest to athletes and consumers with a health-conscious lifestyle. Athletes all know to consume diets which are high in carbohydrates. Raisins naturally contain approximately 36% fructose and 31-33% glucose. Both sugars are readily absorbed by the body and are easily converted into energy. Because of the naturally high sugar content, raisins can be used as a natural sweetening agent in many foods, displacing some of the sugar in yogurts, breads and cereal snacks.

California Raisins in the Food Industry

Today, more than half of all of the raisins produced in California are utilized by the world food processing industry. The most discriminating customers at home and abroad specify the California raisin in a whole range of food products.

Raisins in Breads

Raisin breads and other raisin containing products have become a mainstay in the food industry. In fact, in a typical American store you can find numerous raisin-containing baked foods. Because of the sun dried nature of the product, the California raisin will absorb up to 38% in moisture when conditioned properly.

Research at the American Institute of Baking (AIB) has shown that the addition of California raisins to bakery products actually slows the migration of water to the starch granule which causes staling in breadstuffs. Raisins help to

Table 14.1

World Tour of Raisin Use

- Germanic dishes, such as *sauerbraten*, include raisins as a sweetener and a preservative. Raisins are commonly integrated into breads. For example, the traditional Christmas bread, called *stollen*, often contains twice the amount of raisins as all of the other ingredients combined.
- Russians use raisins in *babkas* and virtually hundreds of their breads and pastries.
- Slavic countries such as Czech and Slovak republics use raisins in Easter breads and their national dish, called *mazanek*, was even produced during communistic times.
- Hungarians integrate raisins into traditional strudel. Yes, they claim to be the home of the strudel, not Germany, and mix raisins into the apple fillings. It takes more than 10 hours to prepare a proper strudel.
- North African countries such as Tunisia and Morocco use raisins in rice and couscous mixes made from grain. Raisins complement fiery spices and savory flavors.
- East Indians use raisins in curries, snack mixes and as a filling in pastries. It is believed that the golden color of the raisin is symbolic of wealth. Raisins are also used in ayurvedic medicine.
- The Dutch have a strong raisin tradition in their cereal and grain based diet. They produce a product called *krentjenbrood*, which is heavy bread with whole grains. Typical Dutch products contain lots of raisins – sometimes with up to 200% raisins compared to flour by weight in the formulation!
- Italians love to use raisins in light fluffy sweet bread called *panetone*. They like using the golden raisin, along with other candied fruits and sweets. They make dozens of different styles of panetone for different religious holidays around the year including the *columba* which is their traditional Easter bread.

keep products in a state of equilibrium, allowing a baker to reduce or eliminate the use of preservatives and additives. In addition, raisins contain high levels of propionic acid (500-600 ppm), which is used as a preservative.

Nobody actually knows when the first commercial raisin bread was produced, but we believe it was in the early 1900s. As early as the 1940s, there were commercial recipes developed. In fact the FDA lists a Standard of Identity, or law, which states that a bakery product called raisin bread must include more than 50% of raisins by weight compared to the flour. Many types of bread in the USA and abroad include far more raisins to give weight, extend shelf life and offer consumer appeal. Bakers know that the pocket of moisture in the raisins inside the loaf will slow down staling of the bread keeping the loaf fresh and tasty. A naturally occurring organic acid in the grape called tartaric acid actually enhances the flavor of the bread. This makes the spices and flavors taste crisper and more flavorful. Some bakers specifically use raisins in breads to enhance flavor while reducing or eliminating the amount of salt normally needed to keep the loaf from tasting bland.

More Raisins in Breads

Did you know that many types of breads, not just raisin bread, include raisins? Raisin juice concentrate, a product of the raisin process and also known as raisin syrup, is used at a level of about 2% or more in breads as a natural preservative in the form of propionic acid. A baker can use raisin juice concentrate and eliminate artificial preservatives and additives that sometimes are used in commercial breads to extend shelf life. Other bakers use ground raisins, or raisin paste, in bread products. When the paste is incorporated into the dough it can provide subtle sweetness and color. Whole grain bread manufacturers use raisin paste.

The Candy Aisle

The average American eats more than 9 lbs (4.05 kg) of candy each year! Raisins have sometimes been called "Nature's Candy." Although they are sweet enough alone, raisins are also found in some of the best selling candies!

"Panned raisins" is the term used in the industry for chocolate or coated raisins. Raisins are added to huge rotating copper kettles which resemble a cement mixer. Expert confectioners carefully add a special blend of chocolate and the coatings adhere to the individual raisins creating a chocolate or yogurt-coated raisin. These raisins even come in a variety of different colors including a purple yogurt coated raisin which is designed to emulate a grape. Did you know that the chocolate-coated raisin is the leading candy sold in movie theaters across the USA? Raisins are also used in bar candies as a tasty addition and in trail mixes and granola snacks, including a favorite called GORP. Can you guess what it stands for? Would you believe "good old raisins and peanuts"?

The Dairy Case

Raisins were probably one of the earliest ingredients used to sweeten yogurts and milks. Today you will find raisins in a variety of different items. Rum raisin ice cream is one of the most popular flavors in Japan, Spain and other countries around the World. You will find this in your freezer case where rum flavor or real rum is normally added to the sweet

- Scandinavians have long integrated raisins into a sour and sturdy bread called *limpa*. The raisins give body and texture to the loaves.
- Finnish people love raisins and integrate them into sauces and breads including the national bread made from reindeer blood. No Finnish meal would be complete without a paste of liver and raisins in this homemade sausage.
- South Africa features a heavy raisin roll served at meals and strangely named, "Spotted Dick." No one seems to know how the name began!
- The Spanish utilize raisins in sweet goods and love the raisin in their national sweet called *turon*. It is a chewy caramel nougat and includes dried raisins or "uva pasas" (dried grapes).
- People in the Philippines use raisins in dozens of rich Spanish-style baked goods including the national bread called the *pan de sal* or salt bread. The day would be incomplete without a visit to the local bakery for a supply of *pan de sal*.
- The English integrate raisins into heavy plum pudding, which is like super fruitcake. The pudding is steamed for more than four hours. This product is found in places settled by the former British Empire. Plum puddings are traditionally exchanged at holiday times. Raisins are a staple item in the English Caribbean islands where each island boasts the very best plum pudding or raisin pastry.
- France is home to the brioche, a light pastry with golden and natural raisins in a miniature cake that can only be made in a real French-style brioche pan. The raisin tradition has also spread to French Canada where raisins are a staple in the French Canadian kitchen.
- The health minded Swiss developed a product called muësli in the 1890s as a spa food for those seeking better health. It is a mixture of raisins and whole grain cereals that is normally consumed with yogurt.

Table 14.2

Nutritional Evaluation of Fruit Samples Compound Quantity per 100 Gram Serving.

Sample 2 (11/20/87)

Element and Compound Content of Fresh Grapes Processed in MIVAC (Puff), Fresh Grapes Sundried into Raisins, and Fresh Fruit Control.

Compound	MIVAC Grapes	Sun-dried Raisin	Fresh Fruit
Protein (NX6.25) (g)	3.63	3.10	1.08
Fat (g)	0.00	0.11	0.18
Carbohydrate (g)	90.99	89.02	24.91
Calories (4-9-4#)	378.00	369.00	106.00
Vitamin A (I.U.)	175.00	n.d.	80.00
Vitamin C (mg)	12.50	8.83	0.30
Thiamine (mg)	0.29	0.17	0.04
Riboflavin (mg)	0.31	0.15	0.06
Niacin (mg)	1.54	2.58	0.50
Calcium (mg)	54.40	54.30	21.10
Iron (mg)	1.38	3.74	1.02
Sodium (mg)	3.90	3.60	3.60
Potassium (mg)	900.00	870.00	200.00
Crude Fiber (g)	0.79	1.18	0.38
Moisture (g)	2.68	5.49	73.27
Ash (g)	2.70	2.28	0.56
Sulfur Dioxide (ppm)	38.00	47.00	16.00
Dietary Fiber (%)	3.90	6.30	1.60

Source: Petrucci, V. E. and C. D. Clary, "Microwave Vacuum Drying of Food Products," EPRI Report CU-6247, Project 2893-7, Final Report, February 1986, Palo Alto, California.

There are no preservatives in the natural sun-dried raisins – they don't need them.

It's hard to imagine a better reason to eat raisins than their naturally sweet taste. However, there is wholesome nutrition packed into each one-quarter cup (40 g) serving. In addition to contributing dietary fiber and essential vitamins and minerals, like iron, potassium, phosphorus and copper, raisins are one of the best fruit sources of antioxidants. It is believed that fruits and vegetables high in antioxidants may protect against diabetes, certain forms of cancer and heart disease.

A study at the Health Research and Studies Center in Los Altos, California has verified that two servings of raisins a day may change some colon cancer risk factors. It confirmed that the combination of dietary fiber and tartaric acid in sun-dried raisins has an important role in colon function and health. Tartaric acid, present in significant amounts only in raisins, grapes and tamarind, plays a significant role in decreasing the time it takes for food and waste to move through the digestive system. The combination of fiber and tartaric acid in raisins helps to speed transit time and decrease bile acids in the colon beyond what would be expected from fiber alone. (**Table 14.2**)

In addition to the fiber and essential vitamins and minerals raisins, contribute to the diet. Further research has determined that raisins are a good source of inulin and several antioxidants, including catechin, plant sterols and flavonoids. Inulin, a fiber-like carbohydrate, ferments in the colon and creates conditions that promote healthy colon cell growth and helps prevent growth of abnormal cells that may lead to disease. Only onion, garlic and wheat, among other commonly eaten foods, are known to be good sources of inulin.

Other studies have shown that catechin and other phytochemicals are effective in helping to inhibit tumor development in animals. Because the environment, including our diets, may account for many of the risks associated with these diseases, it is recommended that a diet with plenty of vegetables, whole grains, nuts, beans and fruits, including raisins, is the best one.

Appendix A: Organizations Servicing the California Raisin Industry

As adapted in part from "Raisin Production Manual," Christensen, 2000.

I. Educational & Research Institutions

California State University, Fresno
Department of Viticulture and Enology
Viticulture and Enology Research Center
2360 East Barstow Avenue, VR 89
Fresno, CA 93740-8003 Phone: (559) 278-2089
Website: *http://cast.csufresno.edu/ve* and *http://cati.csufresno.edu/verc*
A viticulture and enology teaching curriculum and a research center devoted to viticulture and enology, including a reference library and industry education activities.

University of California, Davis
Department of Viticulture and Enology
1 Shields Ave. Davis, CA 95616 Phone: (530) 752-0380
Website: *http://wineserver.ucdavis.edu*
Center for research and teaching faculty and extension specialists serving the public, industry and students.

UC Kearney Agriculture Center
University of California
9240 South Riverbend Avenue
Parlier, CA 93648 Phone: (559) 646-6500
Website: *http://www.uckac.edu*
A Research center and field station emphasizing horticulture production; also, regional administrative center for Cooperative Extension.

University of California Cooperative Extension
UC Cooperative Extension Farm Advisors provide production information and technical assistance via mass media, publications, meeting, short courses, and personal contacts. They also serve as a link to the agricultural research and teaching personnel and facilities of the University of California and conduct local research to solve problems and improve technology.

Fresno County
1720 South Maple Avenue
Fresno, CA 93702
Phone: (559) 456-7285
E-mail: cefresno@ucdavis.edu

Madera County
328 Madera Avenue
Madera, CA 93637
Phone: (559) 675-7879
E-mail: *cemadera@ucdavis.edu*

Kern County
1031 South Mt. Vernon Avenue
Bakersfield, CA 93307
Phone: (661) 868-6200
E-mail: *cekern@ucdavis.edu*

Merced County
2145 West Wardrobe Avenue
Merced, CA 95340-6496
Phone: (209) 385-7403
E-mail: cemerced@ucdavis.edu

Kings County
680 North Campus Drive
Hanford, CA 93230
Phone: (559) 582-3211, ext. 2730
E-mail: *cekings@ucdavis.edu*

Tulare County
4437 S. Lansing St., Suite B
Tulare, CA 93274
Phone: (559) 685-3303
Website: http:*//www.ucce.tulare.ca.us*
E-mail: *cetulare@ucdavis.edu*

USDA Agricultural Research Service
Horticulture Crops Research Laboratory; Postharvest Quality and Genetics; Commodity Protection and Quarantine Insects
Research Unit, 9611 S. Riverbend Ave
Parlier, CA 93648 Phone: (559) 596-2790
Website: *http://pwa.ars.usda.gov/fresno*
A Breeding facility for new table grape and raisin varieties; research facility for genetics and postharvest quality and storage., including insect control in stored raisins.

II. Industry, State & Federal Organizations

California Raisin Marketing Board
3445 North First Street, Suite 101
Fresno, CA 93726 Phone: (559) 248-0287
Website: *http://www.calraisins.org*
E-mail: *info@calraisins.org*
Grower-only supported State Marketing Order. Administers and coordinates raisin promotional activities via education, public relations, and advertising. Supports health, nutrition, production, and raisin quality research; extends information to consumers and institutions.

Raisin Administrative Committee
3445 North First Street, Suite 101
Fresno, CA 93726
Phone: (559) 225-0520
E-mail: *RAC@raisins.org*
Website: *http://www.raisins.org*
Institutes operations and policy of the Federal Raisin Marketing Order in*cluding quality controls, volume controls, research, and development. Tabulates raisin delivery and inventory statistics and publishes Raisin Industry News.*

Raisin Bargaining Association
3425 North First Street, Suite 209
Fresno, CA 93726-6819 Phone: (559) 221-1925
Represents grower members in raisin sales negotiations with consignatary raisin packers.

USDA Agricultural Marketing Service
Fruit and Vegetable Division
Marketing Order Administrative Branch
2202 Monterey Street, Suite 102-B
Fresno, CA 93721 Phone: (559) 487-5901
Website: *http://www.ams.usda.gov/fv*
Administers the Federal Raisin Marketing Order and provides marketing order publications.

California Agriculture Statistics Service
P.O. Box 1258
Sacramento, CA 95812 Phone: (916) 498-5161
E-mail: *nass-ca@nass.usda.gov*
Website: *http://www.cdfa.ca.gov/statistics*
Gathers and disseminates annual crop statistical reports: California Raisin Grape Measurement Survey, California Agriculture Statistics, all crops; Grape Acreage, Grapes for Crushing.

USDA Agricultural Marketing Service
Fruit and Vegetable Division
Processed Products Branch
2202 Monterey Street, Suite 102-A
Fresno, CA 93721 Phone: (559) 487-5210
Website: *http://www.ams.usda.gov/fv/moab*
Conducts incoming and outgoing raisin inspection with standards established under the Federal Raisin Marketing Order.

Appendix B: Raisin Packing Plants, 1998/99 Season.

American Raisin Packers, Incorporated
PO Box 30*
2335 Chandler Street
Selma, CA 93662
John Paboojian, Jr., President
Greg Paboojian, Secretary/Treasurer

Biola Raisin Company
PO Box 277*
12814 West "G" Street
Biola, CA 93606
Tom Gularte, General Manager

Boghosian Raisin Packing Company, Incorporated
P.O. Box 338*
726 8th Street
Fowler, CA 93625
Phil Boghosian, President
Peter Boghosian, Secretary/Treasurer

California Valley Raisin Cooperative
2589 N. Vineland
Kerman, CA 93630
Marvin Horn, President
Mike Jerkovich, Vice President
Pat Torosion, Treasurer

Camara Raisin Packing, Incorporated
21853 Road 24
Madera, CA 93638
Ronald D. Camara, President
Yvonna Malvinni, General Manager

Caruthers Raisin Packing Company, Incorporated
12797 South Elm Avenue
Caruthers, CA 93609-9711
Donald Kizirian, President
Dennis Housepian, Vice President/Marketing

Central California Packing Company
P.O. Box 220*
5316 South Del Rey Avenue
Del Rey, CA 93616
Dan Milinovich, President

Chooljian Bros. Packing Company, Incorporated
P.O. Box 395*
3192 South Indianola Avenue
Sanger, CA 93657

Circle K Ranch
8700 South Leonard Avenue
Fowler, CA 93625
Melvin Kazarian, Partner
Ronald Kazarian, Partner
Skip Karabian, Manager.

Del Rey Packing Company
P.O. Box 160*
5287 South Del Rey Avenue
Del Rey, CA 93616
Carl Chooljian, President
Kenneth Chooljian, Vice President
Gerald Chooljian, Secretary/Treasurer

Enoch Packing Company, Incorporated
P.O. Box 339*
10715 East American Avenue
Del Rey, CA 93616
Dennis Vartan, President
Janice Enoch-Kroger, Vice President
Mona Cade, Office Manager

Fresno Cooperative Raisin Growers, Incorporated.
4466 North Dower Avenue
Fresno, CA 93722
Jerald Rebensdorf, President
Richard Orique, Business Manager

International Raisins, Incorporated
1445 Nebraska Avenue
Selma, CA 93662
John Lidestri, President/C.E.O.
Donna Yanicky, Vice President
Rick Murphy, General Manager

Lion Enterprises, Incorporated
3310 East California Avenue
Fresno, CA 93702
Al Lion, President
Bruce Lion, Vice President
Jeff Lion, Vice President
Leo Chooljian, President
Mehran Chooljian, Vice President
Nic Boghosian, Vice President
Michael Chooljian, Vice President

Mariani Raisin Company
4087 North Howard Avenue
Kerman, CA 93630
Mark Mariani, C.E.O.
George Salwasser, Chief Operating Officer
Charlotte Salwasser, Secretary
Stefan Kaercher, Bulk/Exp. Sales
Scott Burns, Reporting

Mariani Packing Company
320 Jackson Street
San Jose, CA 95112
Mark Mariani, President
George Sousa, Sr., Senior Vice President
George Sousa, Jr., Vice President.

National Raisin Company
P.O. Box 219*
626 South 5th Street
Fowler, CA 93625
Ernest Bedrosian, President
Krikor Y. Bedrosian, Vice President
J. Kenneth Bedrosian, Secretary

Pacific Sources, Inc.
568 S. Temperence Ave.
Fresno, CA 93727
Allen J. Vangelos, Chairman/CEO
Ian Crabtree, Chief Operating Officer
Mark Nesbitt, General Manager

Sun Beam Raisin Company, Incorporated
27421 Avenue 12
Madera, CA 93637
Richard Mosesian, President

Sun-Maid Growers of California
13525 South Bethel Avenue
Kingsburg, CA 93631
Barry F. Kriebel, President

Sunset Raisin & Nut
2544 N. Lake Ave.
Kerman, CA 93630
Mark Mariani, President
George L. Kenneson, President
George M. Kenneson, Vice President.

Victor Packing, Incorporated
11687 Road 27 °
Madera, CA 93637
Victor Sahatdjian, President

**For mail use only*

Source: Raisin Administrative Committee

Appendix C: Raisin Dehydrator Plants, 1999/2000 Season.

A & C Dryers, Incorporated
2210 North Grantland
1256 W. Manning*
Del Rey, CA 93616
Phone: (559) 275-1654
Eddie Wayne Albrecht, President

Albrecht Farms
12400 East Adams
Del Rey, CA 93616
Phone: (559) 888-2071
FAX: (559) 888-2084
Eddie Wayne Albrecht, Partner

Albrecht & Simonian
1420 South Academy
Sanger, CA 93657
Phone: (559) 875-4402/888-2071
Eddie Wayne Albrecht, Partner

Campos Brothers
15082 South Walnut
15516 South Walnut Avenue*
Caruthers, CA 93609
Phone: (559) 864-9488
FAX: (559) 864-3807
Joseph Campos, Partner

Chooljian Enterprises, Incorporated
3192 South Indianola Avenue
P.O. Box 395*
Sanger, CA 93657
Phone: (559) 875-5501
Michael Chooljian, President

Del Rey Packing Company
11272 East Central and Indianola Avenue
P.O. Box 160*
Del Rey, CA 93616
Phone: (559) 888-2177
Kenneth Chooljian, Vice President

Enoch Packing Company, Incorporated
10715 East American Avenue
P.O. Box 339*
Del Rey, CA 93616
Phone: (559) 888-2151
Dennis Vartan, President

Exeter Dehydrator, Incorporated
26783 Road 176
Exeter, CA 93221
P.O. Box 219*
Fowler, CA 93625
Phone: (559) 592-3221 or 834-5981
Ernest A. Bedrosian, President

Farmersville Dehydrator
980 North Farmersville Boulevard
Farmersville, CA 93223
P.O. Box 219*
Fowler, CA 93625
Phone: (559) 747-6011/834-5981
Ernest A. Bedrosian, President

Four Bar C Farms, Incorporated
10616 South West Avenue
10825 South West Avenue*
Fresno, CA 93706
Phone: (559) 266-7965
Carl C. Gunlund, Jr., Vice President

Fowler Dehydrator
726 South 8th Street
P.O. Box 338*
Fowler, CA 93625
Phone: (559) 834-5348
Philip Boghosian, Partner

H & R Gunlund Ranches
3675 West Saginaw
2763 West Mountain View*
Caruthers, CA 93609
Phone: (559) 864-3606
Hans P. Gunlund, Director

Haggmark Dehydrator
12057 South Hughes
12137 South Hughes*
Caruthers, CA 93609
Suzan Haggmark, Owner

International Raisins, Incorporated
1445 Nebraska Avenue
Selma, CA 93662
Phone: (559) 896-2140
Larry Ellis, Plant Manager

Lamanuzzi Pantaleo
No. 1 - 3636 North Grantland Avenue
Fresno, CA 93722*
No. 2 - 11767 Road 27 °
Madera, CA 93637
Phone: (559) 275-6131
FAX. (559) 275-5892
Lamanuzzi & Pantaleo, Owners

Lion Raisins
9400 South DeWolf
P.O. Box 1350*
Selma, CA 93662
Phone: (559) 834-3021
FAX: (559) 834-6622
Al Lion, President

Lone Star Dehydrator
2730 South DeWolf
Sanger, CA 93657
Phone: (559) 485-6191
Mark Melkonian, Vice President

Rosendahl Farms, Incorporated
4624 West Nebraska Avenue
Caruthers, CA 93609
Phone: (559) 864-3501
Don Rosendahl, President

Salwasser, Incorporated
4677 North Howard
4087 North Howard*
Kerman, CA 93630
Phone: (559) 843-2882
George Salwasser, President

Sherman Thomas Dehydrator
25829 Avenue 11
25810 Avenue 11*
Madera, CA 93637
Phone: (559) 661-4026/674-6468*
Mike Braga, Manager

Simone Fruit Company, Incorporated
8008 West Shields
Fresno, CA 93722
Phone: (559) 275-1368
Mauro Simone, Vice President

Six Jewels
6692 South Peach Avenue
7357 East Denett Avenue*
Fresno, CA 93727
Phone: (559) 834-4690/456-4900
Jeff Jue, President

Sun Beam Raisin Company, Incorporated
27421 Avenue 12
Madera, CA 93637
Richard Mosesian, President

Sun-Maid Growers of California
27400 Avenue 6
Madera, CA 93637
13525 SOUTH Bethel Avenue*
Kingsburg, CA 93631
Phone: (559) 888-2101
James M. Henderson, Vice President/ Operations
Harold Hilker, Vice President/Sales

Tri Boro Fruit Company, Incorporated
2500 South Fowler Avenue
Fresno, CA 93725
Phone: (559) 486-4141
Tom Fazio, General Manager

Victor Dehydrator
11687 Road 27 °
Madera, CA 93637
Phone: (559) 673-5908/485-2858
Margaret Sahatdjian, Vice President

Yettem Dehydrator
13990 Ave. 384
Yettem, CA 93670
Phone: (559) 528-6939
Dale Sedao, General Manager

** For mail use only*

Source: Raisin Administrative Committee

Appendix D: Seedless Grape Cultivars

Cultivar	Source	Reference	Attributes	Parentage
Autumn Royal	USDA, 1996	USDA Release Notice, 1996	Berries large, ovoid to elipsoidal, black to purple-black, sweet and neutral, ripens in October, very productive. Medium vigor, medium to small aborted seeds.	Autumn Black x C74-1. Parents of C74-1 include Blackrose, Calmeria, Ribier, Red Malage, Tifafihi Ahmer, Muscat of Alexandria and Sultanina.
Autumn Seedless	USDA, 1984	Ramming, 1987	Berries white, ovoid to 6.0g. Clusters large, loose to med. Late maturing, holds well in storage, sensitive to gibberellic acid. Not recommended.	Calmeria x (Muscat of Alexandria x Sultanina)
Beauty Seedless	UC Davis, 1954	Brooks & Olmo, 1972	Berries black, tender flesh, size of TS. Clusters conical, heavily shouldered. Productive when spur pruned, subject to sun scald if planted on light soils, distinctive blue-green foliage.	Scolokertek kiralynoje x Black Kishmish
Black Emerald	USDA, 1994	Ramming, 1987	Berries medium, round to slightly oval, black and neutral flavor. Clusters are medium, wel-filled, conical with shoulder. The vine has medium vigor and is productive.	A69-J 90 X C84-I 16 parents include Blackrose, Red Malaga, Tifafihi Ahmer, Muscat of Alexandria, Sultanina, Agadia, Muscat Hamburg, Perlette, Italia and Calmeria.
Blush Seedless	UC Davis, 1981	Brooks & Olmo, 1985	Berries bright red, to 3.8g. Clusters large, uniform, conical. Seed traces noticeable but not flinty, good storage and shipping variety for late market.	Emperor x Z 4-87
Bonnet Seedless	L. Bonnet, 1934	Brooks & Olmo, 1972	Berries dull green, ellipsoidal. Clusters med. size, loosely filled. Plant pat. 88. Leaf deeply lobed. Poor productivity, appears inferior to TS.	Muscat of Alexandria x Sultana

Cultivar	Source	Reference	Attributes	Parentage
Bonnet Seedless Muscat	L. Bonnet, 1941	Brooks & Olmo, 1972	Berries small, unknown color, variable in set. Clusters med. size. Seed partially hardened, vine produces irregular crops, dv. now obsolete.	Muscat of Alexandria x Thompson Seedless
Bronx Seedless	A. Stout, 1937	Brooks & Olmo, 1972	Berries red-skinned, oval, and small. Clusters large, long, loosely filled. Susceptible to anthracnose and downy mildew. Hardy vigorous vine, recommended where berry cracking in not a problem.	(Goff x Iona) x Sultanina
Canadice	NY State Ag. Expt. Sta., 1977	Brooks & Olmo, 1978	Berries slip-skin and tender, med. size. Clusters large and compact. Labrusca flavor, resistant to phylloxera, vines winter hardy, tolerant to post harvest fumigation with sulphur dioxide.	Bath x Himrod
Canner	UC Davis, 1958	Brooks & Olmo, 1972	Berries dark green, ovoid elongated, larger than Sultanina, resistant to cracking. Clusters large and conical with prominent shoulders, compact. Berries process very well, susceptible to powdery mildew in late season, very vigorous vine.	Hunis x Sultanina
Centennial Seedless	UC Davis, 1980	Brooks & Olmo, 1982	Berries have yellow green skin, ovoid elongated, mostly parthenocarpic. Clusters med. to large, conical, well-filled to loose, self-thinning at fruit set. Very responsive to gibberellic acid, fruitful when spur pruned, named to commemorate centennial of Dept. of Vit. & Enol.	Gold x (F2 selection of Emperor x Pirovano 75)
Crimson Seedless	USDA, 1989	USDA Release Notice, 1989	Berries medium, cylundrical to oval, bright red, sweet and neutral. Cluster medium, conical, medium to tight. Ripens in September, medium production, very vigorous and very small aborted seeds.	Emperor x C3 3-199 Parents of C33-199 include Italia, Calmeria, Muscat of Alexandria and Sultanina.

Cultivar	Source	Reference	Attributes	Parentage
Dawn Seedless	UC Davis, 1980	Brooks & Olmo, 1982	Berries have yellow green skin, ovoid elongated, mostly parthenocarpic. Clusters large, less compact than Perlette, usually requires no berry thinning. Vine less vigorous, ripens one week later than Perlette.	Gold x Perlette
Delight	UC Davis, 1947	Brooks & Olmo, 1972	Berries green yellow in color, uniform and oval, firm flesh with distinct flavor. Clusters med. to small. Sib of Perlette, slight muscat flavor, sometimes astringent, suitable for table and raisins.	Scolokertek kiralynoje x Sultanina marble
Diamond Muscat	USDA, 2000	USDA Release Notice, 2000	Berries medium-small, oval, white , with a muscat flavor. The cluster are medium to large and conical with shoulders and well-filled. The vine is vigorous and very productive.	A13-2 X 132-11 Parents include Agadia, Muscat Hamburg, Perlette, Red Malaga, Tifafihi Ahmer, Muscat of Alexandris, Sultanina, Calmeria and Autumn Seedless.
Do Vine	USDA, 1995	USDA Release Notice, 1995	Berries white, medium size, oval to round. Cluster medium to larges, conical and shouldered. Ripens 14-17 days before Thomspson Seedless. Candidate for DOV raisins. Airstream sorter grades show 85-100% B or Better, same as Thompson Seedless or better. High vigor and must be cane pruned.	P79-101 x Fresno Seedless
Einset Seedless	NY State Ag. Expt. Sta., 1986	Reisch, Et al., 1986	Berries bright red, ovoid, to 2.3g, tender flesh with slight labrusca flavor. Clusters med. to small in size. Responds well to gibberellin sprays, susceptible to powdery mildew, resistant to bunch rot, named to honor the late Dr. John Einset.	Fredonia x Canner

Cultivar	Source	Reference	Attributes	Parentage
Emerald Seedless	UC Davis, 1968	Brooks & Olmo, 1972	Berries greenish-yellow, med.-large, obovoid, moderately firm flesh. Clusters large, loose to well-filled, uniform. Recommended for fancy raisins, ripens midseason, moderately productive, canes large and few in number.	Emperor x Pirovano 75
Emperatriz	S. Gargiulo, 1986	Brooks & Olmo, 1999	Berries medium pink to redish, ovoid. Clusters loose and large. Two soft seeds. Very productive and vigorous. Plant patent 5,833.	Emperor x Thompson Seedless
Fantasy Seedless	USDA, 1989	USDA Release Notice, 1989	Berries large, oval, black and neutral flavor. The clusters are medium, loose and conica. The vine is very vigorous and medium productive.	B36-27 X c78-68 are siblings and their parents include Blackrose, Red Malaga, Tafafihi Ahmer, Muscat of Alexandria and Sultanina.
Fiesta	USDA, 1973	Brooks & Olmo, 1974	Berries have white skin, rounder than TS, med. size. Clusters med.-large, usually not compact. 10 to 14 days earlier than TS, recommended for raisins, must be cane pruned for maximum yield.	(Calmeria x 43-13N) x [(Carndinal x Sultanina) x 43-13S] (43-13N and 43-13S are: [Red Malaga x Tafafihi Ahmr] x [Musc. Of Alex. X Sultanina])
Flame Seedless	USDA, 1973	Weingberger & Harmon, 1974	Berries bright red, spherical, very crisp flesh. Clusters med., conical, berries are well-spaced. Responds well to girdling and gibberellin, berry color a problem with high temperature, may be spur-pruned.	(Cardinal x Sultanina) x [(Red Malaga x Tafafihi Ahmr) x (Muscat of Alexandria x Sultanina)]
Fresno Seedless	USDA & CSIRO, 1988		Berries white, 2 - 3g, round to ovate, firm and slightly crisp. Clusters med.-sized (.45-0.7kg), conical, med. to tightly-spaced berries. 3 to 5 days earlier than Perlette, very productive when cane pruned, does not respond well to girdling or gibberellin, not recommended for California.	(Cardinal x Sultanina) x [(Red Malaga x Tafafihi Ahmr) x (Muscat of Alexandria x Sultanina)]

Cultivar	Source	Reference	Attributes	Parentage
Glenora	NY State Ag. Expt. Sta., 1952	Brooks & Olmo, 1978	Med. size berries have blue-black skin, non slip-skin, melting flesh. Clusters large, cylindrical, not excessively compact. Responds well to gibberellic acid, does not store well, not fully winter hardy.	Ontario x Black Monukka
Himrod	NY State Ag. Expt. Sta., 1952	Brooks & Olmo, 1972	Small oval berries have greenish-yellow skin, tender and melting flesh, vinous flavored. Clusters large, long, poorly filled. Vigorous vine with low productivity, fairly hardy.	Ontario x Sultanina
Interlaken Seedless	NY State Ag. Expt. Sta., 1946	Brooks & Olmo, 1972	Berries greenish-white, non slip-skin, crisp, sweet flesh. Clusters med. and tapering. Excellent quality, hardy for the eastern U.S., long-cane pruning is needed.	Ontario x Sultanina
Jupiter	Clark and Moore, 1999	Hort Science, 1999	Berries medium large, oval to slightly oblong, reddish-blue and mild muscat flavor. The clusters are small, well-filled and conical. The vine is medium in vigor and medium in productivity.	Ark 1258 X Ark 1672 and their parents include Campbell's Early, Seneca, Galiber 256-28, Gold, Blackrose, Aurelia, Alden and Glenora.
Lakemont	NY State Ag. Expt. Sta., 1946	Brooks & Olmo, 1973	Med. small berries have yellowish green skin, oval, tender flesh. Clusters are med.-large, wedge shaped, med. to compact. Recommended for home garden and roadside market, perfect flowered, less hardy than Himrod but more so than Interlaken Seedless.	Ontario x Sultanina
Marquis	NY State Ag. Expt. Sta., 1996	Brooks & Olmo, 1997	Berries large, spherical, amber, thick skin, medium aborted seeds, Mild Labrusca flavor, ripens middle September. Clusters large, shouldered, moderately loose. The vine is moderately resistanct to Botrytis bunch rot.	Athens X Emerald Seedless

Cultivar	Source	Reference	Attributes	Parentage
Marroo Seedless	CSIRO Div. Of Hort., Merbein, Australia, 1985	Clingeleffer, 1985	Berries 3-4g, black skin, crisp texture, pleasant sweet flavor. Clusters med. to large, conical, med. compactness. Large seed trace (20mg fw), sensitive to gibberellic acid, some disease resistance, produces high quality raisins.	Carolina Blackrose x Ruby Seedless
Mars	Univ. of Arizona, 1984	Moore, 1984	Berries blue-skinned, spherical, average 3.5g, labrusca flavor. Clusters med. Fruit easily held on vine after ripening, resistant to black rot, downy and powdery mildew, anthracnose, recommended for home gardens.	Island Belle x Ark. 1339
Marvel Seedless	H. Folmar, 1983	Brooks & Olmo, 1972	Large ovate berries are yellowish-green, med. texture, sweet flavor. Clusters large, well-filled, less compact than TS or Delight. Responds well to girdling and gibberellic acid, produces short canes. Plant patent 2335.	Tetraploid bud mutation of Delight
Princess	USDA, 1999	USDA Release Notice, 1999	Berries white, 5-6g, cylindrical. Clusters medium to large in size and length, conical with shoulders. Ripens mid-August. Medium production and very vigorous.	Crimson Seedless X B40-208. Parents of B40-208 include Blackrose, Calmeria, Italia, Muscat of Alexandria and Sultanina.
Merbein Seedless	CSIRO Div. Of Hort., Merbein, Australia, 1981	Kerridge & Antcliff, 1983	Berries white, similar to TS, tender skin. Clusters large, conical, with small wings. Primarily developed as a raisin grape, earlier than, and outyielding TS, resistant to rain damage during the harvest period.	Farana x Thompson Seedless
Neptune	Clark and Moore, 1999	Hort Science, 1999	Berries small, elliptic to slightly ovate, yellow-green, with fruity flavor. The cluster are medium small and conical with small shoulders and well-filled. The vine is medium in vigor, with medium productivity.	Ark. 1562 X Ark. 1704 and their parents include Athens, Seneca, Carolina Blackrose, Seibel 14-664, Dattier of St. Vallier, Mills, Triumph, Diamond, Muscat Hamburg, Alden, Seibel 14-514 and Lakemont.

Cultivar	Source	Reference	Attributes	Parentage
Orlando Seedless	Univ. of Florida, 1987	Mortensen & Gray, 1987	Berries light green, spherical, slip-skin, even ripening, average weight of 1.4g. Clusters average 139g, shouldered, with tapering tips, moderately loose. Resistant to Pierce's disease, attractive bunches, early ripening, longevity.	Fla. D4-176 x Fla. F9-68
Perlette	UC Davis, 1946	Brooks & Olmo, 1972	Berries spherical, white to yellow, very tender. Clusters large, compact to very compact. Very early cv., resistant to high temps, requires berry thinning, does not respond well to gibberellic acid.	Scolokertek kiralynoje x Sultanina
Reliance	Univ. of Arkansas, 1982	Brooks & Olmo, 1972	Round berries are light red, to 2.7g, tender flesh. Clusters med., cylindrical, well filled. Delicate labrusca flavor, stores well for up to 3 mos., tolerant of black rot, anthracnose, powdery and downy mildew.	Ontario x Suffolk Red
Remaily Seedless	G. Remaily, 1980	Brooks & Olmo, 1983	Oval berries light green, med., resistant to cracking. Clusters large, tapered, attractive. Resistant to berry cracking, ripens with Concord, must be cluster-thinned, must control excessive vegetative growth to maintain berry size.	Lady Patricia x (Black Monukka x Ontario)
Romulus	NY State Ag. Expt. Sta., 1952	Brooks & Olmo, 1972	Small berries variable in size, greenish-yellow, melting flesh, vinous flavored. Clusters large, shouldered, compact. Resembles Sultanina, ripens with Concord, vigorous, very productive, fairly hardy.	Ontario x Sultanina
Ruby Seedless	UC Davis, 1968	Brooks & Olmo, 1972	Med., ovoid berries, reddish-black to dark red skin, firm flesh. Clusters very large, conical and shouldered, well-filled. Good eating quality, stores well, late midseason, very productive.	Emperor x Pirovano 75

Cultivar	Source	Reference	Attributes	Parentage
Saturn	Univ. of Arkansas, 1989	Moore, 1989; Brooks & Olmo, 1991	Berries large, bright red, oval and crisp. Clusters medium large, conical. Midseason, productive, vigorous, moderate resistance to black rot, powdery mildew and anthracnose. Plant patent 6,703.	Aurore) X NY45791 (Bath X Himrod)
Saturn	Univ. of Arkansas, 1988	Moore, 1988	Berries bright red, firm flesh, 3g, adherent skin. Clusters med. size, compact. Very good storage grape.	Dunstan 210 x NY45791
Seedless Tokay	W. Perrin, 1965	Brooks & Olmo, 1972	Berries white, ovoid-truncate average 9/16" diam. Clusters small, loose to well-filled. Plant pat. 2340, gibberellin treatment necessary to produce adequate berry size, vine similar to parent in all aspects.	Bud mutation of Flame Tokay
Selma Pete	USDA, 2000	Ramming and Tarailo, 2001	Berries white, medium size, oval. Clusters medium size and well-filled. Excellent raisin quality: 95-100% B or better. Neutral flavor. Vine vigorous and highly productive when cane pruned.	(C66-144 = [B36-27 = (Blackrose X 43-13N) X P54-3 = ((C64-80 X Fresno Seedless)] x DOVine)
Skookum Seedless	Summerland, BC, Canada Research Sta., 1996	Brooks & Olmo, 1997	Berries medium, elliptical, and green. Clusters large, moderately well-filled, conical and winged. Tolerant to powdery mildew.	Vineland 37034 (Seneca X Golden Muscat) X Romulus
Sooke Seedless	Summerland, BC, Canada Research Sta., 1996	Brooks & Olmo, 1997	Berreis medium to small, round and green. Cluster medium-large, well-filled, conical and winged. Tolerant to powdery mildew.	Vineland 37022 (Seneca X Golden Muscat) X Romulus
Sovereign Coronation	L. Denby, 1977	Brooks & Olmo, 1998	Berries purple to black skin, crisp flesh, Concord flavor, to 2.8g. Clusters small, approx. 9 per kg, cylindrical in shape, tight. Concord type, vine is med. to weakly vigorous, large leaves, may be successfully spur-pruned.	Lady Patricia x Himrod
Stout Seedless	A. Stout, 1930	Brooks & Olmo, 1973	Berries small, oval, greenish yellow skin. Clusters large, med. compact. Old seedless Eastern seedless cv., not hardy and fails to ripen at Geneva, NY. Not recommended for planting.	(Triumph x Dutchess) x Sultanina

Appendix G: Chapter Tables in Metric

Table 1.1.1

World Raisin and Sultana Production by Country.

Country	(mt)	(%)
United States[1]	402,000	35.8%
Turkey[1]	250,000	22.2%
Iran[1]	102,000	9.1%
China[6]	70,000	6.2%
Greece[1]	53,000	4.7%
Australia[1]	38,000	3.4%
South Africa[1]	35,000	3.1%
Chile[1]	32,000	2.8%
Afghanistan[2]	28,000	2.5%
India[4]	28,000	2.5%
Mexico[1]	17,500	1.6%
Syria[2]	16,000	1.4%
Argentina[5]	9,800	0.9%
Lebanon[2]	8,000	0.7%
Spain[5]	7,500	0.7%
Cyprus[3]	6,000	0.5%
Morocco[2]	5,000	0.4%
Tunisia[3]	4,000	0.4%
Pakistan[2]	4,000	0.4%
Albania[2]	3,000	0.3%
Tajikistan[2]	3,000	0.3%
Uzbekistan[2]	3,000	0.3%
Turkmenistan	1,100	0.1%
Yeman[2]	1,100	0.1%
World	**1,127,000**	**100.0%**

Sources:
[1]International Conference of Sultana and Raisin Producing Countries, Estimates. San Francisco, CA. 1998.
[2]FAO Production Yearbook, Vol. 51-1997. Series No 142, Rome, Italy.
[3]O.I.V. Bulletin Supplement 803-804. January and February, 1998. Paris, France.
[4]S.D. Shikhamany, National Research Center, Pume, India.
[5]Agriculture Attaché Query Detail, 1998. U.S.D.A./F.A.S., Washington, DC
[6]Editorial Board for China Agriculture Yearbook, China Agriculture Yearbook (1993-1996), Beijing: China Agriculture Press, 1993, 1994, 1995, 1996.

Table 1.1.2

World Grape Production by Continent, 1997.

Continent	ha	% World Total	Total t	t/acre
World	7,655,202	100	57,882,300	7.6
Europe	4,743,984	62	29,644,590	6.3
Asia	1,693,708	22	12,694,770	7.6
S. America	418,175	5.5	4,855,950	11.6
N.C. America	365,153	4.8	6,603,300	18.1
Africa	356,149	4.7	3,089,790	8.7
Oceania	78,033	1	996,930	12.8

Source: FAO Production Yearbook, Vol.51, 1997

Table 1.1.3

Total World Grape Acreage Production and Yield per Hectare, 1971-1997.

Years	Hectares	Tonnes	t/ha
1971-75	9,961,135	55,880,395	5.65
1976-80	10,213,140	34,044,682	6.03
1981-85	9,823,135	63,312,667	6.50
1986-90	8,715,119	60,931,584	7.03
1991-95	8,086,111	61,909,848	7.71
1996	7,742,106	59,150,650	7.71
1997	7,655,202	57,882,308	7.62

Table 1.1.5

Utilization of Raisins and per Capita Consumption by the Leading Raisin Producing Countries, 1997-98.

Country	Population (million)	Production (1,000 t)	Domestic (t)	Export (t)	Import (t)	Per Capita (kg)
United States	271.6	361.8[1]	217,800	144,000	10,890[3]	0.86
Turkey	63.5	225.0[1]	36,000	189,000	4,500	0.64
Iran	71.5	91.8[1]	2,295	68,850	0	0.32
China	1243.7	63.0[6]	56,700	6,300	4,680[3]	0.05
Greece	10.5	47.7[1]	7,155	40,545	2,700[3]	0.93
Australia	18.3	34.2[1]	6,840	27,360	8,640[3]	0.86
South Africa	43.3	31.5[1]	4,725	26,775	0	0.11
Chile	14.6	28.8[1]	4,320	24,480	0	0.30
Afghanistan	22.1	25.2[2]	7,200	18,000	0	0.33
India	960.2	25.2[4]	2,520	0	6,030[3]	0.26
Mexico	94.3	15.75[1]	1,575	14,175	1,440[3]	0.19
Syria	14.9	14.4[3]	14,400	0	0	0.97
World Total	**5848.7**	**989,100**				**0.17**

Sources:
[1]International Conference of Sultana and Raisin Producing Countries, Estimates. San Francisco, CA. 1998.
[2]FAO Production Yearbook, Vol. 51-1997.
[3]O.I.V. Bulletin Supplement 803-804. January and February, 1998.
[4]S.D. Shikhamany, National Research Center, Pume, India.
[5]Agriculture Attaché Query Detail, 1998.
[6]China Agricultural Yearbook, 1998.

Table 1.1.1

Total Grape Production by Type and State for the United States, 2000-2001.

State & Type	Total Production 2000	Total Production 2001	Bearing Ha 2000	Bearing Ha 2001	Yield Per Ha[1] 2000	Yield Per Ha[1] 2001
	t	t	ha	ha	t/ha	t/ha
Arizona	18,007	13,939	1,659	1,295	10.9	10.8
Arkansas	3,780	2,430	567	607	6.72	4.0
California	6,356,322	5,338,323	334,687	344,400	19.1	15.6
Wine	3,025,548	2,790,720	185,353	194,256	16.4	14.5
Table	696,870	627,354	36,018	36,328	19.5	17.2
Raisin	2,620,800	1,920,240	113,316	113,316	23.3	17.1
Georgia	3,154	2,880	486	445	6.5	6.5
Michigan	78,525	26,015	5,059	4,978	15.6	5.3
Missouri	2,655	2,067	344	352	7.8	5.9
New York	138,632	134,096	12,748	12,748	11.0	10.6
North Carolina	2,068	1,802	243	283	8.6	6.4
Ohio	6,930	5,400	809	809	8.6	6.7
Oregon	16,524	21,168	3,278	3,400	5.2	6.3
Pennsylvania	56,678	55,286	5,180	5,180	11.0	10.8
South Carolina	468	n/a	162	n/a	2.9	n/a
Texas	n/a	8,561	n/a	1,174	n/a	7.4
Virginia	3,687	n/a	n/a	688	n/a	5.4
Washington	238,392	254,880	17,807	19,426	13.5	13.2
Wine	81,000	90,072	8,094	9,713	10.1	9.3
Juice	157,464	164,808	9,713	9,713	16.3	17.1
US	**6,925,882**	**5,866,857**	**383,029**	**395,785**	**18.1**	**14.8**

Source: USDA Non Citrus and Nuts 1997 Summary, July, 1998.
[1] *Yield is based on total production*

Table 1.1.2

California Raisin Production by County, tons, 1994-1998.

Year	Fresno	Madera	Tulare	Kern	Kings	Merced
1994	285,300	54,523	30,870	8,424	5,220	2,574
1995	249,300	51,824	28,710	14,400	5,040	2,101
1996	209,700	55,761	30,780	2,250	5,850	1,120
1997	288,900	81,844	30,330	5,652	4,950	2,494
1998	185,400	47,312	22,140	2,835	5,850	1,417
Average	243,720	58,307	28,566	6,712	5,382	1,940
Percent	71.0	17.0	8.0	2.0	1.5	0.5

Source: County Agriculture Commissioner, Annual Reports from each county.

Table 1.1.3

Climatic Data for Fresno, California, 1961 – 1990.

Table 1.1.5a: Temperature and Daylight Data

Date	Mean Temp (°C)	Maximum Temp (°C)	Hours of Sunshine per day
August 15	26.9°	36.1°	13.57
September 1	25.5°	34.4°	12.83
September 15	23.6°	32.2°	12.42
October 1	21.4°	30.0°	11.80
October 15	18.6°	26.6°	11.28

Table 1.1.5b: Precipitation During the Drying Season

Date	Average Precipitation (cm)
July 1 to August 15	0.0254
July 1 to August 31	0.0762
July 1 to September 1	0.1016
July 1 to September 15	0.3302
July 1 to October 1	0.7366
July 1 to October 15	1.0922
July 1 to October 18	1.2446
July 1 to October 19	1.2954

Source: Brad Adams, Hydrometeorological Technician, Fresno, California.
Data from National Climatic Data Center, Asheville, North Carolina.

Table 1.1.5c: Relative Humidity

	Relative Humidity at 1600 hrs (4:00 p.m.)	Normal Rainfall
August	25%	0.0762 cm
September	28%	0.6096 cm
October	35%	1.3462 cm

*Average annual rainfall July 31 through June 30 is 26.924 cm.

Table 1.2.1

Production of Seedless Raisins in Turkey, 1994-95 through 1998-99.

Year	Metric Tons
1994-95	163,692
1995-96	198,416
1996-97	218,257
1997-98	231,155
1998-99	248,020

Source: Dr. Niyazi Adali, Deputy Director General, Tekel, Unkapani, Istanbul, Turkey.

Table 1.2.2

Seedless Raisins Exports (World and Turkey), metric tons, 1994-95 through 1998-99.

Years	World Exports	Turkish Exports	Turkey's Share (%)
1994-95	- 443,460	155,388	35.0
1995-96	472,230	169,343	35.8
1996-97	453,380	168,850	37.2
1997-98	448,421	191,763	42.8
1998-99	495,236[1]	147,022[2]	29.7

[1]Estimated
[2]First 5 months
Turkeys export is mostly to Germany and England.
Source: Niyazi Adal, Deputy Director General, TEKEL, Istanbul, Turkey.

Table 1.7.1

Production of Raisin Grapes by Region, dry tons, 1996.

Region	Area	Sultanas	Currants	Raisins
Western Cape	Vredendal	591.39	1,403.91	224.37
	Citrusdal	8.1		3.78
	Tulbagh/Worcester		13.95	
	Robertson/Montagu	0.63	2.52	
	Calitzdorp	0.36	4.41	2.43
	Other		1.26	
Northern Cape	Groblershoop	1,251.0	6.84	0.45
	Karos	2,533.5	1.08	1.08
	Upington	5,502.15	0.0	31.23
	Neilersdrift	2,576.88		
	Kelmoes	5,916.51	0.27	
	Kakamas	4,182.93	0.09	4.32
	Namibia	132.93		1.08
	Augrabies	5,411.88	0.0	0.54
Total		**28,108.44**	**1,461.24**	**269.19**

Table 1.7.2

Production of Raisin Grapes in South Africa, dry tonnes, 1992-98.

Year	Thompson Seedless	Unbleached Sultanas[1]	Golden Sultanas	Currants	Muscat & Monukka	Total
1994	18,813.6	5,084.1	4,393.7	1,136.7	276.3	29,704.5
1995	23,355.9	5,301	6,057.9	1,005.3	295.2	36,015.3
1996	17,627.4	3,233.7	4,638.6	1,325.7	243.9	37,069.3
1997	23,688.9	10,116	5,866.2	1,388.7	297	41,356.8
1998	12,012.3	2,640.6	5,857.2	1,017	187.2	21,714.3
Mean	18,855	5,274.9	5,363.1	1,175.4	260.1	30,928.5

[1]OR and WP unbleached sultanas

Table 1.11.1

Raisin Production in Caborca, Mexico, tonnes, 1985-1998.

Year	Hectares	Tonnes
1985	3,127.93	6,621.3
1986	2,964.02	5,762.7
1987	5,836.99	9,380.7
1988	4,611.96	10,323.9
1989	5,400.32	11,022.3
1990	4,065.21	8,032.5
1991	3,080.17	6,062.4
1992	2,500.24	4,980.6
1993	3,950.28	7,837.2
1994	6,256.26	14,895
1995	6,180.17	14,583.6
1996	5,787.21	13,491.9
1997	5,961.23	18,390.6
1998	4,600.22	16,095.6
Mean	**4,594.56**	**10,531.8**

Source: District of Rural Development No. 139, Dept of Agriculture and Rural Devlopment, Caborca, State of Sonora, Mexico.

Table 1.11.2

Raisin Production in Hermosillo, Mexico, 1998.

Variety	Tonnes
Flame Seedless	468.9
Perlette	615.6
Superior Seedless	35.1
Ruby Seedless	
Thompson Seedless	41.4
Other	
Total	**1,161**

Table 3.1

California Raisin Production by Grape Variety, tonnes, 1994-1998 average.

Variety	Metric Tons
Thompson Seedless	
A) Natural Sundried	281,282
B) Golden Seedless	17,471
C) Dipped Seedless	11,271
D) Oleate Seedless	1,361
Fiesta	15,876
Black Corinth (Zante Currant)	4,287
Flame Seedless	2,722
Ruby Seedless	1,814
Black Monukka	544
Sultana	227
Muscat of Alexandria	181
DoVine[1]	0
Summer Muscat[2]	0
Diamond Muscat[3]	0
Selma Pete[4]	0
Total	**337,035**

NOTES:
**** Of the above production, approximately 90% is naturally sundried.***
**** Not to be confused with the official USDA Classification of "Raisin Varieties" as shown in Chapter 12, Table 12.7.***

Estimated Tonnage:
[1]First commercial production expected 2002
[2]First commercial production expected 2003
[3]First commercial production expected 2004
[4]First commercial production expected 2005

Table 3.2

Net Returns Per Hectare Above Total Costs for Thompson Seedless Raisins.

Price ($/ton)	Yield (metric ton/hectare)							
Raisins	3.14	3.58	4.03	4.48	4.93	(5.15)	5.38	5.82
635.00	-2,448	-2,221	-1,993	-1,764	-1,536	-1,423	-1,307	-1,079
726.00	-2,102	-1,825	-1,549	-1,270	-993	-895	-664	-437
817.00	-1,756	-1,430	-1,104	-775	-450	-287	-129	205
930.00	-1,324	-936	-548	-158	230	**(425)**	620	1,008
998.00	-1,065	-640	-215	212	637	852	1,065	1,488
1,089.00	-719	-245	230	706	1,181	1,420	1,657	2,132
1,180.00	-373	151	151	1,200	1,724	1,988	2,250	2,774

Source: Karen Klonsky, et al., "Sample Costs to Establish a Vineyard and Produce Raisins in the San Joaquin Valley," UC Cooperative Extension, 1997.

NOTE: Numbers for 5.15 metric tons/hectre are extrapolated based on data from 4.93and 5.38 metric tons/hectare.

Table 5.2

Estimated annual N requirements and subsequent losses for 'Thompson Seedless' grapevines[z, y]

Nitrogen Status	Vine Part	Amount (kg per acre[x])
Requirements	Leaves	39.2
	Stems	39.2
	Clusters	39.2
	Total	**84**
Losses	Shoot trimming	
	-Leaves	5.6
	-Stems	3.36
	Fallen leaves	22.4
	Pruning	16.8
	Fruit Harvest	33.6
	Total	**781.76**

[z]Values were obtained by averaging the data collected over a 3-year period in the same vineyard.
[y]Modified after Williams, L.E., J. Amer. Soc. Hort. Sci. 112:330-333. (1987)
[x]Vineyard spacing was 8' by 12'; 454 vines per acre; avg. yield = 25.5 t/ha.

Table 6.1

Capital Budgeting Example Costs and Returns.

Year	Cost/Return for Investment 1	Costs/Return for Investment 2
Year 0*	($100)	($100)
Year 1	$25	$50
Year 2	$25	$40
Year 3	$25	$20
Year 4	$25	$10
Year 5	$25	$5

** This Year*

Table 6.2

Summary of Example Capital Budgeting Calculations.

Measure	Example Investment 1	Example Investment 2
Payback (years)	4.0	2.5
NPV (dollars)	$8.24	$13.32
IRR (%)	7.93%	11.97%
MIRR (%)	6.68%	7.66%

Table 6.3

Projected Per Hectare Revenue and Costs by Production System.

	Production System			
	DOG	Rocca	Simpson	Pitts
Revenue				
Yield (tonnes)	4.9	6.4	12.4	14.8
Price	$931	$931	$931	$931
Total Revenue	$5,063	$6,583	$12,659	$15,191
Operating Costs				
Insecticide, Herbicide, Fungicide	$405	$620	$454	$618
Vine Pruning & Brush Control	$516	$1,502	$2,114	$1,482
Water and Fertilizer	$378	$526	$259	$618
Harvesting and Machinery	$1,457	$1,203	$346	$1,186
Miscellaneous	$417	$106	$289	$247
Total Operating Costs	**$3,174**	**$3,957**	**$3,463**	**$4,150**
Income Above Operating Costs	**$1,890**	**$2,500**	**$9,196**	**$11,041**

Table 6.4

Net Present Value, Payback Period, Internal Rate of Return (IRR), and Modified Internal Rate of Return (MIRR) of Production Systems.

Net Present Value

Discount Rate	DOG	Rocca	Simpson	Pitts
3%	$5,441	$7,387	$50,736	$64,784
6%	$2,127	$3,213	$33,932	$47,383
9%	$73	$667	$23,310	$38,324
12%	($1,245)	($939)	$16,314	$29,672
15%	($2,117)	($1,981)	$11,528	$23,717
18%	($2,709)	($2,673)	$8,139	$19,427
21%	($3,119)	($3,142)	$5,667	$16,217
24%	($3,408)	($3,464)	$3,815	$13,737
27%	($3,614)	($3,688)	$2,395	$11,771
30%	($3,761)	($3,843)	$1,284	$10,176
33%	($3,866)	($3,951)	$401	$8,858
36%	($3,940)	($4,024)	($312)	$7,749
39%	($3,991)	($4,313)	($895)	$6,803
42%	($4,025)	($4,307)	($1,377)	$5,985
45%	($4,046)	($4,294)	($1,778)	$5,267
Payback	10 yrs 7 mos	14 yrs 7 mos	4 yrs 1 mos	3 yrs 1 mos
IRR	9.14%	10.09%	34.61%	46.36%
MIRR	6.48%	6.90%	13.09%	16.03%

Table 6.5

Net Returns Per Hectare by Price Per Ton and Production System.

	Production System			
Price	DOG	Rocca	Simpson	Pitts
$363	-$1198	-$1438	$1477	$1778
$545	-$210	-$178	$3947	$4742
$726	$778	$1082	$6417	$7706
$908	$1766	$2341	$8887	$10670
$1090	$2754	$3601	$11357	$13634
$1271	$3742	$4861	$13827	$16598

Table 6.6

Breakeven Yield by Price Per Tonne and Production System.

	Production System			
Price	DOG	Rocca	Simpson	Pitts
$363	7.9	4.0	8.6	10.4
$454	6.4	3.2	6.9	8.4
$545	5.2	2.7	5.7	6.9
$636	4.4	2.3	4.9	5.9
$726	4.0	2.0	4.4	5.2
$817	3.5	1.8	4.0	4.7
$908	3.2	1.6	3.5	4.2
$999	3.0	1.5	3.2	3.7
$1090	2.7	1.3	3.0	3.5
$1180	2.5	1.2	2.7	3.2
$1271	2.2	2.7	2.5	3.0

Table 6.7

Breakeven Raisin Prices by Yield and Production System.

	Production System			
Yield	DOG	Rocca	Simpson	Pitts
2.2	$3174	$3957	$3463	$4150
4.2	$1588	$1978	$1731	$2075
6.7	$1057	$1319	$1153	$1383
9.0	$793	$988	$865	$1037
11.2	$635	$790	$692	$830
13.5	$529	$659	$578	$692
15.7	$454	$566	$494	$593

Table 12.1

Raisin Type Grape Hectares, 1967-97.

Year	Bearing Hectares	Non-bearing Hectares	Total Raisin Type Hectares
1967	101,765	4,458	106,223
1968	101,138	2,441	103,579
1969	101,445	1,259	102,704
1970	99,332	1,077	100,409
1971	98,704	2,282	100,987
1972	96,866	2,415	99,281
1973	96,866	4,609	101,475
1974	97,714	4,509	102,223
1975	96,316	4,433	100,749
1976	98,429	3,123	101,553
1977	97,595	3,330	100,924
1978	97,347	5,275	102,621
1979	96,973	9,151	106,123
1980	98,768	13,716	112,484
1981	101,135	15,145	116,281
1982	105,579	12,542	118,121
1983	110,857	9,042	119,899
1984	113,763	5,555	119,318
1985	114,757	3,116	117,873
1986	112,098	1,613	113,712
1987	111,134	1,003	112,137
1988	109,595	1,496	111,091
1989	109,717	2,202	111,918
1990	109,312	2,957	112,269
1991	107,692	3,066	110,758
1992	107,692	3,036	110,728
1993	109,717	4,136	113,852
1994	109,328	3,449	112,777
1995	108,685	3,538	112,223
1996	109,148	2,874	112,022
1997	109,067	2,100	111,159

Obtained from the RAC Marketing Policy Tables

Source: California Grape Acreage, California Fruit and Nut Statistics (1973-89), California Agriculture Statistics Service, Sacramento, CA

Table 12.2

California Raisin-Grape Hectares by Variety, 1998-99.

Variety	Bearing	Non-bearing[1]	Total Ha
Black Corinth[2]	996	8	1,003
Fiesta	1,700	317	2,018
Sultana	67	0	67
Thompson Seedless	106,415	1,730	108,145
Other raisin	25	26	51
Total raisin	109,202	2,080	111,283

[1]Non-bearing includes plantings for 1996, 1997 and 1998.

[2]Black Corinth/Zante Currant.

Source: 1998 California Grape Acreage by California Agriculture Statistics Service, June 10, 1999.

Table 12.3

Production & Utilization by Raisin Variety Grapes, metric tons, 1967-68 through 1998-99.

Crop Year	Total Production	Dried	Crushed	Canned	Fresh Sales
1967-68	1,486,364	682,727	559,091	49,091	195,455
1968-69	1,940,909	1,009,091	644,545	58,182	229,091
1969-70	1,959,091	915,455	768,182	60,273	215,182
1970-71	1,700,909	745,455	773,636	48,818	133,000
1971-72	2,101,818	811,818	1,094,545	53,091	142,364
1972-73	1,221,818	396,364	651,818	45,909	127,727
1973-74	2,160,182	879,091	1,100,182	53,636	127,273
1974-75	1,790,909	928,364	685,545	55,636	121,364
1975-76	2,000,909	1,136,000	662,727	47,909	154,273
1976-77	1,779,091	891,818 [1]	686,364	43,636	157,273
1977-78	1,759,091	1,029,091	540,000	49,091	140,909
1978-79	1,518,182	689,091 [2]	638,182	50,000	140,909
1979-80	2,109,091	1,250,909	636,364	54,545	167,273
1980-81	2,447,273	1,465,455	707,273	57,273	217,273
1981-82	1,617,273	930,909	462,727	38,182	185,455
1982-83	2,401,818	1,390,909 [3]	703,636	31,818	275,455
1983-84	2,173,636	1,612,727	300,000	31,818	229,091
1984-85	2,024,545	1,260,909	527,273	27,273	209,091
1985-86	2,301,818	1,456,364	508,182	40,909	296,364
1986-87	1,859,091	1,072,727	459,091	36,364	290,909
1987-88	1,972,727	1,300,000	400,000	36,364	236,364
1988-89	2,336,364	1,645,455	377,273	36,364	277,273
1989-90	2,336,364	1,681,818	336,364	36,364	281,818
1990-91	2,131,818	1,572,727	243,636	36,364	279,091
1991-92	1,968,182	1,420,000	258,182	37,273	252,727
1992-93	2,427,273	1,412,727	713,636	41,818	259,091
1993-94	2,140,000	1,492,727	360,909	40,909	245,455
1994-95	2,171,818	1,712,727	179,091	34,545	245,455
1995-96	2,047,273	1,392,727	392,727	31,818	230,000
1996-97	1,992,727	1,189,091	561,818	32,727	209,091
1997-98	2,620,909	1,616,364	714,545	40,000	250,000
1998-99 [4]	1,781,818	1,140,909	408,182	32,727	200,000

[1] Doesn't include 266,364 metric tons lost due to rain.

[2] Doesn't include 225,455 metric tons lost due to severe weather.

[3] Doesn't include 327,273 metric tons harvested but not sold and 54,545 metric tons not harvested.

[4] Preliminary.

Source: California Fruit and Nut Statistics, 1965-77, 1973-84, 1985-86, 1986-87. California Resource Directory by California Agricultural Statistics Service.

Table 12.4

Raisin Variety Grape Yields, metric tons, 1967-1998.

Crop Year	Metric Tons per Bearing Hectare
1967-68	14.56
1968-69	19.15
1969-70	19.27
1970-71	17.09
1971-72	21.24
1972-73	12.61
1973-74	22.25
1974-75	18.28
1975-76	20.72
1976-77	21.37
1977-78	17.90
1978-79	17.88
1979-80	21.69
1980-81	24.89
1981-82	15.97
1982-83	26.44
1983-84	20.75
1984-85	17.86
1985-86	20.01
1986-87	20.19
1987-88	18.71
1988-89	23.07
1989-90	21.24
1990-91	19.47
1991-92	19.71
1992-93	23.52
1993-94	21.44
1994-95	19.83
1995-96	20.77
1996-97	18.19
1997-98	23.93
1998-99	16.26

Source: California Agriculture Resource Directory by California AgricultureStatistics Service (CASS). California Fruit & Nut Statistics, 1967-86, California Crop & Livestock Reporting Service (now known as CASS).

Table 12.5

Raisin Production, metric tons, dry weight, 1967-68 through 1999-2000.

		RBA Announced		N.S. Season
Crop Year	**All Raisin Variety Production**	**Natural Seedless (N.S.) Production**	**Free Metric Tonnage (N.S.) Field Price**	**Average Producer Price**
1967-68	165,292	146,655	$277	$265
1968-69	239,978	219,045	285	235
1969-70	228,506	206,754	289	237
1970-71	175,397	160,060	291	253
1971-72	177,204	156,679	295	290
1972-73	94,684	82,962	455	455
1973-74	205,062	180,685	636	636
1974-75	220,102	193,082	582	550
1975-76	257,814	230,246	589	552
1976-77	129,022	106,914	45	955
1977-78	226,311	198,921	764	778
1978-79	90,394	67,645	545	1,455
1979-80	275,588	239,189	1,045	1,055
1980-81	285,132	231,506	1,091	1,061
1981-82	233,843	204,057	250	1,118
1982-83	231,391	187,000	1,182	1,182
1983-84	352,122	316,312	1,182	521
1984-85	304,096	272,248	636	535
1985-86	370,100	329,688	745	545
1986-87	345,968	315,404	818	692
1987-88	350,198	320,453	859	735
1988-89	378,104	344,594	23	834
1989-90	393,227	305,001	105	897
1990-31	359,395	324,773	1,014	799
1991-92	352,005	320,599	1,050	875
1992-93	378,255	337,742	1,050	819
1993-94	397,665	351,825	1,050	823
1994-95	383,535	344,025	1,055	844
1995-96	337,053	296,283	1,055	915
1996-97	285,293	247,330	1,109	954
1997-98	393,299	347,680	1,136	*
1998-99	254,993	762,461	1,170	1,170
1999-00	**	**	1,293	*

* All payments for the reserve raisins have not been paid to the grower.

** Not available.

Source: RAC Marketing Policies and final payment statements, Raisin Bargaining Association.

Table 12.6

Domestic and Export California Raisin Shipments by Types of Packages, packed metric tons, 1967-68 through 1998-99.

Crop Year	Domestic shipments				Export shipments		
	Cartons/bags	Bulk	USDA purchases, reserve	Other govt purchases	Cartons/bags	Bulk	Crop year totals
1967-68	60,859	67,406	14,785	912	17,286	39,437	200,686
1968-69	55,891	68,104	11,017	2,405	17,320	39,061	193,798
1969-70	57,456	65,077	18,424	748	16,012	42,473	200,190
1970-71	57,503	64,164	2,670	1,466	15,607	37,101	178,511
1971-72	61,238	67,007	3,762	556	16,856	43,965	193,385
1972-73	41,831	51,399	0	115	7,803	7,511	108,659
1973-74	69,417	61,678	0	77	15,325	21,975	168,473
1974-75	66,366	59,016	0	884	19,019	38,010	183,295
1975-76	67,505	65,525	6,934	461	16,684	31,994	189,101
1976-77	58,945	59,986	0	0	16,626	20,442	156,000
1977-78	63,549	64,214	0	32	20,691	27,727	176,213
1978-79	42,869	54,065	1,890	0	11,865	9,690	120,380
1979-80	64,026	64,615	6,746	187	18,234	45,416	199,225
1980-81	68,215	80,467	3,865	24	20,738	45,379	218,689
1981-82	70,934	82,533	6,423	14	22,181	36,713	218,796
1982-83	73,853	82,308	2,813	23	16,739	36,036	211,772
1983-84	69,595	85,967	4,240	28	18,879	40,768	219,477
1984-85	84,662	102,537	7,165	49	20,801	51,736	266,951
1985-86	81,703	111,836	4,147	24	23,804	59,921	281,435
1986-87	79,509	108,885	4,344	40	25,937	60,965	279,680
1987-88	83,355	115,536	3,207	68	30,502	70,174	302,843
1988-99	88,135	127,712	7,720	144	30,467	67,486	321,664
1989-90	84,385	118,195	3,385	446	29,972	75,994	312,377
1990-91	82,707	118,792	11,120	195	35,815	80,005	328,633
1991-92	79,571	120,468	11,538	31	30,588	81,663	323,859
1992-93	82,045	118,376	9,514	4	29,351	83,286	322,576
1993-94	80,853	122,019	14,697	0	29,326	91,008	337,904
1994-95	75,281	115,630	4,825	0	29,169	92,216	317,121
1995-96	75,064	110,975	5,115	0	29,178	88,100	308,432
1996-97	67,626	116,481	199	0	28,940	86,586	299,833
1997-98	71,823	113,893	686	0	26,696	93,595	306,693
1998-99	66,368	109,846	5,480	0	26,705	82,057	290,456

[1]Crop year: 1967-74 = Sept.1-Aug.31; 1976-present = Aug.1-July 31.

[2]1975 crop year = 11 mo. statistics - Sept. 1, 1975-July 31, 1976.

Source: RAC Final Shipment Report for each crop year.

Table 12.7

California Raisin Shipments by Raisin Variety, packed metric tons, 1967-68 through 1998-99.

Crop Year[1]	Raisin Varieties							Total
	N.S.	G.S.	D.S.	Z.C.	Muscat	Sutana	O.V.	
1967-68[1]	182,975	12,085	54	2,297	2,323	645	307	200,686
1968-69	175,625	13,154	251	2,270	1,745	504	250	193,798
1969-70	182,664	12,452	54	2,505	1,502	388	626	200,190
1970-71	161,726	12,133	15	2,720	1,177	377	363	178,511
1971-72	176,037	12,558	38	2,861	1,018	290	583	193,385
1972-73	95,379	10,295	377	1,232	885	175	317	108,659
1973-74	149,005	12,351	3,687	1,649	1,245	164	371	168,473
1974-75	160,707	12,264	1,089	1,830	1,407	371	5,627	183,295
1975-76[2]	167,630	12,696	4,755	1,721	1,386	318	594	189,101
1976-77	136,490	11,285	4,745	2,201	719	205	356	156,000
1977-78	154,087	12,803	5,541	2,695	467	259	361	176,213
1978-79	99,758	7,710	9,927	2,116	478	133	257	120,380
1979-80	169,355	14,492	10,448	3,525	641	377	385	199,225
1980-81	181,278	14,793	17,895	2,566	1,017	284	855	218,689
1981-82	188,457	12,655	8,251	2,000	1,310	284	5,840	218,796
1982-83	177,751	12,274	13,276	2,245	520	212	5,494	211,772
1983-84	186,391	12,468	10,631	2,057	638	367	6,925	219,477
1984-85	234,562	14,023	9,016	2,829	275	180	6,066	266,951
1985-86	247,776	14,283	9,395	2,612	270	426	6,672	281,435
1986-87	253,025	13,355	6,639	2,839	306	381	3,135	279,680
1987-88	271,853	15,491	8,421	2,783	251	150	3,895	302,843
1988-99	292,057	14,688	9,418	3,220	279	117	1,884	321,664
1989-90	280,656	14,987	10,257	3,382	179	146	2,768	312,376
1990-91	298,710	14,997	9,657	3,093	170	99	1,906	328,633
1991-92	292,559	15,384	9,365	3,165	164	130	3,093	323,859
1992-93	289,387	16,480	11,088	3,217	125	177	2,101	322,576
1993-94	300,174	17,110	14,595	2,568	112	88	3,256	337,904
1994-95	279,076	17,978	12,560	3,892	104	160	3,351	317,121
1995-96	272,884	18,365	9,596	4,443	143	204	2,798	308,432
1996-97	267,841	17,605	8,632	2,607	81	79	2,987	299,833
1997-98	265,554	17,487	15,318	3,329	5	93	4,220	306,006
1998-99	251,567	13,440	13,203	2,865	3	65	3,833	284,976

[1]Crop year: 1967-74 = September 1 - August 31; 1976 to present = August 1 to July 31.

[2]1975 crop year = 11 months statistics--September 1, 1975 to July 31, 1976.

[3]Raisin Varieties NS=natural (sun-dried) seedless; GS=Golden Seedless; DS=Dipped Seedless; ZC=Zante Currants;

OV=Other varieties include the Oleate and related seedless, Monukkas and other seedless raisins.

Source: RAC Final Shipment Reports for each crop year.

Table 12.8

Raisin import inspections in kilograms (pursuant to Section 608[e] of the AMAA of 1937) August 1 - July 31.

	Meeting Raisins							
Crop Year	**Mexico**	**Turkey**	**Greece**	**Chile**	**Other Countries***	**Total Meeting Raisins**	**Total Failing Raisins**	**Total Raisins**
1972-73	0	665,295	3,362,884	0	474,771	4,502,951	4,983,738	9,486,689
1973-74						1,401,791 [4]	843,146	2,244,937
1974-75			69,922	0	19,278	89,200	24,948	114,148
1975-76						39,000 [4]	44,371	83,371
1976-77						13,720,245 [4]	14,952,212 [1]	28,672,457 [1]
1977-78						10,003,625 [4]	997,980	11,001,605
1978-79						13,975,974 [4]	13,599,895 [1]	27,575,869 [1]
1979-80						3,125,414 [2,4]	1,516,217	4,641,631
1980-81	0	0	0	0	14,066	14,066	0	14,066
1981-82	120,117	0	167,033	0	0	287,150	296,541	583,690
1982-83	3,439,368	34,936	323,188	0	10,478	3,807,970	1,775,212	5,583,182
1983-84	699,785	16,765	26,535	0	362,617 [3]	1,105,702	1,547,307	2,653,009
1984-85	0	0	0	0	425,771 [3]	425,771	25,120	450,891
1985-86	2,549,240	0	0	0	63,277	2,612,517	986,154	3,598,671
1986-87	3,535,138	680	0	0	0	3,535,819	2,033,490	5,569,309
1987-88	4,823,019	27	21,092	172,265	133,131	5,149,534	2,331,384	7,480,918
1988-99	3,023,716	100,572	20,412	1,405,309	703,597	5,253,606	2,299,081	7,552,687
1989-90	3,965,401	1,219,428	20,412	2,058,806	366,179	7,630,225	2,390,396	10,020,621
1990-91	2,571,509	19,051	0	3,166,552	7,307	5,764,419	2,321,257	8,085,676
1991-92	3,055,434	234,501	0	2,335,861	165,717	5,791,513	2,073,928	7,865,441
1992-93	2,890,547	1,233,426	0	1,040,509	58,854	5,223,336	1,226,102	6,449,438
1993-94	3,318,289	1,339,096	20,412	841,881	19,418	5,539,096	481,137	6,020,233
1994-95	5,086,728	1,277,516	0	1,653,941	181,006	8,199,190	727,492	8,926,681
1995-96	7,760,576	579,391	0	782,364	112,151	9,234,482	1,653,053	10,887,535

*Includes: Afghanistan, Argentina, Pakistan, South Africa, Sweden, Portugal, and Spain

[1]Some of the failing raisins could have been reconditioned and submitted more than once for inspection.

[2]About one-half the meeting raisins were successfully reconditioned from prior year's failing imports.

[3]Primarily from South Africa.

[4]Country breakdown not available.

Source: Agriculture Marketing Service, Fruit and Vegetable Programs, USDA.

Table 12.9

Raisins and Zante Currants Exports from United States, packed metric tons, 1995-96 through 1998-99.

	Crop year: August 1 - July 31							
	1995-96		1996-97		1997-98		1998-99	
Country of Destination	**Natural Seedless**	**Other***	**Natural Seedless**	**Other***	**Natural Seedless**	**Other***	**Natural Seedless**	**Other***
Austria	73	3			305	33	203	21
Belgium	944	293	1,150		465	349	386	322
Denmark	5,581	180	5,321		4,957		4960	0
Ireland	79		185		129		106	0
Finland	2,631	8	2,257		2,285	6	2349	5
France	309	49	368	2	1,294	134	636	83
Germany	9,585		8,005		9,865	21	6397	13
Israel	143	556	202	54	115	842	113	981
Italy	14	31	32		76	11	91	0
Netherlands	2,725	760	2,539		3,511	565	2751	208
Norway	2,194		2,371		1,999	15	2209	0
Spain	532	7	535		739	6	590	0
Sweden	4,749	50	4,930		5,295	31	5166	0
Switzerland	383		426		54		80	0
U.K.	26,908	1,500	25,171		28,158	613	24851	376
Total Europe	56,848	3,437	53,491	55	59,246	2,625	50886	2021
Brazil	424	83	584		1,357	229	484	27
Columbia	82		141		371	1	101	0
Costa Rica	51	2	80		200	1	73	1
Dominican Rep.	413	1	427		555	1	454	1
Ecuador	27	1	29		45	3	43	4
Mexico	430		598		458	2	20	0
Panama	366	4	407		535	3	436	1
Puerto Rico							354	1
Venzuela	343	11	345		471	20	320	8
Others	1,331	24	1,067		298	35	396	5
Total Latin America	3,466	125	3,678	0	4,290	294	2683	48
Hong Kong	1,193	3,243	1,314		1,877	3,039	893	1091
Iceland	238	1	201		201	1	141	2
Japan	24,004	294	26,202	148	24,570	534	29650	795
Korea	2,297	59	2,558		2,191	16	2288	0
Malaysia	596	475	681		717	592	469	431
New Zealand	1,363	35	1,600		1,407	31	1324	41
NIS (USSR)	485	5	812		1,551	19	327	0
Phillippines	657	18	1,047		1,310	3	1253	1
Singapore	2,494	1,609	2,358		1,902	908	2026	962
Taiwan	3,355	112	3,562		4,155	506	3528	350
Others	2,160	1,302	2,237		3,070	1,233	2174	1115
Total other countries	38,843	7,153	42,572	148	42,951	6,882	44070	4787
Canada	6,217	1,190	5,116	166	3,370	633	3800	479
Grand total	105,374	11,905	104,857	370	109,857	10,434	101439	7335

Varieties include: Golden Seedless, Zante Currants, Dipped Seedless, Oleate and related seedless, Muscats, Monukkas and other seedless.

Source: Raisin Administrative Committee

Table 12.10

1999-2000 Trade Demand, Raisin Administrative Committee.

	Natural	Dipped	Oleate	Golden	Zante Currants	Sultanas	Muscats	Monukkas	Other
Base shipments (packed metric tons)	251,567	13,203	229	13,440	2,865	65	3	607	2,997
%Shrink factor (5 yr avg.)	0.93834	0.92061	0.78869	0.90927	0.88867	0.70193	0.57807	0.76193	0.90853
Shrink %	6.166	7.939	21.131	9.073	11.133	29.807	42.193	23.807	9.147
=Base metric tonnage (sweatbox t.)	267,983	14,367	288	14,866	3,213	93	(5)	732	3,188
x 90% Formula	90%	90%	90%	90%	90%	90%	90%	90%	90%
= Adjusted base	241,185	12,930	260	13,379	2,892	84	(4)	659	2,869
Physical inventory 7/31/98	92,484	1,382	262	4,516	1,729	36	73	72	1,426
- Desirable inventory	66,958	2,436	5	3,974	520	15	0	138	532
= +/- Inventory adjustment	(25,525)	1,054	(257)	(542)	(1,209)	(21)	(73)	66	(894)
=1998/99 Trade demand	230,856	13,277	454	12,838	1,683	454	454	725	1,975

* RAC adopted a policy that any trade demand computed at less than 455 metric tons will be set at 455 metric tons.

Source: Raisin Administrative Committee.

Table 12.11

Raisin Preliminary Percentages, 1999-2000.

Variety	Trade Demand	Estimated Production	Preliminary Free	Percentages Reserve
Natural Seedless	230,856		56%	44% [1]
Dipped Seedless	13,277	17,247	50%	50% [2]
Oleate Seedless	454	458	100%	0%
Golden Seedless	12,838	14,678	100%	0%
Zante Currant	1,683	3,798	29%	71% [3]
Sultana	454	150	100%	0%
Muscat	454	73	100%	0%
Monukka	725	459	100%	0%
Other Seedless	1,975	2,362	100%	0%

*Percentages shown reflect 65% of trade demand

[1]Interim percentages of 84.75% free, 15.25% reserve for natural seedless raisins was announced on Feburary 14, 2000. The interim final rate establishing final percentages of 85% free, 15% reserve was published on April 10, 2000.

[2]On October 25, 1999, the Dipped Seedless percentages were announced 100% free, 0% reserve.

[3]Interim percentages of 50.75% freem 49.25% reserve for Zanter Currant raisins January 13, 2000. The interim final rule establishing final percentages of 51% free, 49% reserve was published on April 10, 2000.

Source: Raisin Administrative Committee Marketing Policy, 1999-2000.

Table 12.13

Raisin Diversion Program, metric tons, 1985-86 through 1999-2000.

Crop Year	Natural Condition
1985-86	53,965 [1]
1986-87	94,187 [1]
1987-88	27,105 [1]
1988-99	44,952 [1]
1989-90	0
1990-91	0
1991-92	36,212 [1]
1992-93	22,496 [1]
1993-94	44,984 [1]
1994-95	0
1995-96	55,861 [3]
1996-97	1,865 [4]
1997-98	0
1998-99	115 [2,4,5]
1999-2000	0

[1]Natural Seedless raisins.

[2]Zante Currants.

[3]Natural Seedless Raisins 61,098 & Zante Currants 478.

[4]Vine removal only.

[5]Zante Currant raisins came from 1997-98 pool.

Source: Table 3 of the 1995-96, 1996-97, 1997-98 and 1999-2000 RAC Market Policy Reports.

Appendix H: Raisin Recipes

Recipes were provided by the California Raisin Marketing Board, P.O. Box 5195, Fresno, CA 93755-5195. Telephone: (559) 248-0287; Fax: (559) 224-7016; E-mail:*info@raisins.org.*;Website:*http://www.calraisins.org.*

Breads and Pies

Raisins can be kneaded into the dough of almost any white, whole wheat or rye bread, and bagels or rolls. Add them to fruit pies or tarts, such as apple or rhubarb, to flavor and sweeten the pie with less sugar. Of course, raisins make a superb pie or tart all by themselves. Bake them into cornbread, biscuits, muffins, scones or nut breads for something a little different.

Raisin Bread

This classic raisin bread recipe is full of tangy, sweet, natural sun-dried raisins.

1-1/2 cups milk
1/2 cup butter or margarine
1/2 cup sugar
2 teaspoons salt
2 packages active dry yeast
1 cup warm water (110° to 115° F)
2 eggs, beaten
7 to 8 cups all-purpose flour
3 cups California raisins, dusted in flour

Heat milk to scalding; add butter, sugar and salt. Cool to lukewarm. Dissolve yeast in warm water. Add the lukewarm milk mixture. Stir in eggs. By hand or using an electric mixer, gradually beat in 5 cups flour. Add raisins. By hand, work in remaining flour to make a medium-firm dough. Place in deep, greased bowl, turning to grease top. Cover; let rise in warm place until doubled in bulk, about 1-1/2 to 2 hours. Punch down dough. Turn dough out onto floured surface; knead slightly. Form into three loaves and place in well-greased 8 x 4-inch loaf pans. Cover; let rise in warm place until doubled, about one hour. Bake at 375° F for 30 to 35 minutes or until golden brown. Remove from pans; brush tops with butter and cover with cloth. Cool on wire rack.

Makes 3 loaves.

Raisin Banana Bread

Your favorite banana bread takes on a new sparkle with either California natural or golden raisins.

1/3 cup butter or margarine, softened
2/3 cup sugar
2 eggs
3 tablespoons milk
2 cups all-purpose flour
1 teaspoon baking powder
1 teaspoon salt
1/2 teaspoon baking soda
1 cup mashed bananas
1 cup chopped California raisins
1/2 cup chopped nuts

Beat together butter, sugar and eggs. Add milk. Sift flour with baking powder, salt and baking soda. Stir into egg mixture just until moistened. Blend in bananas, raisins and nuts. Turn into a greased 9 x 5-inch loaf pan. Bake at 350° F for one hour or until toothpick inserted in center comes out clean. Cool.

Makes 1 loaf.

Hot Cross Buns

Sparkling nuggets of candied fruit make these Easter buns a pretty treat any time of the year.

1 cup milk
2 tablespoons butter or margarine
1 package active dry yeast
1/4 cup warm water (110 to 115° F)
4 cups flour
1/3 cup sugar
1 teaspoon salt
1 teaspoon cinnamon
1 cup California raisins or Zante currants
1/2 cup mixed candied fruit
2 eggs, well beaten
1 egg yolk, diluted with 1 teaspoon water for topping
Lemon Icing (below)

Scald milk; stir in butter and cool to lukewarm. Dissolve yeast in warm water. Sift flour with sugar, salt and cinnamon in a large bowl. Stir in raisins and candied fruit until well coated. Stir in eggs, cooled milk and yeast; blend well. Turn dough out onto lightly floured board and knead until smooth and elastic, five to eight minutes. Place in greased bowl, turning to grease top. Cover; let rise in warm place until doubled in bulk, about 1-1/2 hours.

Punch down dough, pinch off pieces, and form smooth, rounded balls about 1-1/2 inches in diameter. Place balls of dough on greased baking sheet about 2-inches apart. Brush each bun lightly with diluted egg yolk. Snip 1/2 inch deep cross in center of each bun with greased scissors. Let buns rise in warm place until doubled in bulk, about 30 minutes.

Bake at 400° F for about eight to 10 minutes, or until lightly browned. Cool on wire racks about five minutes. Drizzle icing on the cross.

Lemon Icing:

Combine 1 cup powdered sugar, 2 teaspoons lemon juice, and 1 teaspoon water; beat until smooth.

Makes 30 buns.

All-American Raisin Pie

The aroma of cranberries, apples and raisins is scrumptious. Lattice the top crust for a peek-a-boo effect.

3/4 cup sugar
2 tablespoons cornstarch
1/4 teaspoon nutmeg
1/2 teaspoon cinnamon

2 teaspoons vanilla
2 eggs, well beaten
2 cups flour
1-1/4 teaspoons salt
1 teaspoon baking soda
1 teaspoon cinnamon
1/2 cup sour cream

Chop raisins and nuts. Dissolve chopped chocolate and instant coffee in boiling water. Beat butter with half of sugar until light. Beat in remaining sugar and vanilla until fluffy; add eggs, blending thoroughly. Stir in raisins, walnuts and chocolate liquid. Sift flour with salt, soda and cinnamon; add alternately with sour cream to chocolate mixture, stirring only until blended. Turn into well-greased and floured 9-cup bundt pan. Bake at 325° F for 50 minutes, or until toothpick inserted in center comes out clean. Cool in pan for five minutes, then turn out on wire rack to finish cooling. Sift powdered sugar over top if desired.

Makes 10 to 12 servings.

Golden Gate Carrot Cake

Carrot cake wouldn't be carrot cake without raisins!

2 cups flour
2 teaspoons baking soda
2 teaspoons cinnamon
1/2 teaspoon salt
3 eggs
3/4 cup vegetable oil
3/4 cup buttermilk
2 cups sugar
2 teaspoons vanilla
1 small can (8 oz) crushed pineapple, well drained
4 cups grated carrots (4 to 6)
1 cup chopped walnuts
1 cup grated coconut
1-1/2 cups California golden raisins, coarsely chopped
Buttermilk Glaze (below)

Sift first four ingredients together. Beat eggs; add oil, buttermilk, sugar and vanilla. Blend well. Add to dry ingredients, mixing thoroughly. Stir in pineapple, carrots, nuts, coconut and raisins. Pour into greased and floured 13 x 9-inch pan. Bake at 350° F for 55 minutes, or until toothpick inserted in center comes out clean. Slowly pour Buttermilk Glaze over hot cake until it is absorbed.

Makes 10 to 12 servings.

Buttermilk Glaze:

Mix 1 cup sugar, 1/2 teaspoon baking soda, 1/2 cup buttermilk, 1 tablespoon light corn syrup and 1/2 cup butter or margarine together in saucepan. Boil five minutes. Remove from heat and add 1 teaspoon vanilla.

Ice Cream and Yogurt

Stir them into yogurt, custards, or rice, tapioca and Indian puddings. Soak in rum or brandy and serve flaming over ice cream. Combine raisins with chopped nuts and chocolate or maple syrup to serve over ice cream.

Fruit Flambé with Ice Cream

Flaming melds these luscious fruit flavors for an exquisite ice cream topping.

2 tablespoons butter
1/4 cup brown sugar
1/2 teaspoon vanilla
1 ripe mango; peeled, seeded and cubed (about 1-1/2 cups)
1 cup California raisins
1/4 cup dark rum
1 pint vanilla ice cream

In medium skillet over medium heat, melt butter. Add brown sugar; cook until bubbly. Add mango and raisins. Reduce heat; cover. Cook 10 to 15 minutes or until mango is tender, stirring occasionally. Add rum; warm quickly and carefully light with long match. Tilt and rotate pan carefully. When flames have extinguished, pour sauce into each of four dessert bowls. Top with scoop of ice cream.

Makes 4 servings.

Tip: Substitute 1-1/2 cups fresh pineapple or banana chunks.

Two-Ginger Golden Ice

For a snappy, light palate cleanser or dessert, combine two kinds of ginger with golden raisins.

1 cup water
2/3 cup light corn syrup
1/3 cup sugar
1-1/2 cups clear apple juice
1/4 cup lemon juice
2 tablespoons chopped candied ginger
1 tablespoon grated fresh ginger
2/3 cup golden raisins

In medium saucepan, mix water, corn syrup, and sugar. Bring to boil, stirring constantly. Reduce heat; simmer five minutes. Stir in juices and ginger. Pour into shallow bowl; cool. Freeze until firm about 1-inch around edge. Beat with electric mixer to blend evenly. Mix in raisins. Freeze until almost solid. Break up and beat again with electric mixer. Cover and freeze until firm. Serve.

Makes about 3-1/2 cups (6 to 8 servings).

Frothy Yogurt Shake

The perfect tangy pickup with California raisin sweetness.

1/2 cup raisins
1/2 cup milk
1/2 cup yogurt, plain or flavored
Half of medium banana, sliced
1/2 teaspoon vanilla

Puree raisins, milk and yogurt in blender or food processor. Add banana and vanilla; process until smooth.

Makes one serving.

Index

Note: Page numbers in *italic type* indicate photographs or illustrations.

Index

Index

E

F

Index

Index

Index

Index

Q

R

Index